A History of the
World Economy

16
29
49
86
173

A History of the World Economy
International Economic Relations Since 1850

Second Edition

James Foreman-Peck
St Antony's College
University of Oxford

HARVESTER WHEATSHEAF

New York London Toronto Sydney Tokyo Singapore

First published 1983
This edition published 1995 by
Harvester Wheatsheaf
Campus 400, Maylands Avenue
Hemel Hempstead
Hertfordshire, HP2 7EZ
A division of
Simon & Schuster International Group

© James Foreman-Peck 1995

All rights reserved. No part of this publication may be reproduced, stored in a retrieval system, or transmitted, in any form, or by any means, electronic, mechanical, photocopying, recording or otherwise, without prior permission, in writing, from the publisher.

Typeset in 10/12 pt Times and Univers
by Keyset Composition, Colchester

Printed and bound in Great Britain
by Redwood Books, Trowbridge, Wiltshire

British Library Cataloguing in Publication Data

A catalogue record for this book is available from the British Library

ISBN 0-7450-0935-2

2 3 4 5   99 98 97 96 95

Contents

List of figures ix
List of tables x
Preface xii

Introduction xiii

1 International economic relations in the middle of the nineteenth century   **1**
The political impact of mid-Victorian trade   3
Trade based on voluntary exchange   7
Trade without policy independence   9
Trade without law and order   11
Market structure and the gains from trade   13
Justice and fairness   15
Summary and conclusion   16

2 Economic links between Britain, India and the United States at the mid-century   **19**
Measuring international transactions   19
The British balance of payments   21
The balance of payments of the United States   24
The Indian balance of payments   26
Summary and conclusion   29

3 Liberalisation and free trade   **31**
Transport costs and trade   31
Integration   35
Specialisation   36
The Hecksher–Ohlin theory of comparative advantage   40
Commercial policy   43
Summary and conclusion   49

4 International factor mobility, 1850–75 — 50
The cosmopolitan bourgeoisie — 50
Trade, factor price equalisation, and income gaps between countries — 52
Worker migration — 54
Capital movements — 58
Income and welfare — 60
Economic growth and international economic relations — 61
Summary and conclusion — 63

5 The world monetary system, 1850–1875 — 65
Merchant bankers and bills of exchange — 65
Exchange rates — 71
The new gold — 72
The international transmission of price increases — 75
Floating exchange rate regimes — 77
Monetary unions — 79
International fluctuations in economic activity — 81
Summary and conclusion — 86

6 International trade and European domination, 1875–1914 — 90
The changing pattern of comparative advantage in manufactures — 91
The temperate zone primary product exporters — 97
Tropical trade and the less developed countries — 102
The commodity terms of trade — 106
Trade and colonisation — 108
The political response to depression — 113
Summary and conclusion — 118

7 Capital movements, 1875–1914 — 120
The supply and demand for capital — 120
The capital exporters — 123
The political impact of foreign investment — 126
The gains from foreign investment — 133
Direct investment — 136
Summary and conclusion — 138

8 International migration, 1875–1914 — 140
Magnitudes and directions — 140
The causes of international migration — 143
The gains and losses from migration — 147
The political economy of international migration — 148
Summary and conclusion — 152

9 The heyday of the international gold standard, 1875–1914 — 154
Money and the international price level — 155

	The silver standard economies	160
	The working of the gold standard	161
	The gold standard and public finance	164
	Fluctuations in economic activity	166
	Monetary policy under the gold standard	169
	Summary and conclusion	173
10	**International trade in the twilight of liberal capitalism**	**175**
	The First World War and European economic relations	176
	The decline of international trade	180
	Economic performance in the inter-war years	182
	Trade in primary products	187
	Commodity control schemes	192
	Trade in manufactures	194
	Commercial policy and the Depression of 1929	198
	Summary and conclusion	204
11	**The disintegration of the gold standard**	**208**
	Floating exchange rates and post-war inflation	209
	Floating exchange rates and sterling's return to gold	212
	The reconstructed gold standard	215
	Liquidity and the onset of international depression	222
	The managed exchange rates of the 1930s	228
	Summary and conclusion	232
12	**The redirection of the international economy, 1939–53**	**235**
	International economic relations during the war	236
	The post-war international system	239
	The dollar gap	242
	The Marshall Plan and west European reconstruction	245
	The Cold War and the reconstruction of eastern Europe	248
	The pattern of trade and finance	251
	Summary and conclusion	255
13	**The new liberal trade order**	**258**
	The new technology: generation and international transfer	259
	Trade policy of the industrial world	268
	Trade policy of non-industrial countries	274
	Trade patterns and spheres of influence	278
	The terms of trade between primary and manufactured products	280
	Communist-system trade policy	281
	International labour mobility	283
	Welfare and economic growth	287
	Summary and conclusion	289

Contents

14 The Bretton Woods system and its transformation — 293
- The growth of international reserves and liquidity — 294
- Balance of payments adjustment under Bretton Woods — 299
- The end of the fixed exchange rate regime — 305
- The crash of 1974 — 307
- International private capital mobility — 311
- International official capital flows — 315
- International debt — 318
- Summary and conclusion — 319

15 The search for a new regime: the world economy of the 1980s — 323
- Trade and trade policy — 324
- Convergence and productivity slowdown — 330
- The non-convergers: Latin America and Africa — 332
- Catching-up in Asia — 333
- OPEC and oil — 335
- Centrally planned economies and the transition to the market — 338
- European economic integration — 340
- Monetary disorder — 346
- International resource policy — 351
- Summary and conclusion — 353

Postscript — 355

Glossary — 358

References — 365

Index — 396

Figures

Figure 1.1	World population in 1850	2
Figure 1.2	World trade in 1850	2
Figure 5.1	Telegraph routes to India in 1870	69
Figure 5.2	Siberian telegraph route to China and Japan in 1871	70
Figure 5.3	National price levels, 1850–75	73
Figure 6.1	Simplified world pattern of settlements in 1910	93
Figure 6.2	The convergence of British and US wheat and beef prices 1865–1913	98
Figure 6.3	World trade in 1913	105
Figure 6.4	The scramble for Africa	109
Figure 7.1	International capital flows as a percentage of gross national product, selected countries, 1870–1910	121
Figure 8.1	International migration, selected countries, 1870–1920 (in thousands)	142
Figure 9.1	National price levels, 1875–1914	156
Figure 9.2	Ratio of the price of gold to the price of silver, 1800–1914 (Friedman, 1990)	159
Figure 10.1	European frontiers after Versailles	179
Figure 10.2	Flow chart of Birnberg and Resnick's (1975) model of colonial development	189
Figure 13.1	Product composition of the trade of industrial and oil-importing developing countries	264
Figure 13.2	Direct contribution of exports and other factors to economic growth, 1960–70	288
Figure 14.1	Price levels under the Bretton Woods system and after	308
Figure 14.2	International debts rescheduled, 1979–82	320
Figure 15.1	Yen–dollar exchange rate, 1970–92	326
Figure 15.2	Oil prices and the world price level, 1970–92	337
Figure 15.3	Impact of the completion of the European internal market	341

Tables

Table 2.1	The British balance of payments in 1858	22
Table 2.2	The balance of payments of the United States in 1855 (financial year)	25
Table 2.3(a)	An estimate of the Indian balance of payments in 1858/9	27
Table 2.3(b)	Banerji's estimate of the Indian balance of payments in 1858/9	29
Table 3.1	The international spread of railways to 1875	33
Table 3.2	Labour productivity and trade in United States manufactures, 1860	38
Table 3.3	Population densities and wheat exports	42
Table 4.1	Intercontinental migration rates, 1851–80	55
Table 5.1	Specie points for the sterling–franc exchange rate	72
Table 5.2	The relative value of production and market price of gold and silver, 1801–80	73
Table 5.3	The growth of money, national income and prices in Britain, 1846–80	74
Table 5.4	Average annual exchange rates against sterling, 1845–59	77
Table 6.1	The pit price of coal in various locations, 1861–1901	93
Table 6.2	Changes in the volume of export of manufactures attributable to changes in the world market, in the pattern of trade and in market shares, 1899–1913	95
Table 6.3	European revealed comparative advantage, 1899 and 1913	97
Table 6.4	Growth in the volume of world trade, 1850–1913	104
Table 6.5	Total grade compared with colonial trade, for selected countries, 1892–6	110
Table 7.1	Foreign investment in China and Chinese trade	129
Table 7.2	Foreign investment in Japan and Japanese foreign trade	130
Table 7.3	Return on state bonds issued 1850–1914 or outstanding in 1850	134
Table 7.4	Railway loans and exports to China, 1898–1912	136
Table 8.1	International migration regimes: a first approximation	148
Table 9.1	Bondholders' interest on loans to various nations at market prices in 1889	173

Table 10.1	Annual growth rates of real gross national product per head, 1913–50	184
Table 10.2	Growth rates per annum in the dollar value of exports, 1913–37	185
Table 10.3	The composition of world exports, 1913–37	186
Table 10.4	Changes in the volume of exports of manufactures, 1913–37	195
Table 11.1	Wholesale price levels in 1928	218
Table 11.2	Cost of the French franc and sterling as a percentage of purchasing power parity cost in 1934	229
Table 12.1	Total net ERP aid after utilisation of drawing rights, as a percentage of 1949 GDP	247
Table 12.2	East–West trade as percentages of the trade of the developed West and of the East, 1938–53	249
Table 12.3	Changes in the volume of exports of manufactures, 1937–50	253
Table 12.4	World trade, 1937–55	253
Table 13.1	The network of world trade, 1963–78	263
Table 13.2	US net exports of manufactures by production characteristics, 1960–70	266
Table 13.3	Changes in the volume of exports of manufactures attributable to changes in the size of the world market, the pattern of world trade, and market shares, 1963–7	267
Table 13.4	Rates of protection of manufactured imports	275
Table 13.5	Approximate net international migration, 1960–70	284
Table 14.1	Percentage composition of international reserves, 1956–73	295
Table 14.2	Growth in world reserves and world trade, 1950–76	299
Table 14.3	The effect of the 1967 devaluation on the British balance of payments	301
Table 14.4	Percentage changes for the United States economy in some macroeconomic variables, 1929–31 and 1973–5	310
Table 14.5	The main recipients of British aid, 1974	317
Table 15.1	Population and merchandise trade by region in 1990	325
Table 15.2	Income transfer from an $11 per barrel fall in oil prices in 1986	338

Preface

Were international economic history not so instructive for contemporary policy and attitudes, the vast scope of the enterprise might justly deter authors and students. But national economies have been and are interdependent, so that one country's progress and policies cannot be understood in isolation from others'. Like the first edition this history of the world economy is a story of economic relations between nations. Trade flows, commercial and exchange rate policies, foreign investment and international migration not only have economic causes and effects but also political consequences, which often give them their main significance. Economic relations are equally likely to be subject to political causes of great interest. Although I try to take these political influences into account, the central focus is on the actions traditionally described as economic.

As in the first edition an objective is to facilitate the understanding of history by means of economic concepts, when they are relevant, in a non-technical way. The book is intended to interpret, to synthesise, and in some instances, to make more accessible, a substantial body of published research by a large number of scholars. The greater number of the theoretical concepts are outlined in the earlier chapters, where the historical context should encourage the reader's interest. Inevitably as the world economy changes new concepts, such as technology-gap and product-cycle theories of trade become relevant and are introduced when appropriate. Rather belatedly, I should like to thank Edward Elgar for encouraging the original project. In revising this edition I have taken advantage of suggestions by readers of the earlier version, especially Larry Neal. I am particularly grateful to Tim Hatton and Liam Halligan for their last-minute assistance, without in any way implicating them in the deficiencies of the final product.

Introduction

People on different sides of the world found their fortunes linked by mid-nineteenth century trade. A shock to one economy could quickly spread to others, as did the ending of the Crimean War, releasing Russian grain on to the world market. In 1857 grain prices dropped with the increased supply, and financial institutions that had lent on the security of high prices were in difficulty. Financial panic and collapse spread from New York to Liverpool, to London and Paris, to Hamburg, Oslo and Stockholm. From London the crisis spread to South America, South Africa and the Far East. Less than a decade later the ending of another war, this time in the United States, pulled down world cotton prices. Indian banks that had lent to expand cotton cultivation especially for Lancashire, were left with liabilities greater than assets and failed in large numbers during 1866.

Long before the mid-nineteenth century world markets had partly integrated some national or regional economies. Land prices in Bordeaux varied with foreign demands for wine and gold mining in Latin America raised European prices. Equally, foreign supplies and demands were considerably less pervasive in the 1860s than they were to become. By 1973 the Organisation of Petroleum Exporting Countries' quadrupling of the posted price of crude oil slowed the growth of the industrial countries and precipitated balance of payments crises for many poor oil importing economies. Transport and communications improvements from the mid-nineteenth century combined with state policies increasingly favourable to free markets, to accelerate the growth of international economic relations. In so doing the foundations of today's interdependent global economy began to be laid.

The task of this book is to explain the pattern of relations between national economies, to assess their consequences, and to evaluate the costs and benefits. Because the objective is to analyse and discriminate between possible explanations, not merely to tell a story, it is necessary to employ some theory as well and to appraise systematically a range of evidence, often statistical.

Some of the questions addressed are directly relevant to recent problems of economic policy. The appropriate choice of exchange rates at which economies should link to the European Exchange Rate Mechanism or to the US dollar, and the costs of setting a one-for-one exchange rate between the East and West

German marks on unification find analogies in debates over national returns to the gold standard during the 1920s. The apparent success of the United States' Marshall Aid of the 1940s is sometimes invoked as an argument for western aid to the states of the former Soviet empire. For many decades poorer countries after 1945 often justified their economic policies with claims about previous exploitation by colonial powers and companies and by allegations that international specialisation had stunted their economic development. Free trade was and is an appealing policy for those who believe that it fuelled Britain's economic precocity in the nineteenth century. Assessments of these positions should be possible on the basis of evidence and tested theories, rather than emotionally or politically convenient assertion.

At the most abstract level economic relations generally involve voluntary exchange (although theft and coercion have economic aspects as well). A swap between only two parties is a bilateral exchange which is comparatively difficult to negotiate because the deal requires each party to want what the other is offering and to agree the terms. Easier to arrange are exchanges intermediated by money or financial instruments, where some commodity such as gold, and/or a commitment to pay in the future, is accepted, instead of barter. Such transactions include exchange of labour services for wages through migration, borrowing or lending money, as well as buying or selling goods and services for consumption. Exchange is greatly facilitated by organised markets in which prices and other terms of exchange are quoted, (and by traders or brokers specialising in particular goods or services). There must be some agreement or convention as to the commodity or currency in which prices are quoted. Definitions of what is bought and sold, both legal and physical measures, are also necessary. Within the territories settled by Europeans, by the mid-nineteenth century the state assumed responsibility for such definitions and for the enforcement of rights. In much of the rest of the world, especially where land was concerned, the European notion of 'fee simple', that land ownership could be passed by will to anybody, was foreign. Property rights in people persisted even in mid-nineteenth century Europe; within the Russian empire, serfs were still sold with land. In the United States, southern agriculture was organised principally around slaves of African descent.

The nation was a relatively novel concept in the nineteenth century. Empires continued to be widespread, though they were a diverse collection, ranging from the British empire with democracy for white males, more liberal in practice than most nation states, and allowing self-government for many territories, to the despotic Chinese and Ottoman empires. In between the Austro-Hungarian empire by the end of the nineteenth century provided a customs union that maintained peace and order among a variety of often antagonistic peoples sufficiently for them to experience sustained economic growth. The term international economic relations must therefore be only loosely applied. What matters is that the character of trans-border economic activities distinctively depend on the nature of the different political jurisdictions. Potentially traders must cope with two currencies, two legal systems and two sets of weights and measures, when they are clearly specified. But a great deal of political friction also arises when for one or both parties rights are not clearly defined. This is as true of the former Soviet empire in the 1990s as in Africa of the 1880s.

Both for the individual and the nation, exchange, or equivalently trade, gives rise to specialisation. If a person or an economy is able to sell say, silk, profitably then they will grow and spin more than is needed for their own consumption and with the earnings buy more of other goods and services that they would otherwise have provided for themselves. In so doing they are able to consume more at the same time as rendering themselves more vulnerable to falls in the demand for silk (an acute problem for the Japanese after 1929). Their pattern of specialisation may also influence the future evolution of the economy for at least two reasons. First industries acquire and require specific and distinctive institutions that are likely to become powerful if the industry has grown large supplying foreign as well as domestic markets. These institutions then gain both the capacity and the motive to block certain future developments that might have benefited the national economy. That has been alleged of the atomised British cotton industry in the second and third quarters of the twentieth century. Closely related is the second source of influence over future development, that of learning by doing. Writing in Germany during the 1840s, overshadowed by British industry, Friedrich List was particularly conscious of the role of experience in creating competitiveness, and so favoured tariff protection for 'infant industries'.

Nineteenth-century (and later) economies may be classified according to their natural and human endowments:

1. First there are those relatively high population density industrialised or industrialising countries – much of western Europe, north eastern United States, and at the very end of the nineteenth century, Japan. In the twentieth century industrialisation was no longer restricted to high population density countries or even to temperate zones. South east Asia, Brazil, and Mexico among others, joined the group.
2. A second category is the temperate zone regions of recent European settlement; the mid United States, Canada, New Zealand, Australia, South Africa, Argentina, Chile, Uruguay and possibly parts of Siberia.
3. Third are the tropical (broadly defined) economies. They usually generate only low incomes and typically were colonised in the nineteenth century, formally or informally. Oil slightly upsets the classification in the twentieth century. The Middle East anyway fits only awkwardly into the third category.

Different combinations of natural and human endowments characteristic of these groupings are associated with the predominance of different activities, which typically deploy very different technologies:

1. Natural resource intensive agriculture provided major bases for trade and specialisation in category 2 economies.

2. Extractive industries, such as coal mining, rarely dominate a national economy to the same extent as agriculture, though they may strongly influence it. Oil in Saudi Arabia, Venezuela and elsewhere is the exception not matched by the impact of nitrate mining in nineteenth-century Chile, gold from the 1880s in South Africa, and copper from the interwar years in the Belgian Congo (Zaire). Indeed the extractive industries have often been accused of promoting harmful 'enclave' development.
3. Manufacturing is widely billed as the source of modern economic growth. The nature of trade in manufactures changes in the twentieth century, with consumer durables and narrow specialisation of capital equipment. The western world after 1945 increasingly traded manufactures for manufactures. National endowments of formally trained manpower seemed to become more important as determinants of competitive success after this shift.
4. Services were almost always exchanged between advanced economies in the nineteenth century. But in the later twentieth century, this diverse group, that included insurance, banking and financial services, clearly faced income elastic demands. Faster productivity growth in manufacturing allowed more resources to be released for their supply.

As well as endowments and technology, institutions, doctrines, policies and powers determine international economic relations. An international economic regime or order can be defined as the rules and norms which provide the framework in which international economic relations take place. The regime influences the policies chosen by national or imperial powers through the international institutions, conventions and beliefs. Regimes, or orders, themselves are influenced by dominant states, by technology and the distribution of endowments. The combination of international economic policies, production patterns, institutions, and predominant beliefs constitute epochs or eras. As a simplification it can be helpful sometimes to classify predominant beliefs as combinations of, or selections from, liberalism, nationalism and socialism.

Five distinctive epochs structure the present study of international economic relations. The first from around the middle of the nineteenth century is the era of free trade and the spread of international liberalism, strongly influenced by the British example. This was a period of the 'cosmopolitan bourgeoisie', in which families such as the Rothschilds and the Siemens could pursue their business in the different states where they had settled, assimilating into the life of the host nation without losing their family identification. The nation state did not dominate economic life; neither Germany nor Italy existed at the beginning of the epoch.

In the last quarter of the century increasing market integration brought a political response. Although economic liberalism was the prevailing doctrine in the western world, the state became more interventionist both at home and in relations with other powers. Tariff barriers increased, a new wave of imperialism broke, particulary in Africa, and industrialisation, especially in Germany and the

United States, eroded British pre-eminence in the world economy. Some formal co-operative institutions emerged however. Already in 1869 an International Office of Telegraphy was set up at Berne, the forerunner of the ITU. The Universal Postal Union was established in 1874 and in 1883 an agreement was reached on trade marks and patents. European subsidies to beet sugar exporters and tariffs on imported sugar were ended at the 1902 Brussels convention. Although no agreement could be reached on a common coinage, the gold standard emerged as the international monetary system more or less spontaneously.

The First World War ended the great nineteenth-century expansion of the world economy in the slaughter of millions of Europeans. A third epoch began when fighting ceased, not only because old empires had crumbled and new independent states emerged but because of the international hostility, mistrust and ineffective co-operation that marked the period. At first some co-operative economic institutions were created. The new League of Nations contributed to post-war reconstruction and in 1920 IATA was formed to regulate international air traffic. But when in 1931 an international financial crisis struck in the depths of an already severe world depression, the ramshackle international economic framework collapsed, and political extremism fed on the rising mass unemployment. The Bank for International Settlements, founded in 1929 to manage and depoliticise German war reparations payments, explicitly set itself the goal of achieving co-operation between central banks, but proved impotent in the face of the disaster.

World war once more marked the end of one era and the beginning of another. A new set of economic institutions, the IMF, the World Bank and the GATT, governing exchange rates, investment in poorer countries and international trade, effectively established by the United States underwrote the great post-war boom. Other economies, most spectacularly Japan, were empowered to catch up the US technological lead because the new liberal order maintained the conditions in which transfers of ideas, products, personnel and capital could be maintained. Instruments of these transfers, huge multinational companies, with turnovers greater than some sovereign states, replaced family enterprises.

The fifth epoch is generally dated to the oil crisis and embargo of 1973/4. World economic growth slowed and international debt held by many newly industrialising countries accumulated in an inflationary environment. As the US tightened monetary policy, while increasing defence spending, real interest rates rose and by 1982 for a considerable number of states these debts became too expensive to service. A decade of retrenchment set in during which Latin America showed no growth and African output per head actually contracted. The inefficiencies of central planning proved too great for the Soviet empire, which disintegrated in 1989. Newly independent states were obliged both to learn how to construct and manage domestic market economies and to conduct external economic relations as well.

The western market economies desultorily experimented with policy co-ordination as an alternative to the adjustable peg exchange rate regime that had already broken down by the first oil crisis. Regional arrangements, such as the European Exchange-Rate Mechanism and the North American Free Trade Area, testified to the continued hankering for a more stable basis for international economic relations. Whatever the shortcomings of the international economic regime that evolved after 1972/3, it provided a tolerable if not ideal structure. Ordinary people could prosper as their national economies participated in international exchange and specialisation, given appropriate domestic policies. Unlike international power politics, the world economy allowed all nations these chances regardless of their size or political power, as the 'four little dragons', Hong Kong, Singapore, Taiwan and South Korea, amply demonstrated.

1 International economic relations in the middle of the nineteenth century

Karl Marx and Friedrich Engels flamboyantly described the international impact of Western trade and technology in 1847:

> by the rapid improvement of all instruments of production, by the immensely facilitated means of communication, [they] draw all, even the most barbarian, nations into civilisation. The cheap prices of ... commodities are the heavy artillery with which [they] batter down all Chinese walls, with which [they] force the barbarians' intensely obstinate hatred of foreigners to capitulate. (1893, p. 47)

Novel machinery, including mechanical reapers, steam-driven mills and blast furnaces, destroyed traditional occupations and created new. Domestic economic transformations were reflected in international relations, with increased specialisation in production and trade between nations. Steam-ships, railways and the electric telegraph stimulated the international movement of capital, labour and technology. Bursts of investment in the new techniques were followed periodically by crises which contracted exports and imports. As foreign sales fell, domestic businesses shed workers, tried to cut costs and sometimes defaulted. In turn, domestic demand for foreign goods declined, transmitting depression to enterprises in other economies.

These themes form the core of international economic relations. This and subsequent chapters trace out the ideas, institutions and economic conditions that encouraged the expanding international division of labour, and also explore the political and economic effects. The military metaphors in Marx and Engels' quotation above turn out to be particularly appropriate in many cases.

Viewed from western Europe, Europe dominated world trade and income, but not world population (see Figures 1.1 and 1.2). Europe accounted for somewhere between one-sixth and one-quarter of the world's population, but for almost 70 per cent of world trade at the mid-century.[1] One 'guesstimate' gives north-western Europe almost one-third of world income earned by little over one-tenth

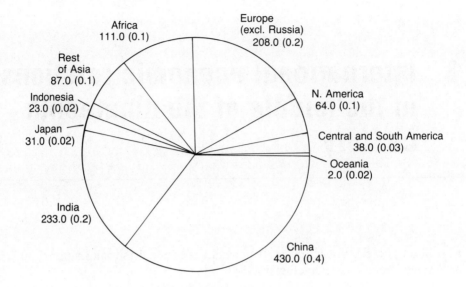

Figure 1.1 World population in 1850 (millions (percentages))

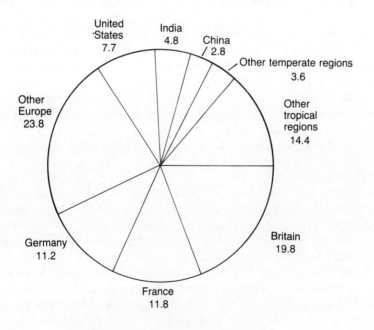

Figure 1.2 World trade in 1850 (%)

of the people on the earth.[2] Within Europe, Britain achieved the highest national income per head with one-third of total fixed steam power installed in the world's factories. Britain also disproportionately engaged in international trade. Perhaps one-fifth of this trade by European measurements originated in, was destined for, or passed through Britain, even though those islands were occupied by less than 2 per cent of the world's population in the middle of the nineteenth century. French trade was the second largest national total, but amounted to little more than half of Britain's.

The present chapter begins the task of explaining this remarkable European economic pre-eminence. The international economy is introduced by discussing the political transformations triggered by international trade at the mid-century and earlier. Then the chapter analyses the economic impact of trade and the political preconditions. Was success in trading the result of successful violence and coercion? Did trade create the economic strength of western Europe by 1850? Or was it a response? After considering contemporary doctrines of the gains from economic freedom which stimulated trade, the focus shifts to historical experience as a means of appraising the strengths and weaknesses of these ideas.

The political impact of mid-Victorian trade

In most of the world, governments claimed a monopoly of physical coercion, and therefore of violence, within their borders. That conferred upon their legal and police systems the ability to enforce and therefore effectively to adjudicate decisions to resolve disputes and enforce rights. One source of political change induced by international economic relations arose from disputes *between* nationals or residents of different jurisdictions, for how they should be resolved was less clear.

A second source was more purely economic. The power of government, the ability to maintain a monopoly of violence or physical coercion within a given area, depends on control of sufficient resources, usually tax revenues, to prevent unauthorised coercion. Where power is disputed, either within the region or from without, more resources are needed. Typically, the group that can command greater economic power will ultimately win, unless the less well-endowed coalition is much more effective. International trade changes the value of resources and so potentially alters the distribution of power within a region.

A third source of political change was the determination of one state to extend its borders so as to gain economic advantage deriving from new access to world markets. The international economy may increase the profitability of imperialism. Extensions of government authority to peoples, natural resources or tax revenues in other regions long antedated the world economy, however, and therefore should not necessarily be attributed to it exclusively.

The European, and especially British, rise to dominate the international economy went hand in hand with the extension of political control or influence.

The exact nature of the relationship is controversial, some maintaining political action caused economic pre-eminence, and others, the view that will be argued here, that economic relations stimulated political influence.[3] Even so, not all European political acquisitions of this period were primarily by-products of international trade. Non-economic motives were fundamental in the creation of the nineteenth-century French empire in particular. Algeria was annexed in 1830 initially because the Restoration monarchy wanted a spectacular political success (Fieldhouse, 1982, pp. 304–5; Doyle, 1986, pp. 307–8). Napoleon III's imperial aspirations were key contributors to France's initial foray into Indo-Chinese colonies when a minor crisis arose over missionaries in the early 1860s. By contrast, British acquisitions and exercise of influence abroad were much more a consequence, direct or indirect, of commercial activity.

Gallagher and Robinson (1953) asserted that mid-nineteenth-century British merchants preferred informal control, which allowed much less taxation and interference than if a formal empire was imposed. But if informal control was impossible, then direct British rule was imposed to secure trade. Between 1841 and 1851 New Zealand, the Gold Coast, Labuan, Natal, the Punjab, Sind and Hong Kong were occupied or annexed. In the twenty years after the mid-century, Berar, Oudh, Lower Burma, Kowloon, Lagos, the hinterland of Sierra Leone, Basutoland, Griqualand, Transvaal, Queensland and British Columbia were subject to the same process.

Despite the length of this list, empire building did not begin in the middle of the nineteenth century. Spain lost most of its huge Latin American empire, dating from the sixteenth century, in a series of uprisings after the Napoleonic Wars. The thirteen British North American colonies had been founded and lost by then. Indeed, they were engaged in commercial expansion overseas of their own. Most of the Indian sub-continent had fallen under British rule by 1818 as an indirect consequence of trade with Europe and because of wars between the French and British trading companies, on orders from their respective home countries during European hostilities. The impact of Europe on India is both the most striking example of the 'imperialism of trade' and in many respects the archetype of a large proportion of later nineteenth-century European acquisitions. Hence the process by which India came under British rule is worth elaborating briefly.

During the eighteenth century, the old Moghul Empire at Delhi collapsed, opening the way both for a Hindu revival and for many adventurers who won their thrones by force. In order to obtain security and the stable conditions which were necessary for their trade to make a profit, the British East India Company, and sometimes its servants independently, intrigued with native princes (each of the three Company stations at Bombay, Madras and Calcutta separately ran its own foreign policies). The Company was drawn into war and the government of Indian provinces, which generated the revenue necessary to support the Company's establishment (Spear, 1978; Majumdar, *et al.* 1958; Jones, 1987; Raychaudhuri, 1983).

Continued British success in the eighteenth-century wars with such a vast and ancient civilisation is more problematic than that the wars should have been fought. Sea power, built up over centuries of long-distance trade, was a necessary but not sufficient condition, for eventually land battles had to be won. Superior weapons cannot explain the British victories because military technology was quite easily imported from Europe. Under Ranjit Singh (died 1839), the Sikhs did just this, building up a regular army that included European officers and employed French and Italian generals to train the troops. The Sikh infantry were as effective as any available to the British in India, and the Sikh artillery was superior. Yet the Sikhs were defeated, and the Punjab annexed (by Lord Dalhousie on his own initiative in 1849), because of the lack of an orderly succession and the absence of civil discipline after Singh's death. The British ultimately won India because of the habitual loyalty of their military and civil officers and because, being interested primarily in trade, they did not usually threaten the social and religious objectives of Indian society – as had the Portuguese.

The Indian mutiny and revolt of 1857 shows the limits of this generalisation. The use of pig and cow fat in Enfield rifle cartridges (which troops were obliged to bite) succeeded in offending the beliefs of both Muslim and Hindu soldiers. Progressive territorial annexation, of which the latest acquisition had been Oudh, reduced the area in which the old ruling class could exercise their prerogatives and benefit from political office. Yet a full-scale revolt did not occur, otherwise the British could not have survived. Brutally suppressed by the Moghuls, the Sikhs had nothing to gain from a restoration of the Moghul Empire. By and large the Hindus also preferred to be ruled by the British than by the Muslims, and similarly the Muslims preferred the British to the Hindus. The British most of the time supplied more acceptable law and order than the perceived alternatives.

A relatively equitable and efficient public administration was costly, however, as is suggested by the British government's continuing concern with the finances of the East India Company until it was discredited by the Mutiny. Company profits from the Indian trade were barely adequate. Annexations of additional territories were therefore not regarded favourably by a government that would not receive the profits but might bear the costs. Where there was a more or less effective central government, the 'imperialism of trade' led to concessions rather than acquisitions.

Mid-century European trade with China followed this route. The Imperial Chinese government was unwilling to concede extra-territorial rights for resident European traders, even though it allowed restricted commercial contacts. Chinese attempts to impose their laws on foreign nationals could easily explode into a war in which superior European military technology and organisation would be brought to bear (Welsh, 1993).

The proximate cause of British military and naval intervention in China from 1839 to 1842 was Indian opium exports. China banned opium smoking in 1729, and in 1800 imports of opium were forbidden altogether (Bourne, 1970, pp. 44–5, 81–3; Fieldhouse, 1973, pp. 210–23; Wakeman, 1980; Fairbank, 1980; Hurd,

1967; Wong, 1974). By the early 1830s Indian trade with China therefore depended on the ability of British merchants to smuggle opium, which was usually no great problem until 1839, when the law was enforced at last. The Chinese government allowed foreigners only to trade in Canton, too far for convenience from the main markets in central China. Lancashire manufacturers, shipowners and advocates of international free trade therefore exerted strong pressure on the British government to negotiate better conditions for British trade in China, along the lines of commercial treaties negotiated with the states of Europe and America in this period. In 1839 the Chinese insisted that someone (not necessarily the culprit) should be handed over to be executed for the murder of a Chinese at Kowloon. The Chinese also ordered foreign ships which had left Canton to avoid confiscation of their opium, and which had taken refuge at (Portuguese) Macao and then Hong Kong, either to return to Canton or to leave Chinese waters. When these demands were ignored, Chinese warjunks fired on two British naval vessels, who retaliated by sinking four junks. The resulting fighting ended with the Treaty of Nanking in 1842, which specified the payment of an indemnity, the opening of five ports to British traders, and the ceding of Hong Kong in perpetuity. The trade concessions were extended to the United States and France in 1844.

In 1856 the boarding of an allegedly British-registered ship at Canton and the removal of several of its crew triggered a second war in which Anglo-French forces captured Canton and eventually occupied Peking. The new treaty opened eleven additional treaty ports and the Yangtse river to foreign trade, permitted permanent diplomatic missions at Peking, allowed Europeans to travel throughout China and conceded full toleration of Christianity. Unlike in India, direct rule was not attempted because there was already a government which could maintain law and order, or could be helped to do so in the case of the Taiping Rebellion.

The political impact of western trade on Japan stands at the opposite end of the spectrum from India, leaving China as an intermediate case. Japan possessed a strong sense of political unity and military discipline. It was therefore able to absorb rapidly what the West had to offer and, before the century was out, had itself embarked on a policy of colonial acquisition. But like China, the opening of Japan was forced by the West, although Japanese seclusion had been more total and the process began later. American whalers frequently sailed into Japanese waters in the 1850s and after such long voyages hoped to be granted docking rights. For a time the Shogun (the hereditary commander-in-chief and virtual ruler until 1868) was able to evade the demands from the United States government for permission to shelter, or to trade. But in 1854 Commodore Perry of the US Navy extracted an agreement which meant in effect the end of Japan's seclusion (Allen, 1981, pp. 22–3).[4] This intervention triggered a political turning-point in the Meiji Restoration of 1868, which speeded up a nationalistic drive to modernisation. As a much resented component of the 'unequal' agreement extracted from Japan in 1857, its tariffs on imports were not to exceed 5 per cent for forty years from 1858.

More than Japan's, Thai adjustment to the world economy was initiated, and controlled, by established domestic political authority. Most of the political friction experienced by India, China and even Japan was thus avoided. Siam (now Thailand) in 1851 gained an absolute monarch sympathetic to western and British economic achievements with the ascent of King Rama IV (or Mongkut) to the throne (Ingram, 1971; Feeny, 1989).[5] He combined nominal royal trade and land monopolies with the Bowring Treaty of 1855, giving the British extra-territorial status and restricting export and import duties to 3 per cent. Rice exports, primarily to China and Malaya, grew rapidly (4.4 per cent per annum by volume and 5.6 per cent by value between 1864 and 1910). Living standards probably began a modest but sustained advance.[6] And slavery, which may have included one-quarter or more of the population around 1850, was gradually eliminated partly through moral pressure from the West and partly through the self-interest of the King.

Elsewhere western humanitarian motives (or according to another view, 'moral imperialism') combined with trade to extend European control, as did the British attempt to abolish the international trade in African slaves. Kosoko, the ruler of Lagos, refused to co-operate with the British Consul by ending the export of slaves (Hopkins, 1980; R.S. Smith, 1979). The British navy therefore bombarded the town in 1851 and installed in power Kosoko's more pliant rival, Akitoye. The ending of the external slave trade greatly reduced the revenue available to the rulers of Lagos, and low export duties on the nascent palm oil trade were not an adequate compensation. This trade was not helped by the unwillingness of the dependent rulers after 1851 to confer alienable property rights in the land of Lagos. Foreign and domestic merchants therefore agitated for reform. Matters came to a head when the arrival of French warships signalled a possible pre-emptive French annexation. Supplementary support was a British desire to find alternative sources of cotton supply to the United States. All these elements encouraged the British in 1861 to persuade Akitoye's son and successor to cede his territory in exchange for a pension to compensate him for his lost palm oil trade revenues. In Lagos, then, European economic interests became increasingly important after the first humanitarian interventions, and they eventually prompted more political control, a pattern similar to that in China and India. In Japan and Thailand, the process was limited by stronger government and an awareness of what had happened in China and India.

Trade based on voluntary exchange

By the middle of the nineteenth century, in those countries where beliefs in the advantages of international trade were most strongly held, governments enforced well-defined property rights and laws for exchanging them. Under these conditions, and as long as the distribution of income and wealth within the country was acceptable, there was considerable justification for the belief of free-traders and economic liberals that allowing people to buy and sell freely would maximise national income and welfare. Liberals defined welfare as the sum of the personal fulfilments of individuals. Liberals generally assumed that, if people were allowed to do what they wanted, especially where buying and (usually) selling were concerned, then they would naturally make themselves as well off as possible. Individual pursuit of self-interest, constrained only by other people's fulfilment and the scarcity of resources, was necessarily the way to maximise society's well-being.

These doctrines were by no means universally accepted then, or now. Governments generally favour doctrines that bolster their authority rather than undermine it. Identification of the state with the nation or society is particularly helpful. And individualist ideologies must actually be predicated upon other values, such as respect for the truth and for others, which themselves are generally taken for granted. An appeal of liberalism to many merchants and others in mercantile countries was that, unlike nationalism, it offered a justification for the forcible opening of foreign countries to trade, and the dependence of Indian exports on the opium trade with China. In his *Essay on Liberty*, John Stuart Mill who had worked for the British East India Company for thirty-five years, becoming chief administrator in 1856, explicitly referred to the Chinese ban on the importation of Indian opium. This, he wrote, was objectionable not as an infringement upon the liberty of the producer or seller, for that was legitimate, but upon the liberty of the Chinese buyer:

> He who lets the world, or his own portion of it, choose his plan of life for him has no need of any other faculty than the ape-like one of imitation ... It is possible that he might be guided in some good path and kept out of harm's way ... But what will be his comparative worth as a human being? It really is of importance, not only what men do, but also what manner of men they are that do it. (Mill, 1972, pp. 117, 151)

Since opium was the second-largest source of Indian state revenue, the convergence of Mill's doctrine and expediency was particularly fortunate for the East India Company. The extension of European political influence clearly allowed a greater volume of trade because of the minimal tariff policies imposed. According to the prevailing liberal ethic, the countries opened up to trade benefited, as well as the instigators of economic relations. Governments that declined to let their people participate were, in liberal eyes, tyrannical and unrepresentative. The whole world, liberals thought, would gain from being able to buy in the cheapest market and sell in the dearest. Not only would people become better-off in a material sense, but according to advocates such as Richard Cobden and John Stuart Mill, the intellectual and moral gains from trade were even greater. The diffusion of new ideas and opportunities, and the interlocking of national interests, were among the greatest benefits of increasing international trade, making war obsolete, they thought (Mill, 1871, pp. 122–3).[7]

The weaker economies of western Europe and the United States did not embrace the free-trade ethic of liberalism as wholeheartedly as did Britain. The ideas of Friedrich List, the German nationalist economist, who died in 1849, attracted a considerable following. List agreed with liberals that liberty was a fundamental cause of industrial progress, maintaining for instance that national success in navigation was directly proportional to the initiative and courage fostered by national conditions (List, 1856, pp. 181, 183). But List believed in the primacy of the nation over the individual. Protection against foreign competition was necessary for the industrial education of the nation, to integrate new ideas, rights, duties and institutions into national life. List was interested in maximising the attainment of goals other than the sum of individuals' self-interest. Germany, List asserted in 1846, deprived of an energetic commercial policy to restrict international specialisation, unlike the United States, became 'like a colony' in the face of the superior productive power of Britain.

Trade without policy independence

India was actually a colony and was unable to pursue an independent tariff policy. Under British rule during the first half of the nineteenth century, the composition of Indian trade was transformed in a manner that would probably not have been permitted had India been truly independent. Imports of cotton goods from Britain increased massively to become the largest single category of imports, while cotton piece goods, formerly a major export, dwindled into insignificance (Chaudhuri, 1971; Maddison, 1971). According to Marx, the plains of India were covered with the bleached bones of starved hand-loom weavers (cited in Macpherson, 1972). India was subject to a number of famines, but foreign competition was not necessarily the source of tragedy. The predominance of agriculture may have enabled some of the extensive manufacturing population of the countryside to continue spinning and weaving during the hot dry season of the agricultural year, albeit in diminished numbers (Morris, 1963, 1968; Desai, 1971; Twomey, 1983).[8] The real issue, however, is not so much what happened to the numbers of hand-loom weavers because of increasing international trade, but what happened to former hand-loom weavers.

Some probably increased their specialisation in agriculture, supplying exports of opium, raw cotton and rice. Others may have found employment working up imported cotton twist and yarn to make more satisfactory cloth than the old hand-loom textiles (the first Indian cotton mill opened in Bombay in 1851). An independent and probably protectionist Indian tariff policy (similar to that of the United States described in Chapter 3) would have reduced the incentive to introduce the new, mechanised textile industry and, by keeping workers in low-productivity hand-loom weaving, may well have lowered Indian production and consumption below what was actually achieved.[9] That the Indian economy averaged a growth of income per head of about 0.5 per cent per annum for the remainder of the nineteenth century, and that earlier in the mid-eighteenth century technology had long been unprogressive, suggests that the impact of nearly free trade was not disastrous and probably on balance was beneficial (Heston, 1983; Raychaudhuri, 1983).

India was the most important market for the British cotton textile exporters, but Lancashire never supplied more than a fraction of the Indian consumption of coarse cloth and, during the second half of the nineteenth century, India took little more than one-fifth of the total value of the exports of the industry (Farnie, 1979, pp. 79, 119). Thus it would be wrong to attribute Britain's pre-eminence in the cotton industry to its access to the colonial Indian market. The volume and composition of trade would have been different without British political supremacy, but such a change would not have made the British much worse off.

Although it was not a western colony, a low import tariff was imposed on China, similar to that of India. As in India, the gains in China to the instigators of the expansion of trade beyond the western world were disproportionately small compared with the political consequences. Free-traders expected that China would provide a massive market. Sheffield cutlery imports, they thought, would be greeted avidly as far superior to chopsticks, and consignments of pianos would be welcomed as means of increasing the marriageability of girls, as they were in Victorian England (Allen and Donnithorne, 1954, pp. 17–18). To the surprise of many British manufacturers the Chinese preferred their traditional ways, and for many years unsold cutlery, bent into a variety of ornamental shapes, was to be seen on display in Chinese shops. But if the gains to the West from the opening up of trade with China were small, the costs to the West had not been very great either, in contrast to the Indian trade, because no formal empire had been declared.

In a similar political position to China, Japan by contrast imported capital goods and foreign technicians after the Meiji Restoration, paid for with raw silk and tea exports (Sugiyama, 1988; Baba and Tatemoto, 1968). As the only nineteenth-century Asian economy to industrialise, the gains to Japan soon became obvious. As to the advantages for the Chinese and the Indians, perhaps the most significant imports that they took from the West were ideas. The massive Taiping Rebellion (1851–64) was inspired by Christian doctrines, and, unlike the rulers of China, the rebels welcomed foreign influences. The rulers of China were, however, sufficiently capable of ensuring their own survival to enlist the military aid of western governments in eventually putting down the rebellion after 20–30 million people had been killed (Hookham, 1972, pp. 277–86). A successful revolt may have taken China on a path closer to Japan's. The British occupation of India transplanted western ideas of nationalism and liberalism and, by providing a communications network, helped to weld the sub-continent into a nation that had not previously existed.

Trade without law and order

Influential historians and politicians have judged the debate about tariff policy or the spread of western ideas as of marginal importance. Mercantilists and many later writers thought that international trade was central to national economic growth and development (Coleman, 1969; Hecksher, 1955; O'Brien, 1988, 1989, 1991). They held a doctrine popular in France and Britain during the late seventeenth and eighteenth centuries which emphasised that national strength required bullion imports to create employment. Exports were necessary to obtain gold, but imports, especially basic foods, were dangerous, for they undermined self-sufficiency in times of war. Success in war also turned on bullion, to pay for naval stores, for mercenaries and for subsidies to foreign allies. Not only did foreign trade offer the lure of gold, and supply the prerequisites for naval security, but it also generated government revenue from taxes on imports (those that were not smuggled). Success in war was vital to capture and defend trade.

The beneficial impact of trade, they asserted, was entirely one-sided. International economic relations allowed the rise of western Europe to its powerful nineteenth-century position by despoiling and correspondingly reducing the income of the rest of the world. Physical violence, confiscation and theft in Asia, Africa and America by the superior European military and political organisations caused divergent paths of economic development in the four centuries before 1850 (Barrett Brown, 1974, ch. 4; Wallerstein, 1979). The employment of capital acquired in world markets conferred a monopoly which, allegedly, was used to exploit foreign consumers by charging high prices, and to exploit foreign producers by buying at low prices. So Great Britain's 'industrial monopoly' allegedly distorted Chilean economic development in the nineteenth century. This monopoly furthered Britain's development under free trade, but forced Chile to specialise in the production of primary products supposedly unconducive to economic development (Frank, 1967, p. 37). Mercantilists maintained that foreign demand pulled up exports without any change in transport costs or tariffs. Where smaller economies supply larger ones, the pattern may be common. In the eighteenth century outside Europe, only the Americas were an important industrial export market. African imports were restricted to firearms and linen, while Asia was impenetrable.

An alternative view to mercantilism, favoured here, is that growing economic power at home created the tax base enabling British governments to pursue mercantilist policies successfully. Trade, as well as victory in the imperial struggle, was a reflection of the productive capacity of the domestic economy, which could regularly pay and equip warships, troops and foreign allies, while at the same time providing goods wanted by the rest of the world. Without the same economic support, French governments were handicapped over the long term. Causation runs in the opposite direction according to this second hypothesis (Kennedy, 1988).

The slave trade was probably the most harmful manner in which the international economy impinged upon lower-income economies. Britain declared the international trade in slaves illegal in 1807. It maintained a squadron off the African coasts to intercept slave ships, and paid foreign powers for putting down slave trading, which, with other anti-slaving expenditures, amounted to £29 million between 1808 and 1865. Even so, the trade continued because the strength of western demand for slave-produced goods, in particular for cotton, sugar, and coffee, guaranteed its profitability (Sheridan, 1976). From the nominal abolition of the international trade, Brazil until 1850, and Cuba until the 1860s, imported the largest number of slaves annually in their histories (Klein, 1978, p. 18). However, the slave trade was certainly less than it would have been in the absence of suppression policies. One estimate was that slave exports from Africa would have been 820,000 greater without such policies – 54 per cent of the nearly 1.5 million slaves who were actually imported to Cuba and Brazil between 1821 and 1865 (Phillip Leveen, 1975).

The United States also prohibited the trade in 1807. By 1821 the institution of slavery had been ended north of the Mason–Dixon line, although the legality of slavery in the newly admitted states to the Union remained a bone of contention until the Civil War. Slaves continued to be smuggled into the cotton-growing, slave-based economies of the Southern states and, as the price of slaves rose in the two decades before the Civil War, slave smuggling increased (Mannix and Cowley, 1976, ch. 12). Probably not less than 250,000 slaves were brought in from abroad between 1800 and 1860.

The costs to the African economies were first, those of a country experiencing massive emigration, the use of resources to support and educate young people without a corresponding return on that investment during the productive years; and second, and probably much more important, a lack of security that impeded economic development because of slave raiding parties. Some states, such as Dahomey, the African Sparta, devoted a large proportion of their resources to capturing slaves. The city states of Yorubaland, frequently warring with each other, sold off their captives as a lucrative by-product. By contrast, the feudal kingdom of Buganda, away from the slave-trading zones, experienced substantial economic growth and trade (Rodney, 1972, ch. 4).[10]

As a result of the slavery to which the expansion of trade had given such a boost, the ultimate consumers of goods embodying slave-produced materials paid less than would have been required in a world where all exchanges were voluntary. These consumers throughout the world exploited the slaves, whereas those actually involved in the slave trade and slave plantations are properly described as thieves. Even after the slaves had been freed, the tropical products they supplied were still artificially cheap, because the ex-slaves usually had few alternative sources of employment (outside the British West Indies), and the majority probably would not have chosen to migrate voluntarily from Africa to work at the market wages in the plantations. Hence the substitution of resources induced by slavery to some extent also persisted after abolition. The cotton produced by free labour in India and the Middle East was less profitable and was therefore produced in smaller quantities, even after the American Civil War, than would have been the case if the slaves had never been transported across the Atlantic. Indirectly, because of the slave trade, India and the Middle East suffered a reduction in their opportunities to gain from international trade.

Despite the advantages to some, slavery was impermissible in liberal ideology, and in English Common Law after the Sommersett Case of 1772.[11] The French Revolution abolished slavery, although Napoleon brought it back. Even strong vested interests failed to prevent abolition in the British West Indies in 1834, where there was no evidence that slavery was unprofitable. Significantly, by the 1870s, the only major area of slave trading was between East Africa and the Arabian peninsula, under the control of Arabs who remained untouched by the doctrines of liberalism (Engerman, 1986; Beachey, 1976).[12]

If the slave trade was fundamental to the rise of western Europe, then the success of the abolitionists is truly remarkable. In fact the gains from the trade seem to have been too small a proportion of British investment or income to account for British economic pre-eminence, and if this was true of Britain, it was even more true of western Europe. There are two components necessary for a quick calculation of the gains. The first is the higher price of British or European imports relative to the price of exports that would have been paid in a world in which slavery had never been transplanted across the Atlantic. The second is the share of foreign trade in British or European national income. Multiplying a plausible range of guesses for these two magnitudes cannot yield a large figure for the gains from slavery, on any reasonable assumptions.[13]

In the wider context of all extra-European trade, the magnitudes were still not big enough even for all possible 'thefts', such as the extraction of bullion from Latin America, to have greatly raised the national incomes of the beneficiaries. In 1800 total extra-European trade was in value only around 4 per cent of the aggregate GNP of western Europe. The maximum likely impact of such trade for Britain was only 15 per cent of gross investment during the Industrial Revolution (O'Brien, 1982).

Market structure and the gains from trade

The slave trade was in any case atypical of international transactions by the 1850s. International economic relations in the main were conducted within the framework assumed by nineteenth-century liberals, and there were therefore gains to both parties despite continuous friction about the proper distribution of these gains. In Latin America and elsewhere, European manufacturers did not co-operate to maintain prices merely because so many of them held a common nationality. Instead, they competed and so did not earn large profits (although exclusive contracts and concessions were contentious in shipping).[14]

The guano trade is a case in point. Sea fowls' droppings had no value without access to the world's agricultural demand for natural fertilisers. This vital source of Peruvian government revenue in the 1850s therefore depended on the enterprise of Antony Gibbs & Sons. The division of the gains from this trade between the contractor and the owner of the resource was a matter of controversy both at the time and later, particularly because the Peruvian economy appeared to benefit so little (Albert, 1983, pp. 69–73). The returns to the contractor from the Peruvian birds' droppings, or guano, trade came from commissions on the costs of arranging the shipping of the guano on behalf of the guano owners, the Peruvian government, and from a commission on the total value of guano sales. The controversy centred on whether Gibbs, the contractor, charged too much or too little for the guano in order to increase its commission (Levin, 1960, pp. 68–71). There was no general tendency for prices to be reduced, and the inelasticity of guano supply removed any likelihood that Gibbs could increase its commission by reducing them. Successive Peruvian governments did well out of the trade, taking an average of 65 per cent of gross sales proceeds from the Gibbs contracts. Most of the remainder went to the shippers (Mathew, 1977).[15] Gibbs was, however, guilty of claiming unearned commissions, of excessive centralisation through London of British and European imports, and of heavy exporting in 1860–1 just before its contract was to expire for the last time. Gibbs certainly made ample profits, but although it undertook virtually all the practical managerial responsibilities of the trade, its commission was still only a fraction of the royalties taken by the government.

Competition between merchants prevented persistent monopoly profits in most spheres of international trade, and guano was no exception. Before Gibbs had entered the guano trade in 1842, the Peruvian government drew up contracts with a Peruvian capitalist and with a group backed by a Liverpool firm. Those who needed finance and the associated services under competition were obliged to pay a price sufficient to prevent it being used in the next most profitable use, while those with access to the funds had to charge a price low enough to prevent the customer going to a competitor. The Peruvian government could not itself provide services similar to those of Gibbs, and did not think it could buy them cheaper elsewhere. That the government probably did not use the proceeds from the trade wisely cannot be blamed on the British merchants, or on international economic relations generally.

One instance does not, of course, prove the general point that the distributions of the gains from trade were fair, although the observation of widespread competition between traders gives added force to it. Had trade been monopolistic then nineteenth-century British merchants should have accumulated massive fortunes. Until about 1880 more than half of all the really wealthy men in Britain were landowners and, during the first half of the nineteenth century, the non-landed wealthholders were a virtually insignificant percentage of the entire wealthy class (Rubinstein, 1981). It is true that non-landed fortunes were earned disproportionately in commerce and finance – by merchants, bankers, shipowners, merchant bankers, and stock and insurance agents and brokers – rather than in manufacturing or industry. But this wealth belonged mainly to merchant bankers and came almost exclusively from the organisation of foreign or government loans. Henry Gibbs (1819–1907), of guano fame, became only a half-millionaire, and that in the course of a long and successful career in merchant banking after he had given up guano. Any exploitation of monopoly positions in international trade by the largest trading nation cannot therefore have been substantial at the mid-century. Ironically, it was Peru that held a virtual world monopoly on nitrogenous fertilisers in the form of guano at the mid-century. Any judgement of the outcome of particular or general international economic relations must depend upon the perceived ends of economic activity. These are quite clear to nineteenth-century liberals and to the majority of Anglo-American economists in the twentieth century. Those who place a greater weight on state economic and political independence or power than on individual or family goals will often reach different conclusions. From the liberal economist's position, apart from the slave trade, and possibly some caveats stemming from imposed tariff policies, the assessment seems warranted that both parties usually gained and the distributions were fair.

Justice and fairness

They were fair in the sense that the procedure by which the distribution was brought about was fair: that is, competition prevented the use of bargaining power to raise market prices above competitive levels. But competitive prices and the distribution of purchasing power depend on resource ownership. Prices and incomes might have been unjust if a given distribution of rights or resources was unjust. The issue is whether, as a matter of historical fact, just principles were followed first, for the acquisition of those holdings, and second, for their transfer.[16] Market exchanges could still be exploitative, as in the case of transactions deriving from slave labour, if they were based on a violation of those principles. Exploitation may be defined as a relationship in which one party gains at the expense of another. The exploited group is therefore better off withdrawing from the relationship, whereas the exploiter will be worse off. What a party is entitled to take with them in withdrawing from the relationship is the key to whether it is exploitative. If the holdings, labour or land for instance, were stolen then the outcome of subsequent voluntary exchanges is not necessarily fair; they may entail exploiting someone. In many cases, the legitimacy of claims to resources or rights is questionable depending upon how far into the past the claim is to be tested. Only in the case of 'inalienable rights', such as to the products of one's own labour, can clear results be obtained. For nineteenth-century liberals, the 'state of nature' was competitive markets with the existing distribution of property rights that were acceptable in western Europe (Nozick, 1974, ch. 7; Roemer, 1982; Foreman-Peck, 1989).

Neither the procedural definition of fairness nor any particular rules need be accepted in making judgements about international economic relations. Justice and fairness could alternatively be defined by the characteristics of distributions (Rawls, 1971; Amdur, 1977; Cooper, 1977). A commonly accepted principle is a base position of equality in the allocation of resources and power, deviations from which require further justification. One additional principle might be that only inequalities are acceptable that make the poorer parties better off than they would be under a more equal distribution of power or resources. A practical difficulty is that the fairness of a distribution can be judged only in the light of the knowledge of what would happen if the distribution were different. How would India have developed without European intervention? This is not a question of value, but such counterfactuals are often tainted with wishful thinking, as well as with considerable uncertainty. In short, whatever doctrine of fairness or justice is chosen, general evaluations of international economic relations face severe difficulties.

Summary and conclusion

Economic development in lands inhabited primarily by western Europeans or their descendants disproportionately concentrated world trade, income and political power in the 1850s. Western trade with Asia and Africa was a catalyst for political change, or ultimately triggered direct western political control, then and earlier. The political pre-eminence of the West was due to the lower political and social cohesiveness of other societies at the time of contact with the West. This meant that jurisdictional disputes were likely to trigger international conflicts which they would lose. Thailand and Japan are the major exceptions that prove the rule.

Trade expanded partly because European and American political influence ensured low tariffs. Liberals at the time put forward good reasons why all trading nations should benefit from such a policy. In western economies less advanced than Britain, these arguments were less wholeheartedly embraced. In formal colonies such as India, and other regions, states and empires, such as China and Japan, low-tariff and open-door trade policies were imposed. The foreign trade structure of India was thereby radically changed. This transformation has frequently been represented as harmful, yet it must have provided a stimulus to the more productive use of labour and to the development of an advanced-technology cotton textile industry that was eventually sufficiently competitive to cut back British imports, even under British rule. Imposed tariff policies were not vital to maintain European exports, judging by the magnitude of Asian trade with Europe.

The slave trade, continuing in 1850 despite suppression policies, demonstrates a necessary condition for nineteenth-century optimism about the gains from trade to be warranted. The maintenance of basic rights to security of property and life and limb was essential but not widespread in the world at that time. In addition to the direct losses entailed by the mortality and deprivation of freedom intrinsic to the slave trade, the African economies suffered from the drain of prime human capital. Moreover, the lack of security and the direct destruction caused by slave raiding parties also hampered their economic development. Nonetheless, what statistics there are, when properly analysed, do not confirm the view that Africa's loss was the cause of Europe's economic supremacy, that the capital for European industrial growth was substantially supplied by the profits of the slave trade, or by other acts of theft perpetrated against the rest of the world.

Economic relations between Latin America and the more advanced economies were stormy, but because Latin America had developed institutions and organisations capable of coping with trade on a large scale, and because competition rather than monopoly was the rule, generally the mutual benefits expected by the contemporary liberals were achieved. An analysis of the guano trade provides an instance of these mutual gains, and the absence of large British fortunes earned from nineteenth-century foreign trade suggests that competition was prevalent.

Notes

1. The trade figures are from Lewis (1981). Mulhall (1899) probably overestimated the steampower in Britain in 1850 by two-thirds (see Musson, 1976). There has been a considerable range of estimates of European and world populations. The population figures for 1845 given by Mulhall (1899) are 245 million for Europe and 1,009 million for the world. Durand (1967) gives a lower European population of 208 million and a higher world population of 1,262 million. A.M. Carr-Saunders (1936, p. 42) offers a European population of 266 million and a world total of 1,171 million for 1850.
2. Zimmerman (1962) gives the 'guesstimate' for 1860. See also Maddison (1983).
3. Cain and Hopkins (1993, chapter 1) succinctly summarise some of the more prominent arguments and present their own for Britain.
4. Beasley (1955, p. 99) gives the text of the American President's letter to the Japanese Emperor.
5. King Mongkut expressed his satisfaction in 1858 at the opening of a branch of the (British) Borneo Company in Bangkok two years earlier. The company's representative appointed the school mistress to the royal children and advised the King in 1868 on how to maintain the stability of the kingdom in the event of the King's death while his 29 sons and 33 daughters were still minors (Jones, 1986, pp. 20, 206).
6. Deflated by rice prices, wages seem to have declined, because the price of rice rose relative to textiles, the principal import. However, the ratio of wages to rice prices is unlikely to be a good indicator of general Thai living standards.
7. Cain (1980) believes this element has been overstated because of the idealistic views of Cobden, the most prominent member of the Anti-Corn Law League. See also Cain (1979). Kindleberger (1975) concluded that both ideology and self-interest convinced the English that cosmopolitan as well as national interests would be served by free trade.
8. The cotton textile industry of the Ottoman Empire declined under the impact of foreign competition at the same time. Pamuk's estimates of jobs lost are considerably lower than Twomey's when adjusted for the differences in the populations of the two empires (Pamuk, 1986).

9. This scenario is entirely consistent with the view that Indian output per head suffered a massive fall between 1689 and 1813, and had not recovered former levels by the mid-century (Raychaudhuri, 1969).
10. Eltis (1987, 1991) maintains that the costs to Africa from the slave trade are generally overestimated.
11. Slaves were still traded illegally in the ports of Bristol and Liverpool for some years afterwards (Macinnes, 1934, pp. 143–5).
12. Slave imports to the Middle East in the 1870s may have averaged 6,000–8,000 a year. Two-thirds of the population of Zanzibar and Pemba were slaves in 1858.
13. R.P. Thomas and D.N. McCloskey use this method to calculate the total British gains from all foreign trade and obtain fairly small numbers (Floud and McCloskey, 1981). For an alternative view, see Williams (1964), Solow and Engerman (1987) and O'Brien and Engerman (1991).
14. On shipping, see for example Sturmey (1962) and Greenhill (1977). On the prevalence of competition, see Platt (1977, pp. 6–7).
15. The poor conditions of the labourers extracting the guano casts doubt on whether their share in the guano from the trade was fair. Although they were nominally free, they were treated like slaves.
16. This is the basis of 'entitlement' theories.

2 Economic links between Britain, India and the United States at the mid-century

From the conditions influencing the economic and political consequences of trade, and their distribution, we turn to the pattern of trade. International trade in the mid-nineteenth century was less complex than it was to become, with a great proportion comprising approximately balanced exchanges between economies. Transactions between Britain and the United States were typical except for their sheer magnitude. The United States sold raw cotton to Britain and bought cotton textiles. British demands for Chinese tea could not be satisfied so simply, for the Chinese peasant was impervious to what the West had to offer. India came to play a key role in a more complex and disreputable equilibrating of international exchanges. This chapter describes in detail how external economic relations of Britain, India and the United States were balanced. It explores the different patterns of specialisation and methods of finance of imports, the political repercussions of which were discussed in Chapter 1.

Measuring international transactions

The statements of international transactions of three very different but closely related economies of the mid-century, Great Britain, the United States and India, both provide a snapshot of a major portion of international economic relations of the time, and show the different ways in which economies participated in the burgeoning of world trade. The majority of the North American colonies had thrown off the restrictions imposed upon their trade, industry and agriculture by Britain in the previous century. The merchandise trade of these former colonies, the United States, was in the middle of the nineteenth century only about one-third of that of Britain. Yet this rapidly expanding and independent economy was the United Kingdom's most important trading partner and the principal destination for British emigrants and international investment. Reciprocally, Britain was America's largest trading partner. Second only to the United States in importance to British trade was India, which, despite a population eight times as large as Britain's, conducted merchandise (including bullion) trade of only one-sixth the size. India's international economic relations offer a valuable contrast to those of the other two countries. India was a poor country, both in natural resources and in average standards of living relative to those of western Europe or the New World. The United States was rich in natural resources, with a small population who were well-off compared with the rest of the world. Great Britain was one of the most densely populated countries, yet was relatively industrialised and rich.

Contemporaries could most easily judge these international differences from the statistics of exports and imports. These were among the most reliable national data available and had been collected for many years (which is not to say that they were perfectly accurate). Export and import statistics alone, however, do not give a complete description of how a nation solved the problem of what to buy abroad and how to pay for it. Historical research has remedied this deficiency by constructing complete balance of payments statements for some countries, admittedly with many figures subject to wide margins of error. The solutions adopted to the international economic problem are now therefore better understood. First, contemporary statistics did not measure trade in 'invisibles' – services such as shipping, insurance and tourism. Second, they did not measure other means of paying for foreign goods, such as the selling of national assets to foreigners, gifts or other transfers of money across national boundaries, which did not involve a corresponding current exchange of goods or services.

The calculation of any one of these items may take advantage of the principle that 'the balance of payments must balance'. If something has been bought from abroad and thereby entered as a debit (a minus) in the balance of payments statement, then the finance must have been available to complete the transaction. The means by which this finance was obtained appears in the statement as a credit (a plus). In this sense, the balance of payments is an 'identity', a relationship that is true by definition; any discrepancy between the debits and credits must be the result of errors of measurement. If errors are ignored and, say, all credit items and all but one debit category are known by direct measurement, the missing component can be inferred.

Foreign investment has proved one of the more difficult international transactions in the nineteenth century to estimate directly, so in some cases it has been computed by this method. Since the balance of payments must balance, the difference between the two sums of credits and debits, the 'residual', is attributed to foreign investment. If the discrepancy is negative, foreigners have been investing in the country and providing the foreign currency to buy more imports than would otherwise have been possible. A positive difference is taken to mean that the domestic economy has been investing abroad, and exports have been higher or imports lower than in the absence of such investment.

Although the balance of payments must balance, it does not have to be in equilibrium. An equilibrium exists if the international transactions that residents can undertake correspond with those they would like to carry out under the prevailing circumstances: in particular, given their limited incomes, productive capacity and prices. A state of disequilibrium occurs in the balance of payments if unplanned credits or debits are accumulated in the course of the year (or other accounting period). Unplanned transactions which take place solely to settle the balance and not because of trading or investment opportunities, or for reasons of foreign policy or altruism, are defined as 'accommodating', while planned transactions are called 'autonomous'. Hence a measure of the imbalance or disequilibrium in a balance of payments is the size of accommodating components.

In practice, it is difficult to determine under which heading to classify certain transactions. The 'residual method' cannot distinguish between short- and long-term investment, or between planned and unplanned short-term investment. But the distinction is vital for assessing whether a balance of payments is in equilibrium, and thus whether adjustment is needed in the domestic economy. One result of the incomplete balance of payments statistics available to the nineteenth century is that the nature and timing of the adjustment made to a balance of payments disequilibrium differed from responses in the second half of the twentieth cenury, simply because the information available to judge whether a disequilibrium existed was so much more inadequate. The classification of precious metal flows is particularly problematic in the nineteenth century. Most were probably autonomous rather than accommodating. Gold and silver were the ultimate means of settling imbalances of exports and imports between countries (accommodating transactions), but gold was also an important export commodity for the United States. Together with silver, gold was a major import commodity for India, where there was a strong demand for jewellery, coinage and hoarding (autonomous transactions). Britain intermediated these flows. The considerable magnitude of autonomous precious metal transactions probably eased the adjustment to balance of payments disequilibria, because the proportionate changes required by accommodating transactions were reduced.

The British balance of payments

The balance of British merchandise trade in 1858 (Table 2.1) shows a very substantial excess of imports over exports, amounting to 15 per cent of total imports. But this balance was more than covered by 'invisibles', by earnings from shipping, insurance and interest on foreign investments. The largest single import was raw cotton and the largest export was cotton manufactures. Thus Britain's impact on the international economy owed much to its supremacy in the cotton textile industry in 1858. The countries of origin of imports show the geographical pattern of this impact, with one-fifth of total British imports coming from the United States, the world's largest producer of raw cotton, supplying the world's largest manufacturer of cotton textiles. Even so the cotton industry employed

22 A History of the World Economy

Table 2.1 The British balance of payments in 1858

Imports by country of origin (%)		Merchandise imports (£m)		Merchandise exports (£m)		Exports by country of destination (%)	
USA	20.8	Raw cotton	30.107	Re-exports	23.174	India	12.4
India	9.0	Corn	20.152	Cotton		USA	11.3
France	8.0	Raw sugar	12.322	manufactures	33.422	Australia	8.3
Russia	7.0	Wool	8.970	Cotton yarn	9.579	Hanse Towns	8.3
China	4.0	Raw silk	5.880	Iron and steel	11.197	Others	59.7
Others	51.2	Tea	5.207	Wool and worsted			
		Timber	8.206	manufactures	12.742		
		Flax	3.020	Linen	5.868		
		Others	70.720	Machinery	3.600		
		Total	164.584	Others	40.201		
				Total	139.783		
		Balance of merchandise trade −24.801					
		Gold	22.793	Gold	12.567		
		Silver	6.700	Silver	7.061		
		Balance of bullion and specie flow −9.865					
		Balance of ship sales		+1.0			
		Balance of foreign trade and services		+15.2			
		Balance of insurance etc.		+7.6			
		Balance of shipping credits		+24.2			
		Balance of emigrants' funds		−1.0			
		Balance of tourism		−5.8			
		Balance of interest and dividends		+15.9			
		Balance of foreign investment		−22.4			

Sources: Imlah (1958); *Abstract of UK Statistics*.

only about 4 per cent of the British labour force and generated roughly 7 per cent of national income (Farnie, 1979, p. 24). India was the second greatest supplier of British imports, sending tea, raw cotton, rice, indigo and sugar; sugar and tea were by now well established in British diets and culture.

After the repeal of the Corn Laws in 1846, imports of corn by the 1850s rose to become second in value only to raw cotton. France in 1858 was the largest supplier of British wheat imports, with the United States (exporting wheat flour), Prussia and Russia close behind. Wheat sales were a major reason for France providing as much as 8 per cent of British imports, the third largest share, followed by Russia and, because of the tea trade, by China. Sugar was Britain's third largest import commodity by value, supplied by the British West Indies ('free' sugar) and also to some extend by Cuba and Brazil ('slave' sugar). India was Britain's largest market in 1858, although this position was usually taken by the United States. The United States was forced into second place by the slump of 1857, while Indian demand was beginning to recover from the ending of the Mutiny early in 1858. Britain's role as provider of the capital goods and intermediate products that it had innovated accounts for the second most important category of exports, iron and steel; one-third of this category consisted of railway products. Essentially Britain exported manufactured and semi-manufactured goods, whether they were made of cotton, wool, flax, iron or steel, and imported raw materials and food, with some high-quality consumer goods.

Bearing in mind that the entries in Table 2.1 show the most important traded goods and trading partners, the general impression is of a very diversified foreign trade structure; there was no great dependence on any one commodity or country, expect on the United States as a source of imports, and on cotton exports which accounted for 30 per cent of exports. In this respect, British trade differed from that of the contemporary United States and India, and from the trade of poor countries in the twentieth century. The size of the re-export trade also distinguished British trade from that of most other countries. Raw cotton dominated all other re-exported commodities, amounting to 17.6 per cent of the total value between 1854 and 1859.

Australia, where Britain had transported so many of its convicts, was Britain's third-largest export market.[1] It was no coincidence that Australia was also the main source of the massive gold inflows which, together with raw wool exports, provided the revenue to buy British goods. More gold went to France than came in from Australia in 1858 because of France's bimetallic standard, discussed in Chapter 5. The value of this gold was 2.5 times the visible trade deficit (the excess of visible imports over exports) with France, and equal to five-sixths of the total exports of gold bullion and specie.

The other components of the British balance of payments in 1858 can be estimated with much less accuracy, but there are nevertheless substantial sums of money involved. The largest items are those earnings which Britain received by virtue of its naval supremacy and its high level of financial development. Imlah (1958) estimated the net income from foreign trade and services as 5 per cent of the value of imports and exports. The intermediation of Gibbs in the guano trade is one instance of these service earnings. Insurance revenues were a less important contributor to the British balance of payments than foreign trade services, but nevertheless Imlah's estimate for 1858 suggests that this balance was more than double the value of machinery exports. Shipping earnings in 1858 were apparently very sizeable, being second only to exports of cotton manufactures as a source of foreign exchange.

The 114,000 British residents who emigrated in 1858, almost 60,000 of them bound for the United States, probably contributed a small negative item to the balance of payments because of the excess of the money they took with them over remittances to Britain from relatives and friends abroad.

The balances shown in the bottom two lines of Table 2.1 are perhaps the most conjectural of all. Income from interests and dividends in Table 2.1 was calculated assuming that capital losses balanced capital gains, and the net return on investment overseas was 5 per cent.[2] The measure of the stock of British overseas investment is derived by accumulating the balancing item, the measure of foreign investments, from the balance of payments calculations for each year since 1815, by using the principle that the balance of payments must balance. The resulting interest and dividend figure of £15.9 million is consistent with estimates derived from other sources, and is considerably less than the new foreign investment estimate (Jenks, 1971). The foreign investment figure derived from Imlah's balance is £22.4 million, of which the bulk in normal years during the 1850s went to India and the United States. This method of calculating British foreign investment cannot be expected to give accurate results for individual years, which may be dominated by short-term influences, such as the prosperity of the source or destination regions.

The balance of payments of the United States

The depression of 1858 was particularly severe in the United States, as indicated by the outflow of capital in that year. The balance of payments for 1855 (fiscal year ending 30 June) is analysed because more detailed figures are available for that year and because it was a more typical year in the sense that there was net foreign investment in the United States.

The goods of first importance in American trade were close counterparts of those in British trade. Raw cotton accounted for 45 per cent of United States exports and most of this was sent to the mills of Lancashire (Table 2.2). Britain took 42 per cent of United States merchandise exports, which, apart from raw cotton, included leaf tobacco, meat products and wheat flour as the most important items in 1855. Some 18 per cent of US agricultural output was exported, although total exports were less than 6 per cent of GNP (Davis *et al.*, 1972, ch. 14). Mirroring Britain's international specialisation, the United States exported food and raw materials and imported manufactured goods: iron and steel, cotton manufactures and wool products. Import duties raised the cost of imports by $54 million, or 23 per cent of imports for immediate consumption. The duties also offered an incentive to undervalue or under-report imports: official values perhaps should be increased by 4 per cent as a consequence (North, 1960). Even without the import duties, the United States was running a deficit on the balance of trade. Cuba, the third most important source of imports, suggests an exception to the generalisation that the United States imported manufactured goods. Like Britain, the United States demanded tropical products, especially sugar, although the sources of supply were often different.

The flows of goods are the most easily identified transactions; the others are more a matter of conjecture. It is, however, clear that gold exports and the inflows of foreign capital financed the balance of trade deficit in normal years. United States gold exports, originating mainly in the Californian discoveries of 1849, largely went initially to Britain. In 1857 this was true of 60 per cent of United States gold exports (Hughes, 1960, p. 13). Alone these gold exports only

Table 2.2 The balance of payments of the United States in 1855 (financial year)

Country of origin (%)		Merchandise imports ($m)		Merchandise exports ($m)		Country of destination (%)	
Britain	41.1	Iron and steel		Raw cotton	88	Britain	42
France	12.4	manufactures	29	Meat products	16	British North	
Cuba	7.0	Wool manufactures	28	Leaf tobacco	15	America	28
Others	39.5	Cotton manufactures and		Weat and wheat flour	12	France	13
		semi-manufactures	18	Others	62	Others	17
		Others	183	(Total crude materials	109)		
		(Total finished		Total merchandise	193		
		manufactures	129)				
		Total merchandise	258 (268)^a	Re-exports	26		
		Balance of merchandise trade −39 (−49)^a					
		Gold	1	Gold	55		
		Silver	3	Silver	1		
		Balance of bullion and specie flow +52					
		Balance of ship sales			+3		
		Balance of transport			+14		
		Balance of foreign travel			−21		
		Balance of interest and dividends			−22		
		Balance of private unilateral transfers			+10		
		Balance of government unilateral transfers			−2		
		Balance of foreign investment			+15		

^{a}North's adjusted figure allowing for undervaluation. These figures are used for the calculation of the balance of foreign investment.
Sources: North (1960); *Historical Statistics of the United States: Colonial times to 1970* (US Department of Commerce, 1975).

partially covered the merchandise trade balance; net gold exports were $54 million and the trade balance was $65 million in 1855.

The size of foreign investment in the United States in 1855 ($15 million) is computed as a residual, as in Imlah's estimates, on the assumption that the balance of payments must balance, and hence includes errors and omissions from the other components of the balance.

Immigration to the United States, mainly from Britain and Germany, amounted to 201,000 and is thought to have improved the US account by a sum shown under the heading 'balance of private unilateral transfers'. Immigrants brought funds with them and paid fares to United States shipping companies to a greater extent than earlier immigrants in the United States remitted money back to friends and relatives in Europe. The credit from immigrant funds increased with the number of non-Irish, who were generally much better off than the Irish. The debit from immigrant funds is the converse: the Irish appear to have remitted the most money on a per head basis, almost exclusively for passage money to bring people out of Ireland. An Irish labourer in the United States needed approximately three years to save up for the passenger fare, and the remittance

26 A History of the World Economy

values correlate closely with the number of immigrants from Britain three years earlier. The government unilateral transfer is part of the payment for the Gadsden Purchase (the acquisition of southern Arizona and New Mexico from Mexico in 1853).

Foreign investment imposed costs upon the balance of payments as well as conferring benefits. The payment of interest and dividends on foreign capital in the United States (net of payments to United States capital abroad) required a larger sum in 1855 than was invested in the country. These payments were mainly servicing loans to states, many of which had defaulted at the beginning of the 1840s. By 1853 most had resumed interest payments, but not Mississippi, Arkansas or Florida (Wilkins, 1989). That record must have encouraged foreign investors during the 1850s to prefer the privately owned railways. As in the British case, Americans travelling abroad spent more than foreigners in the United States, but the net negative figure was only partly counterbalanced by receipts from shipping. With the rise of the iron steam-ship, the American mercantile marine was beginning to lose the supremacy formerly based upon abundant supplies of timber on the Atlantic coast.

The Indian balance of payments

Information concerning a country's international economic relations is typically less reliable, or less likely to exist, the lower that country's level of economic development. Since the greater part of the world consisted of poor regions, nations or empires in the 1850s, our knowledge of international transactions then is seriously deficient. The two most populous regions, India and China, were probably among the most poverty stricken of those for which output per head indices might be constructed.[3] Of these two, India, with a population of 233 million, greater than all of Europe's if smaller than China's, was a major force in international economic relations (Visaria and Visaria, 1983, table 5.20). Indian statistics are believed to contain considerable errors and biases. But they are perhaps the most complete of all the poor countries in the middle of the nineteenth century. Investigating what inferences may be drawn from them casts light on the reliability of some judgements about the nineteenth-century world economy.

India's principal trading partners were Britain and China. Not only cotton products, railway materials and arms, shown in Table 2.3(a), were imported from Britain, but also gold. Indian merchandise exports to China were almost as great as those to Britain and closely matched the value of opium exported. Together with the rise of the China trade, the export of American silver to the East, from Acapulco to the Philippines, earlier created the possibility for multilateral trade. This was accentuated by India's transfers to Britain, which necessitated an excess of Indian exports over imports (Chaudhuri, 1983, p. 862). However, in the later 1850s, the positive balance of merchandise trade was more than offset by a

Table 2.3(a) An estimate of the Indian balance of payments in 1858/9

Country of origin (%)		Merchandise imports (Rm)		Merchandise exports (Rm)		Country of destination (%)	
Britain	57	Cotton goods	80.9	Opium	108.3	Britain	38
China	11	Military equipment	6.7	Raw cotton	40.9	China	33
Others	32	Railway material	12.3	Rice	24.3	France	5
		Cotton twist and yarn	17.1	Cotton twist and yarn	8.1	Others	24
		Iron	11.1	Indigo	21.2		
		Others	89.2	Sugar	14.5		
		Total	217.3	Seeds	20.6		
				Others	60.7		
				Total	298.6		

Balance of merchandise trade +83.1

Gold and silver 128.2   Gold and silver 7.0
(by sea)   (by sea)

Balance of bullion and specie flow −121.2

Balance of foreign investment^a +55^b

Balance of interest and dividends, of unilateral government transfers, private unilateral transfers and 'invisibles' trade +55^b
(Interest payment on Indian government debt held in UK −3.7)
(Indian government expenditure in the UK −74.7)

^{a}Converted from sterling back to rupees at the official rate of 1R = £0.1.
^{b}The estimate is from Jenks (1971).
The balance of interest and dividends, of unilateral transfers and invisibles, is calculated as a residual, assuming the correctness of all the other figures. The balance of foreign investment is likely to be an underestimate because of the exclusion of direct investment. Financial year ending 30 April 1859.
Source: Calculated from *Statistical Abstract Relating to British India from 1840 to 1865* (1867).

massive inflow of precious metals, over half the value of exports. The 'drain' of bullion to India, much remarked upon since Roman times, has been explained by the peasants' desire for hoarding. They wanted a store of value that could be drawn upon if their crops failed. In addition, some import of precious metals, probably between 30 and 60 million rupees every year, was necessary merely to maintain the value of the currency against wear and loss. During the late 1850s and 1860s, these volumes were greatly exceeded (Barbour, 1886, ch. 20). The contemporary British economist Thomas Tooke attributed it to an increase in the desire in the West for Asian goods (Tooke and Newmarch, 1928, appendix 23). He also emphasised the investment in the construction of Indian railways of the time. Although only 332 miles were in operation in 1858, already £23.5 million had been spent on the construction of the system (BPP, 1867). Jenks (1971, p. 219) estimated that £5.5 million (R55 million) was invested in 1858, virtually all by British savers. Some of this demand for rupees by the British must have been offset by an increased demand for imports of railway material into India, but less than one-fifth of Jenks' figure in 1858 was so balanced according to Table 2.3(a). The major portion of the money was spent on labour. Hence payment for the investments in India was largely made in precious metals.

The invisible current account items are perhaps the most controversial elements in the study of the Indian balance of payments. Chaudhuri estimated that between 1814 and 1858 India needed to export capital amounting to between £5 and £6 million a year. Payment for this transfer required India to export more than was imported (Chaudhuri, 1971, pp. 35–6; see also Frank, 1976). India's economic structure consequently became 'biased excessively' towards export industries. Table 2.3(a) makes it clear that India's balance of trade surplus in 1858 cannot be understood solely or mainly in terms of the invisible current account transactions, because the bullion inflow exceeded this surplus (Latham, 1978a). The largest single element of the transfer from India to Britain was the 'Home Charges', the payment of pensions to former Indian civil servants and army officers retired to Great Britain, savings of British officials remitted home, and India Offices expenses in Britain. These Chaudhuri estimated at £3.5 million, on average. In addition, there was a private export of capital of around £1 million and occasional debt repayments for the Indian government.

In 1858 the government debt was held mainly in India, with the consequence that most interest payments were paid to Indian residents; £60.7 million was held in India compared with £8.7 million in Britain. Interest payments to British holders of the debt amounted to £0.16 million, or R1.6 million, compared with £2.2 million to holders of debt in India. Though debt payments were low, total Indian government expenditure in Britain amounted to £6.1 million. As far as the balance of payments is concerned, some of this money is already recorded in the merchandise account as military equipment. This second 'drain' of funds from India, mainly the Home Charges, was an extremely small proportion of national income (about 0.33 per cent in 1872). The 'unnecessary' component of these charges, 13 per cent of the total, was even smaller, so that it would be hard to blame any substantial distortion of Indian economic structure on this flow (Mukerjee, 1972). In Table 2.3(a), the balance of measured merchandise and bullion trade amounts to a deficit of R39.9 million. If, for the sake of argument, Jenks' figure was accepted as an accurate measure of the foreign investment balance, then the balance of interest and dividends, invisibles trade and government transfers was R55 − 39.9 million, a surplus of R15.1 million, quite contrary to received interpretations. This probably suggests that foreign investment was higher than Jenks' figure.

Banerji has reconstructed the Indian balance of payments (Table 2.3(b)) with much more detail about non-merchandise and bullion current account transactions, and capital account flows, and without using the constraint that the accounts must balance (Banerji, 1982). That method can be justified by errors and omissions in the data, even after the adjustments he makes. Nevertheless the implied figure for errors and omissions in 1858/9 is rather large, R58.5 million, almost one-third of the estimated capital account balance. Banerji's estimated

Table 2.3(b) Banerji's estimate of the Indian balance of payments in 1858/9

Balance of merchandise trade	+97.1
Balance of bullion and specie	−130.2
Government railway net dividends	−11.0
Interest and dividends on other foreign capital	−32.1
Profits from foreign trade	−18.6
Remittances of services	−37.2
Home Charges less government railway dividends	−93.5
Interest on rupee loans	−6.7
Miscellaneous	−8.6
Balance of total current transactions (including interest etc., and adjusted merchandise and bullion balances)	−240.8
Guaranteed railways net capital receipts	+67.6
Net sterling loans to the government	+62.3
Rupee and sterling company investment	+17.1
Foreign subscriptions to rupee companies	+35.3
Balance of foreign investment	+182.3
Net errors and omissions	−58.5

Source: Banerji (1982).

capital inflow is much larger than assumed in the simple calculation of Table 2.3(a), but his estimated non-merchandise current account deficit is even greater. If errors and omissions were assumed zero, and the non-merchandise current balance were instead treated as an item that ensures the accounts balance, then that balance would instead be 97.1 − 130.2 + 182.3 = 149.3, a deficit of R149.3 million. This is almost R100 million less than the actual value in Table 2.3(b) and less than the capital inflow. This alternative figure for the balance of interest dividends, invisibles and government transfers presents a rather different picture of India's external economic relations, though a conjectural one. Yet another approach, followed in the estimates of the British and United States balances of payments discussed already, is to estimate foreign investment as the balancing item.

Once the railways had been built and interest and dividends fell due, the balance of payments reverted back to the position of the pre-railway age. Indian government purchases of sterling in London were only 13 million rupees in 1857/8, and zero in the following three years. But by 1871/2 these sterling purchases to pay for the 'drain' reached 100 million rupees (Barbour, 1886, p. 108). Hence there must have been an export surplus, unless offset by incoming investment, in the 1870s.

Summary and conclusion

The different resource endowments of Britain, India and the United States influenced their economic links in sometimes surprisingly similar ways. Both the

USA and India exported primary products, imported manufactures and acquired highly specialised foreign trade, despite the higher standard of living in North America. Both countries at the mid-century were importers of capital from Britain, and both accordingly needed to pay dividends and interest abroad. How much capital they imported may be inferred from the balance of payments identity. India in addition paid for British administrative expenses in India, although these were small in relation to the volume of Indian trade.

Britain, unlike its two most important trading partners, earned large foreign balances from managing, financing and shipping foreign trade. This pre-eminence originated in Britain's naval supremacy and greater commitment to trade because of its population density and island location. Precious metals were the ultimate means of settling imbalances in the economic relations between states. However, the balance of payments of Britain, India and the United States imply that the bulk of the bullion movements from the USA to Britain, and from Britain to France and thence to India, were traded in their own right; they were undertaken as autonomous, rather than as accommodating, transactions. Gold and silver flows facilitated international transactions as commodities as well as means of exchange. Opium was the keystone in the multilateral balancing of transactions between China, India and Britain. This should not be taken to imply necessarily that, without India's specialisation in this addictive drug, some other way of balancing accounts might not have been found. Chinese demands for sandalwood were already being met from the islands of the South Pacific, but more probably a substantial increase in Chinese rice imports could have been achieved, as it was later in the century (Wolf, 1982, p. 258).

Notes

1. Germany was in fact a more important market for Britain than Australia, but it was still a collection of independent states.
2. The implications of alternative returns are explored in Foreman-Peck (1989).
3. That seems to be true for 1900 and 1913, although whether China was more or less productive than India is unsettled (Hanson, 1991, table 2; Maddison, 1989).

3 Liberalisation and free trade

The years between the European revolutions of 1848 and the depression of 1873 offer virtually a laboratory experiment for testing theories of international markets. Britain survived 1848 without a revolution and announced its commitment to free trade with the Great Exhibition of 1851. With 14,000 exhibitors and six million visitors, this exercise in internationalism inaugurated the new era. Almost all the 1848 uprisings ultimately failed to establish radically new styles of government, but the rulers of Europe, and elsewhere, during the third quarter of the nineteenth century recognised that they must abandon traditional policies. They turned with varying degrees of enthusiasm and success to liberalising their economies. Austro-Hungarian peasants were emancipated in 1848 and serfdom was abolished in Russia in 1861. Slavery was abolished by France in 1848 and by the Netherlands in 1863, and was eliminated from the United States in 1865. Official recognition of the power of the market warrants describing the rules and implicit assumptions governing states, businesses and enterprises as at least a move towards a free trade regime. These were also years of sustained economic growth in Europe and in the United States. Elsewhere the parallel march of growth and freer trade was much less apparent. The following sections first examine the principal influences on the pattern of international trade under this regime. Then some of the implications of the trade theories are tested, explaining the discrepancies between prediction and reality in these years. Finally the chapter looks at the reasons for governmental restrictions on the free play of international market forces, and the consequences of these policies.

Transport costs and trade

Among the greatest technological improvements of the period was the application of steam to land and sea transport. For most of history, the great cost of moving any considerable distance meant that only goods with a high market value in relation to their bulk were traded internationally, or indeed inter-regionally. Silk, spices and precious metals fulfilled this requirement, but being luxuries, far too expensive for the majority of the population in even the richest countries, trade in such commodities never assumed much economic importance. The extension of the railway network altered this position, not only by reducing land transport costs, but more importantly by allowing relatively cheap travel over routes where it had been virtually impossible to offer any before. The diffusion of the railway therefore offers one indicator of the extent to which international markets could pervade economies. Countries were not homogeneous and therefore comparisons between economies of the spread of railways must somehow correct for land area and population differences. The US railway system by 1875 was greater than the combined networks of Britain, France, Germany, Italy and Russia. But then the United States was so large that not until 1869 did the Union Pacific railway, climbing to 8,600 feet, link the east and west coasts. As Table 3.1 shows, the length of railway per head of population was much greater in the United States than elsewhere. But the other major regions of recent European settlement, Canada and Australia, show a similar pattern. In Europe, where population densities were higher, a rail mile could typically serve more passengers and freight. There, the spread of railways by the end of the third quarter of the nineteenth century apparently depended on when railway building began. First the major trunk routes would be built and then the networks would be filled in with lesser trunks and feeder lines. At the mid-century, much of the European periphery (Norway, Sweden, Portugal, Greece, Romania) was still without railways. With the exception of Sweden these and three other periphery economies – Italy, Russia and Hungary – apparently lagged in 1875. Sweden, Switzerland and, to a lesser extent, Denmark experienced very rapid paces of development after slow starts. The European railway core was clearly located in north-west and central Europe by 1875.

Latin America lacked railways at the mid-century, which Peru began to remedy with guano money in 1851. That country, perhaps together with Egypt, showed that railways could be a mixed financial blessing when mismanaged. Measured by their absorption of railways, the Latin American economies in 1875 fell into two groups, with large population Mexico and Brazil appearing backward in comparison with the temperate zone states.

In Asia, British India stands out as a consequence of the early state support for railways. China remained without mechanical land transport in 1875 and Japan had only just begun building a system. The military early appreciated the value of the railway system for troop movements. The Belgian railway system, one of the earliest completed, and subsequently the densest in terms of length per square mile, was originally planned with military considerations in mind. Napoleon III demonstrated how effective railways could be in his war with Austria in 1859 (Bierman, 1989, p. 195). The British introduced state-subsidised Indian railways after the mutiny as a cost-effective form of military control.

The enormous impact on economic development originally ascribed to railways has been reassessed by many static 'social savings' studies. These studies usually attempt to calculate the costs of shipping the freight and passengers by the next best alternative means: by postcoach for passengers and canals for freight. The social saving is then the difference between this and the actual cost. Only in

Table 3.1 The international spread of railways to 1875

	Date of opening of first line	Rail length/population in 1875 (miles per million)	Rail/area (miles/1000 sq. miles)
Britain	1825	527	137
France	1828	367	66
Bavaria	1835	490	84
Belgium	1835	407	191
Austria	1837	314	55
Saxony	1837	397	190
Prussia	1838	383	74
Russia (Europe)	1838	185	6
Italy	1839	178	42
Netherlands	1839	281	79
Switzerland	1844	471	79
Hungary	1846	26	32
Denmark	1847	386	45
Spain	1848	226	19
Norway	1854	190	0.3
Portugal	1856	160	19
Sweden	1856	495	13
Greece	1869	5	0.4
Romania	1869	129[a]	15
United States	1827	1922	20
Canada	1836	1159	1
Australia	1854	998	1
Peru	1851	362	1
Brazil	1854	72	0.3
Argentina	1857	520	1
Chile	185?	342	4
Mexico	1867	27[a]	0.5
Uruguay	1869	425	3
India (British)	1854	34	7
Japan	1872	1	2
China	1876	0	0
Egypt	1855	90[a]	67

[a] Denotes population census figures considerably later than 1875 and therefore a possible substantial downward bias in the ratio.
Sources: Calculated from *Statistical Abstract for the Principal and Other Foreign Countries* (HMSO, various numbers); *Statistical Abstract for the Several British Colonies, Possessions and Protectorates* (HMSO, 1880).

economies where water was so scarce that canals were not a viable alternative was the static social saving from railways large. A study of the impact of railways on the Mexican economy found a social saving of 30 per cent of national income, and the largest figure estimated for Europe was in Spain (Fogel, 1979; O'Brien, 1977, 1983). Much depends upon date and method of measurement. Dynamic estimates, with rather different alternative worlds postulated, generate larger results (Williamson, 1980; Foreman-Peck, 1991a). In continental Europe the railway system boosted international trade directly by reducing the costs of moving goods across frontiers, but for Britain, the United States and India their effect was indirect, cutting the costs of carrying goods to or from the ports.

The transport of goods by sea was subject to a revolution similar to land transport. British freight rates declined only modestly before 1850 and then rapidly as the steam-ship came into use (Harley, 1988). Improvements in ship operation together with lower ship prices drove freights by 1910 down to just over 40 per cent of their 1855 value, despite a slight rise in general prices. Shipping benefited from the more or less continuous reduction in fuel consumption of marine engines throughout the second half of the nineteenth century.

Isambard Brunel's iron-hulled passenger-carrying *Great Britain* made the first screw steamer crossing of the Atlantic in 1845. Over the next quarter century the two technological innovations that Brunel employed became universal. The Cunard company ordered four iron screw ships for the Atlantic route in 1852. Early in the 1860s the compound engine, with first two, and later three, cylinders of different diameters, was introduced. The steam began its expansion from boiler pressure to atmospheric pressure by moving the piston in the smaller cylinder before passing on to the larger. By using steam more efficiently, these engines saved on coal.

Steam could provide lower transport costs than sail on short routes, while the coal consumption of marine engines was still quite high, but could not successfully compete with sailing ships for bulk cargoes on long routes. The steam-ship carried its own fuel and thus the longer the voyage, the greater the proportion of the ship's capacity devoted to coal, rather than to cargo. One of the first transatlantic steamers, the 178-foot *Sirius*, in 1838 was obliged to burn its furniture and bulkheads before reaching New York, after all coal fuel was exhausted. Fuel efficiency then largely determined the ship's range.

Throughout the nineteenth century, Britain was the world's primary coal source. Consequently, steam-ship costs increased with distance from Britain. By 1855 almost all the cargo entering Britain in the 'home trade' (from Brest to the mouth of the Elbe) was carried in steam-ships, although the large export trade in coal from the north-east coast was still shipped under sail in wooden colliers which returned in ballast. Nearly one-third of the Mediterranean trade was also carried by steam at this date. At the same time, steam was extending into the grain trades of the Baltic and the Black Sea, although the Baltic timber trade remained largely the preserve of sailing ships (Harley, 1971).

Sailing ships were obliged to improve their productivity as well if they were to maintain their competitiveness on a given trade. They attained such a peak of development that the clipper *Thermopylae*, built for the China tea run, was reputed to be able to sail at 7 knots in a breeze so light that a candle could remain alight on deck. The opening of the Suez Canal in 1869 radically shortened distances from Europe to the East. The distance between Liverpool and Bombay was halved, and the route to China was also reduced. The Canal was unsuitable for sailing ships, primarily because of unsatisfactory wind conditions in the Red Sea. Steam-ship, but not sailing ship, route distances were reduced after 1869.

Hence the tea clippers, unable to compete in the China trade, moved to the Australian wool trade, the distance for which had not been affected by the Canal. More importantly for international trade, the Canal allowed a marked reduction in transport costs on the India and China runs.

The volume and distribution of European trade responded to this reduction of transport costs. Domestically produced European exports grew much faster than income, during the 1860s (Bairoch, 1974, 1976). The European average rate was 5.2 per cent per annum, pulled down by the comparatively slow growth of British exports at 3.8 per cent per annum. During the decade of the 1870s, European average trade expansion fell to 3.1 per cent per annum but remained well above income growth, so that openness of national economies continued to rise. The greater part of this expansion was trade between European countries, which in 1850 had accounted for two-thirds of European exports. During the free trade era, the trend towards geographical diversification was broken, with a decline in the share of non-western countries in European trade from a little over one-fifth. The decline is attributable to the greater distances, which could not benefit from the reduced transport costs of the steam-ship.

Integration

Trade links between national economies are connections between individual markets. Either markets (trading identical products) are integrated or the transport, tariff or other barriers are too high. Changes in demand, in supply or in barriers may connect or disconnect a pair of markets in different states. If they are linked, what affects one market will affect the other; the ending of wheat exports from Russia during the Crimean War tended to raise the price of wheat in Britain and increase the British demand for wheat from France. As transport costs fell, land which formerly only supported subsistence farming, or nothing at all, could provide cash crops for the world market. On the other hand, the Argentine beef market at the mid-century was independent of the beef market in Europe because of the costs of transport. Very substantial shocks to the supply of Argentine beef would have left the European beef market quite unaffected.

Because economies are composed of a number of markets, the degree of integration of an economy into a world or regional economy depends upon how many markets are linked and how they are connected to the domestic economy. A limiting case is the enclave economy, the mine or plantation which is entirely self-sufficient and has no effects upon the host economy. In general, even the prices of goods that are not traded internationally, such as housing, will be affected by the world economy if sufficiently numerous or pervasive other domestic markets are internationally linked. Both will be bidding for the same labour. Thus, while an individual national or regional market either is, or is not, linked or integrated with the world or a larger economy, an economy may be more, or less, integrated (McInnis, 1986; Harley, 1992b).

36 A History of the World Economy

The polar case is given by the 'law of one price'. If all prices (of homogeneous products) are not equalised when exchange rates are taken into consideration and transport, tariff and information costs are negligible, then there are profit opportunities to be had from importing or exporting. Taking advantage of these opportunities tends to reduce price discrepancies. What can be bought with a sum in one currency is then equivalent to the goods and services that can be purchased when that sum is converted (at the prevailing exchange rate) into another currency. This is the law of one price. But markets can be integrated even in the presence of transport costs and tariffs. Different prices will prevail in the different national markets, but their changes will be comparable when there is a shock to one market. 'Purchasing power parity' obtains when all prices in two economies move similarly, when corrected by the exchange rate.[1]

Specialisation

Absolute advantage

Though the falling cost of freight may account for much of the increased volume and distribution of international trade, it does not explain why various countries specialised in supplying their particular goods and services to the international market. Nor does it offer much insight into the distribution of the gains from trade. If a product can be made in one location more cheaply – that is, using fewer resources – than in a second, we might suppose that the first would be the site favoured for production. The first place has an absolute advantage over the second in manufacturing that good. Why might there be spatial differences in efficiency? The scale of production might reduce costs by allowing longer production runs and/or more specialised equipment. Then it would make sense to export one type of product and buy whatever imports are needed for consumption. As Adam Smith maintained, the extent of specialisation is limited by the extent of the market, which in turn is constrained by transport costs.

Comparative advantage

David Ricardo (1772–1823) assumed instead that there were no cost advantages or disadvantages from larger output volumes; that there were constant returns to scale (1817, ch. 7). Costs were cut not by producing greater output, but by specialising in production of goods and services that could be made relatively more efficiently. The produce that could be manufactured relatively less efficiently was imported in exchange. Trading goods can accomplish the same results as moving factors. The market value of the immobile factors, such as climate, depends upon their most productive use. Hence profit-oriented entrepreneurs will aim to make in one spot whatever the assets there can do best. In Ricardo's example, Portuguese entrepreneurs could take advantage of Portuguese sunshine by exporting wine to England. Rainy England could best exploit its climate by specialising in cotton textile manufacture. Each country could then buy in the other products wanted for consumption. Because of the natural resource immobility, absolute advantage is irrelevant; sun and rain could not be moved to their absolutely most productive locations. Without trade, prices in Portugal and England would have differed according to their respective combinations of resources and their suitability for making wine and cloth. When trade was then opened up between them, merchants compared prices and tried to export the cheaper goods and import substitutes for the more expensive. The less expensive goods (wine in Portugal, cloth in England) were cheaper in comparison with the similar products of the other location (wine in England, cloth in Portugal). The first location possessed a comparative advantage in those goods.

In the mid-nineteenth century, specialisation between Britain and the United States in food and manufactures was of greater concern to the international economy. Ricardo's theory of trade demonstrates that, if the cost of manufactured goods in terms of food in Britain was lower than in the United States, Britain had a comparative advantage in manufactured goods and the United States had a comparative advantage in food. The cost of manufactures in terms of food is the reduction of, say, food output necessitated by transferring resources to increase the production of manufactured goods, holding constant the level of resource utilisation.

The gains from trade stem from the specialisation of each country in the production of the goods in which it has a comparative advantage. The export of these goods allows imports and greater consumption than in the absence of international trade. These benefits accrue even if one economy is more efficient in the production of both goods than the other. In this case there is a wage differential between the two economies to compensate, although the persistence of that efficiency gap would need to be explained. In the absence of knowledge of demand in the two hypothetical economies, little can be said about which gains more from the exchanges. However, if one economy is much larger than the other, it is reasonable to suppose that the equilibrium terms of trade, which determine the distribution of the gains, will be much closer to the pre-trade prices of the large economy than to those of the small. The greater proportionate gains will then accrue to the small economy.

Although mainly an agricultural country, the United States in 1860 employed 1.25 million in manufacturing industries. The census of that year provides a detailed statistical account of industry which can be used to illustrate Ricardo's theory. Table 3.2 shows that productivity per worker in manufacturing in flour and meal was greater than in all other industries; if Ricardo is correct, and Britain can be represented as the US's principal trading partner, then the observation that this was the largest manufactured export category indicates that Britain's inter-industry ranking of opportunity costs placed its flour-milling industry much lower down its ordering. Of the other main categories traded, the greater net import of cotton goods than of iron manufactures is consistent with the productivity ranking. British productivity in iron manufactures should have

38 A History of the World Economy

Table 3.2 Labour productivity and trade in United States manufactures, 1860

	Value added per worker ($000)	US trade (net exports + net imports −) ($m)
Flour	1.45	+15.4
Leather	1.00	−4.3
Printing and publishing	0.90	−0.8
Machinery	0.81	0
Lumber (sawed)	0.70	+5.14[a]
Iron	0.68	−16.35
Cast	0.78	
Forged, rolled, wrought	0.66	
Pig	0.54	
Furniture	0.64	+1.03
Carriages	0.64	+0.82
Woollen goods	0.62	−38.3
Blacksmithing	0.52	—
Cotton goods	0.47	−21.62
Tobacco and snuff	0.47	−1.41
Boots and shoes	0.40	+0.6
Brick	0.40	+0.15[b]
Men's clothing	0.32	−1.577[c]
Silk manufactures	n.a.	−32.07

[a] Boards, plants, staves, etc. Forest product exports were $13.7m.
[b] Brick, lime and cement.
[c] 'Wearing apparel' exports; 'clothing' imports.
Sources: *Commerce and Navigation of the United States, Year Ending June 30 1860*; *United States Census 1860*, vol. 3.

exceeded that in cotton. The greater net import of woollen goods relative to cotton suggests that Britain's productivity in woollen goods was higher than in cotton. Lumber and forest products were a major export, and productivity in this industry was high by comparison with other industries (the average was $670 value added per worker). In Britain, Ricardian theory predicts that it must have been low.

Discrepancies between the prediction of the theory and the data of Table 3.2 may arise from the neglect by the theory of distance from potential markets and the costs of moving various products. The weight of bricks in relation to their value made unprofitable the import of these products on a large scale. A second reason for discrepancies is tariffs in the domestic economy and abroad offsetting comparative advantage. Between 1857 and the Civil War, American tariffs were lower than at any other time in the nineteenth century, and Table 3.2 shows that the 24 per cent tariff on woollen goods did not prevent these being the largest single category of manufactured imports.

Ricardian theory does not explain why comparative advantages should exist. Much of eighteenth-century trade was in products such as sugar, tobacco, tea and raw cotton – or, as in Ricardo's example, wine – that required climates different from those of the importing country. Even the great nineteenth-century export industry, cotton textiles, depended to a considerable extent upon climate for locational advantage. The influence of Lancashire's natural humidity was estimated as the economic equivalent of a protective tariff of 10 per cent because cotton fibres became more pliable, less brittle and easier to process as humidity increased. The abundant supply of river water was essential to the industries whose demands absorbed up to half of the water used in Manchester and Oldham, and the finishing industries required enormous quantities of soft lime-free water that would neither waste soap nor resist dye (Farnie, 1979, ch. 2). Other natural advantages that generated trading possibilities included temperature. Within Europe, January temperatures in 1860 ranged from −4.6°C in Helsinki to 7.9°C in Rome, while January rainfall in the same year was 172 cm in Berlin but only 1.8 cm in Madrid (Smithsonian Institute, 1927).

Yet another source of comparative advantage is technology differences between nations. An influential view of European economic development in these years places great emphasis on these differences (Landes, 1969, chs. 3 and 4). Britain, according to this account, exported its new technology, which was only slowly absorbed by European industry. The dates of construction of the first national railways in Table 3.1 lend impressionistic support to this position. The question then arises as to why new technologies should spread so slowly over an area in which people and machines were free to move across national boundaries by the middle of the century.

Institutional obstacles are one reason for delay. Vested interests might be able to block for longer in some countries the new technologies that threaten their economic and political power.

A second possibility is that the size of the market accounts for different national technologies. Profitable employment of the new techniques may require a larger number of higher-income customers within an economic distance of the manufacturer. If this were so, the improvement of transport facilities would be a prerequisite for the profitability of the other innovations. In Germany, the establishment of a free-trade area, the Zollverein, from 1834 reduced the economic distance between buyer and seller, and contributed to a stronger subsequent growth of the eventually politically unified economy than was experienced by unified Italy, which lacked an earlier customs union.

A third reason why a technological lead may persist is the elapse of time necessary for the discovery and exploitation in the 'follower' economies of the natural resources on which the new technology is based. Until the middle of the nineteenth century, the coal and iron ore deposits of the Ruhr were unexploited. Then between 1850 and 1855 coke production trebled, and iron ore output rose by fifteen times in the decade after 1852 (Milward and Saul, 1973, pp. 406–9).

British railway exports and construction in mid-nineteenth-century Europe are consistent with the (market size) explanation above. British contractors were heavily engaged in building the French railway system as Napoleon III pursued a vigorous policy to develop the French economy (Jenks, 1971, pp. 164–7). At the same time British engineers, navvies and operatives were working in Piedmont, Switzerland, Austria, Spain and, after the Crimean War (1854–6), Russia and Turkey. In Denmark and Scandinavia, Morton Peto had virtually a monopoly of railway construction in various projects intended to improve the transport of Swedish iron and Danish butter to the English market.

British contractors in Europe tended to order capital goods from Great Britain. Countries that were investing most heavily in their transport systems increased their British purchases by the greatest amounts. Exports of iron and steel from Britain doubled in volume during the years 1850–3. Britain early created the largest market for railways, and those supplying it acquired experience and capacity that allowed them to sell more cheaply than nationals of other countries. These observations are also consistent with the British advantage stemming from coal-intensive industries, which included railways, as well as iron and steel.

In contrast to the means of diffusing advanced technology in the twentieth century, the multinational company, enterprise in railways was generally cosmopolitan. A network of railways in the Papal States was conceded in 1857 to a Paris banker, Jules Mires, who sold stock mainly in clerical circles in France and Italy. (Mires was arrested in 1861 for stock manipulation (Redlich, 1967).) The rails, however, were bought from Newcastle upon Tyne, the engines from Paris and the wheels from Belgium, and the carriages were manufactured for the contractor in Italy. This wide distribution of purchases must have been due in part to the comparative lack of restrictions on international trade in the 1850s compared to the twentieth century, and in part to political realism. Nevertheless, doubt must be cast on the diffusion model if the techniques of production were already so widely spread.

Technologies may have been common across Europe, but the efficiency with which they were employed may have differed. So even if engineering technology differences seem a less plausible reason for the emergence of comparative advantages, 'social technology' differences remain a possibility.

The Hecksher–Ohlin theory of comparative advantage

An alternative theory of trade traces the sources of comparative advantage to the relative scarcities of the factors of production in different countries. This theory, due to the Swedes Eli Hecksher (1879–1952) and Bertil Ohlin (1899–1979), and published in the years between the two world wars, is of considerable value in explaining much of nineteenth-century trade (Hecksher, 1949; Ohlin, 1933). Ricardian theory could account for trade if there were differences in production technology between countries, but the persistence of the differences themselves would then require explanation, as already argued. The simple Hecksher–Ohlin (HO) theory in contrast assumes that the same (efficiently employed) technology is available to all countries.[2] The relative scarcity of productive inputs differs from region to region and from country to country. In the absence of trade, these different relative scarcities lead to different relative prices within the countries, even though countries are assumed to have access to the same production technologies; wheat was cheaper in the United States, compared with cotton textiles, than in Britain. Wheat was cheaper because land was so abundant in the United States relative to labour, in comparison to Britain, and because wheat is a product which needs a substantial land input. Similar considerations apply to the production of wool in Australia to a greater extent. Cotton textiles, in contrast, required little land but relatively large amounts of labour in their manufacture, both in the factory and indirectly, in the construction of the equipment used in the production process.

The implication of the theory is that a nation will export products that are intensive in the relatively abundant factor of production. In the above example, the United States therefore gained from exporting wheat to Britain, which in turn benefited from exporting cotton textiles to the USA.[3]

HO theory implies that, when the nineteenth-century reduction of transport costs opened up vast regions, virtually unpopulated, to the international market, the pattern of international specialisation adjusted to take account of this effective increase in the supply of land. Where land was cheap, more was used per worker and the output of agricultural commodities per worker tended to be higher. Land was more likely to be cheaper, the lower the population density, although of course the quality and accessibility of land differed between countries. Table 3.3 indicates that the HO theory did correspond with the facts for Britain, the United States and India. The US and Indian ratios of labour to land were lower than Britain's and therefore, as shown in Chapter 2, they exported land-intensive agricultural products (wheat in Table 3.3) to Britain. In return, Britain exported labour-intensive manufactured products.

If national economies specialised as HO theory predicts, then although they all faced a common set of technological possibilities, each economy was obliged to choose those techniques of production most appropriate to its factor endowment. The abundant timber of the United States made the adoption of coal-fired steam engines, and for many years the coal smelting of iron, on the British model uneconomical (Hyde, 1991). The fast-flowing waters of Switzerland allowed the Swiss textile industry to use water-powered machinery as efficiently as the British textile mills used steam power. The land-abundant United States not only possessed a comparative advantage in land-intensive agriculture, but also adopted labour-saving machinery in all sectors (Habbakuk, 1962; Temin, 1966; Fogel, 1969; Temin, 1971; David, 1975, ch. 1; Musson, 1980). Because there was no scarcity of land, the productivity of labour in agriculture was high and so were earnings. Moreover, workers in manufacturing industry were paid similarly high wages. It was therefore more worth while developing complex and expensive machinery to save labour in the United States than in Britain, as a group of Englishmen discovered when they went to see the Industrial Exhibition in New York in 1853. They arrived to find that the exhibition was not ready, so instead they visited a number of American firms. Two of them, George Wallis,

Table 3.3 Population densities and wheat exports

	Population densities c. 1860 (person per sq mile)	% of world wheat exports 1854-8[a]
United States	9	24.9
Canada	9	6.4
Russia (in Europe excl. Poland)	31	12.0
India	117	3.2
(British India)	(151)	3.2
France	183	—[b]
Britain	240	(−33)
(England and Wales)	(310)	

[a] A minus sign denotes imports
[b] French grain exports were relatively small from 1854 to 1857, but increased greatly in 1858.
Sources: Stern (1960); *UK Statistical Abstract of Foreign Countries; Official Abstract of British India*, HMSO; Urquhart and Buckley (1965).

headmaster of the School of Art at Birmingham, and Joseph Whitworth, one of the most prominent British engineers, wrote reports concerned with the manufactures and machinery they saw. What matters here is less which country developed the most advanced machine tools, than in which was the use of labour-saving devices, such as the 'system' of interchangeable parts, most widespread.

By the middle of the nineteenth century coal, like agricultural land, was a vital natural factor of production. Some argued that the British owed their prosperity entirely to Britain's abundant coal resources, which gave them cheap motive power for railways and steamships. Among them the Victorian economist W. Stanley Jevons thought that because coal was a non-renewable resource, in contrast to climate, Britain's iron trade was destructive of national economic prospects (1906, pp. 369, 415). In 1865 the iron industry took between one-third and one-quarter of the whole national yield of coal, and at such a rate of consumption he forecasted that Britain's stock of coal would soon be run down and its comparative advantage eliminated.

Like Britain's, the French pattern of trade and technology reflected the central importance of coal. French engine drivers were given a bonus according to their fuel saving, while British and German engine drivers were merely required to arrive on time (Milward and Saul, 1973, p. 173). French ironmasters used less fuel per ton of pig iron smelted than British ironmasters throughout the second half of the nineteenth century. France's most important trading partners were Britain, the United States and Belgium. In trade with the USA, France like Britain bought products that required a great deal of the abundant land in America, but because of the paucity of France's coal deposits, its imports from Belgium and Britain were coal-intensive products. Over 60 per cent of British exports to France in 1854 consisted of raw materials or intermediate goods, including coal, wool, iron, copper and semi-finished textiles, particularly woollen products; while over half of French exports to Britain took the form of finished manufactured goods and processed foodstuffs, such as silk, leather and cotton manufactures, wine, spirits, refined sugar and flour (O'Brien and Keyder, 1978, p. 162). British industry sent machinery and intermediate goods to French industry, which returned finished manufactured goods to British consumers. The French, with a comparative disadvantage in coal-intensive products, possessed a comparative advantage in skill-intensive processes (despite a lower literacy rate) which tended to involve finishing, rather than basic manufacture. The opportunity to trade in coal meant that, although trade patterns reflected coal endowments, an economy without those endowments would not be badly handicapped.

Commercial policy

It was often the desire to buy cheap coal that motivated demands for lower tariffs in continental Europe. A corollary of the Hecksher–Ohlin theory, the Stolper–Samuelson theorem, draws attention to some more general forces behind tariff making. The theorem demonstrates how, for example, tariff-induced changes in domestic prices can change relative factor rewards. If tariffs on labour-intensive goods are cut, wage rates fall relative to land rents. Such a policy might be easier to implement in a legislature where landowners held power than in one where workers' representatives were in control. The relatively low US tariff on manufactures during the 1850s, when the Southern plantation owners were powerful, is consistent with the theorem.

Judged by tariff policies, the twenty-five years after 1850 may be described as the era of increasing free trade. But tariff cuts were rarely guided primarily, or even substantially, by a desire to reap the maximum gains from trade that economic liberals maintained would be forthcoming. For governments, the use of tariffs as a source of revenue often vied with their potential as instruments of foreign policy. In some cases the government merely responded to pressure groups who stood to gain, or thought they did, from the imposition of a particular tax on imports. Beneficiaries were producers expecting to sell at higher prices behind the tariff barrier. The losers, including the ultimate consumers and the buyers of intermediate goods and raw materials, were obliged to pay the higher prices. In addition, mercantile interests suffered from the reduction in the carrying trade.

Where the numbers of producers expecting to gain are small, they will find it in their interests to co-operate in influencing the government to raise the relevant tariff (Frey, 1984, chs. 2 and 3). Where the costs of the tariff are spread over a large number of losers, none of them will find it worth while resisting a change which has such a small individual effect. Other considerations might include the state's need for revenue, and the competitiveness of the industry.

Britain was the first country to move to free trade and adhered to it most tenaciously. Here the reductions were probably undertaken primarily for reasons of budgetary policy rather than because of an ideological commitment to free trade or political lobbying. Moreover, it is arguable that the impact of the reductions on the British economy were not ideal even though they were beneficial. Holding the composition of British imports constant, the reduction of tariff rates from 1841 to 1881 was 21 per cent (McCloskey, 1980). The absence of major wars and the associated accumulation of debt, together with the Victorian views about the proper role of the state, combined to reduce the government's share of the growing national income. This cut the share of national income that had to be financed by the tariff revenue, and anyway, since the ratio of imports to income was rising, a given rate of tariff generated an increasing ratio of revenue to national income.

Although comparative advantage theory shows that free trade is better than no trade, free trade is not necessarily better than reduced trade. For a country large enough to affect world prices there is an optimum tariff which gives more benefit to the country which imposes it, by reducing the price at which foreigners can sell, than is cost by the fall in imports consumed by the domestic country. McCloskey argues that the 5.8 per cent average tariff rate in Britain, on any plausible assumptions, was lower than ideal. He dismisses the likelihood of retaliation against an optimum tariff as no defence, on the grounds that other European countries raised their tariffs anyway during the 1880s. This argument, however, neglects the 'most favoured nation' (MFN) clause in the treaties negotiated to reduce tariffs after 1860. The two signatories agreed to grant each other any advantage already conferred at the time of signing, or in the future, to any third nation. This clause prevented discrimination in trade policy, because the reduction of duties to one country meant that they were automatically reduced to all other countries receiving MFN treatment from the country cutting tariffs. The indirect consequences of an MFN treaty for trade expansion often exceeded the direct effects. Had Britain not reduced its average tariff to 5.8 per cent, other national duties would not have fallen by so much.

The move to virtually free trade, brought cheaper food for the British consumer, and deferred the raising of tariffs until after the First World War. How the gains from opening the British market were distributed between foreign agricultural producers and British consumers depended on the proportion of British demand in total market demand. Inspecting the course of prices over time suggests, possibly misleadingly, that expanding demand tended to raise world prices of cereals and of most transportable foodstuffs to British levels, rather than reducing British prices down to former levels abroad (Imlah, 1958, ch. 6). As to the impact on national security, Britain's increased economic strength outweighed its greater dependence on the rest of the world, which was, in any case, being forced on it by population growth. Increased dependence upon food imports left Britain less vulnerable in the event of emergency such as harvest failure, because it could call upon a wide variety of supplying areas unlikely all to suffer harvest failures at the same time. The principal risk, as became apparent in 1917 at the peak of the German U-boat campaign, was that the Royal Navy would be unable to guarantee safe passage of food imports (Jenks, 1971, pp. 158–63).

British duties were not obviously lower than French duties in the 1850s, even taking into account French prohibitions: that is, when the prohibitions are added into the average tariff by the amount that domestic prices were raised by the ban. Of course, this measures the effect only on the importer, not on the exporter. The French placed heavy tariffs on those products – textiles, iron and steel – in which Britain held a comparative advantage, but they were a very small proportion of total imports. Britain reciprocated, with heavy duties on French wine. It was only towards the end of the nineteenth century that Britain was clearly less protected than France.

During the 1850s, the smaller economies of Europe – the Netherlands, Scandinavia, Portugal and Switzerland – moved towards free trade. But they accounted for a rather small proportion of European economic activity. It was the French liberalisation of trade in 1860 that really inaugurated the shift in continental European policy. French free trade came about for reasons quite different from the earlier British move. Ideas played some part, and Michel Chevalier was an influential intellectual advocate of French economic liberalism. Most significant in overcoming the weight of vested interests in protection was Napoleon III's foreign policy. Only the Emperor was empowered to make treaties, including those dealing with trade. The British (and the rest of Europe) disapproved of his Italian policy, which culminated in the 1859 war with Austria. They were intended to be mollified by the Anglo-French commercial treaty of 1860. Thereafter France, the Zollverein states, Italy and Britain negotiated a whole series of reciprocal trade treaties with 'most favoured nation' clauses.

The French policy shift was also easier because French government revenue was much less dependent on customs duties than Britain's; the proportions were respectively 10–12 per cent and 35–40 per cent. To facilitate the transition and make it more palatable to the losers, Napoleon III granted an Imperial loan in 1860 of 38.4 million francs to modernise French industry. The impact of free trade on the French economy has been hard to disentangle from other retarding influences operating simultaneously. Higher imports clearly damaged some industries, but stronger foreign demand stimulated others (Levy-Leboyer and Bourgignon, 1990, p. 24; Caron, 1979; Nye, 1991a, 1991b).

In Germany, freer trade united authoritarian eastern Prussian landed aristocrats, selling agricultural exports, with the liberal western and urban classes who had unsuccessfully challenged their political control in the 1848 uprisings. The Zollverein (customs union) tariff was reduced after 1850 and cuts continued at intervals until 1879. The objectives of Prussian foreign policy also coincided with Junker interests. Foreign policy was directed to establishing a unified Germany that excluded Austria-Hungary. A trade treaty with France served to isolate Austria. Considerable financial concessions were made in order to achieve unification: Hanover, for instance, was bribed to enter the Zollverein customs union by being offered more revenue from the customs than it would get on the basis of population, the usual method of allocating receipts. In addition to the policy objective of unification, the importance of the British market for German food grains required that some attention be paid to the sensibilities of the British government about the taxation of British exports to Germany. The British managed to foil proposals for a higher external tariff on textiles by referring to the benefits conferred on Germany by the repeal of the Corn Laws, and indicating that they might be threatened if British interests were (Kindleberger, 1975, pp. 38–41; Henderson, 1939, ch. 6).

Tariff protection did not play a major role in the rapid industrial development of Germany during these years, for the Zollverein drew most of its revenue (more than two-thirds in 1871) from the taxation of popular foods, drinks and tobacco. Duties protective of domestic industry contributed little to the budget, although textile revenue was 10 per cent of the 1871 total customs revenue.

Independent states outside Europe were less affected by free-trade doctrines, but in the United States the political strength of the trade-oriented Southern cotton and tobacco interests lowered tariffs in the late 1840s, culminating in 1857 in the lowest tariff since 1816. The antebellum tariff had no important terms-of-trade effects because US marginal exports were food not cotton. The USA exported only a little and was a small player in world markets. With elastic supply and demand, food exports adjusted to the tariff and international prices were unaffected. Competition between imports and domestic production shows that much of American manufacturing was dependent upon tariff protection (Harley, 1992a).

The geographical position of the United States gave little incentive for the government to pay attention to foreigners in setting tariffs, and domestic votes mattered a good deal. Thus the forces at work on American tariff making could be to a greater extent than in other countries those of interest groups. Revenue needs were sometimes the justification for tariff hikes, especially in view of the far greater proportion (than in Britain) of government revenue brought in by tariffs. The financial crisis of 1857 and the associated downturn in economic activity caused a falling-off in the revenue from the duties (Taussig, 1931). The reaction was embodied in the Morrell Tariff Act 1861, nominally intended to restore the rates of the 1846 Act, but the substitution of specific for *ad valorem* duties allowed rather higher rates. The Civil War required an enormous government revenue; duties were increased every session until 1865, a huge debt was accumulated, and an inconvertible paper currency was introduced.

Because the new balance of power favoured the more home-market-centred North, after the war increases of duties demanded by domestic producers were more readily granted. A case in point was the wool producers, who in 1864 were protected by a 40 per cent *ad valorem* tariff which had allowed them to expand output greatly. With the end of the war and the re-emergence of the Southern cotton economy, wool growers and manufacturers successfully applied for continued assistance. Manufacturers in 1867 were awarded 35 per cent *ad valorem* protection to offset the higher, protected, wool prices. The temporarily higher profits and increased sales allowed by this legislation did not serve the manufacturers' interests for very long because new entrants were attracted by the profits. Increased competition soon eliminated the excessive returns, and the industry found a new equilibrium, employing more of the nation's resources than in the absence of protection.

The iron and steel industry also lobbied successfully. By 1877 the duty on steel rails was effectively 100 per cent; when in 1880 the price in Great Britain was $36 a ton, the American price was almost double at $67. The consumers who paid the higher prices of protected goods only gathered enough collective political impetus to object during years of agricultural depression. By 1872 all internal taxes raised during the Civil War had been abolished, yet there was no corresponding reduction of the wartime tariffs. In that year, agrarian unrest against the high price of manufactures, combined with the government budget surplus generated by the great volume of taxed imports, was judged to warrant a tariff reduction of 10 per cent across the board. After the downturn of 1873, imports declined and so did customs revenue. The reaction was to repeal the 1872 Act after only three years to restore the lost revenue.

Despite this evidence of successful lobbying, the height of the 1870 tariff in an industry cross-section is consistent with revenue maximisation. The heaviest tariffs were placed on products with inelastic demands. Liquor, tobacco and consumer goods were significant causes of inter-industry tariff rate variation in 1870. But the share of value added in Republican-dominated states, the level of fabrication, the capital–labour ratio and skill intensity made no significant contribution, contrary to expectations if pressure groups were dominant influences (Baack and Ray, 1983).

Unlike the United States, many other countries outside Europe were unable to pursue an independent tariff policy in these years. Both China and Japan signed treaties placing a low upper limit on their tariff rates, as did the Ottoman Empire. The Chinese customs were managed by British Inspector-Generals, the second of whom, R. Hart, was in post from 1863 until 1908. In Latin America the most protectionist states were Brazil, Mexico and Venezuela with an average tariff of around 30 per cent in 1857. Argentina favoured relatively free trade with an average rate of about 15 per cent. The 1854 tariff did not anyway include the much more liberal Buenos Aires (Bairoch, 1989). In the British Empire, an Act of 1846 gave colonies freedom to pursue their own tariff policies. A treaty of 1854 established free trade between Canada and the USA in agricultural products. Canada raised duties on manufactures to around 20 per cent in 1859, and reduced duties on sugar, tea and coffee, despite the protests of Sheffield manufacturers, in order to raise money for the new railway system. The Australian colonies reacted to their freedom differently: New South Wales opted for free trade and Victoria for protection. South African colonists introduced a protective tariff in 1866/7. Because it had not been a colony in 1846, India was not given the same freedom in these years. This caused such indignation that the formation of policy has received much more intensive study than the effects.

Indian commercial policy was dictated by a combination of revenue requirements and the political pressures on the British government to maintain India as a market for British cotton textiles. When the British government took control from the East India Company in 1858, the import duties were 3.5 per cent *ad valorem* on cotton twist and yarns, and 5 per cent on other articles of British produce and manufacture, including cotton piece goods (Dutt, 1970, book 2, chs. 10 and 12). The duties were double on foreign articles. Heavy financial demands after the Mutiny required the abolition of all differential tariffs the following year, and some rates were pushed higher. Thereafter, throughout the 1860s, as in the British case, a rising revenue from trade allowed tariff cuts. But as a proportion of government revenue, the duties remained small. In 1856/7, customs revenue was £1.19 million of a total revenue of £31.92 million. By 1870/1 customs revenue was providing £2.61 million out of a total government revenue of £49.38 million.

To prevent Indian mills competing in the finer cotton goods, a 5 per cent duty on long-staple cotton imports was imposed in 1875. This was not deemed sufficient, however, and Lord Salisbury, the Secretary of State for India, insisted on the repeal of the 5 per cent import duties on cotton manufactures. The Indian Viceroy, Lord Northbrook (one of the Baring family, see Chapter 5), resigned in protest, and subsequent financial difficulties prevented compliance for some years. In 1879 the duties were eventually repealed, but the electoral advantage that the British government might have hoped to gain in Lancashire was not adequate, for it lost the General Election of 1880.

The consequences of the loss of tariff autonomy for Indian economic development are usually overstated. In other countries, protective duties often merely allowed higher profits and prices, less effort, and the diversion of more resources than ideal to the protected industries. If there were economies of scale in cotton textile manufacture, such that the larger output induced by tariff protection reduced unit costs and prices, then some benefit may have been forgone. Similarly, the existence of cost reductions from 'learning by doing' could result in a net gain from the tariff protection that the Indian industry was denied. The evidence for the pre-Civil War American cotton textile industry suggests, however, that there were only small benefits at stake (David, 1975, ch. 2). Despite, or because of, minimal tariff protection, a large modern Indian cotton textile industry did emerge.

Other dynamic arguments for net benefits from a high level of protection can be found, although their relevance to the Indian case is doubtful. In a growing economy, free trade is not the optimal policy for a large country because the growth of exports will cause an excessive deterioration in the terms of trade (Smith, 1977). There is an optimum tariff which offsets this deterioration (in the absence of retaliation). More important, perhaps, were the revenue limitations placed on the Indian government by free trade. Restricted principally to taxes, such as those on land that hardly increased their yields, (unlike import duties), public expenditure was tightly constrained.

Summary and conclusion

The rapid expansion of international trade and economic development from 1850 to 1875 owed much to the application of steam technology to land and sea transport. Those countries whose incomes were growing experienced an increase in the 'openness' of their economies – in the ratio of trade to national income – largely because of transport cost and tariff reductions. These reductions expanded the demand for exports and for imports more than proportionately to the growth of income, which was itself enhanced by the increased productivity allowed by the extension of international specialisation. The pattern of this specialisation was largely determined by national endowments of factors of production, especially climatic factors, arable or pastoral land, coal or labour, rather than by the gradual diffusion of technical knowledge.

International specialisation and the gains from trade transformed the value of national resources and shifted political power. The regulation of international trade was a response partly to the relocation of political power, or the changed objective of that power, partly to the need for revenue to finance state activity, and partly to political beliefs. Theoretical arguments suggest that economic liberalisation was a great benefit to the European economies. The United States participated for only a few years in this movement, but foreign trade was sufficiently unimportant in national income for the gains or losses from protection in America to have been similarly small.

Notes

1. Some evidence of the international association of nineteenth-century prices is in Hatton (1992).
2. As a logical construction, the HO theory is more limited than the Ricardian theory because the predictions are not unambiguous.
3. Unfortunately, once there are more than two countries and two traded goods in the international economy, this implication is not always valid; it is not necessarily true, only probable, that a traded good will be exported if its relative price exceeds its pre-trade relative price (Dixit and Norman, 1980, p. 8).

4 International factor mobility, 1850–75

The previous chapter explained the general determinants of the pattern of trade and trade policy. Trade also needed merchants to spot the opportunities for profits and to organise the shipping, finance and marketing. Their organisation could influence the gains from, and directions of, trade. For these people at the mid-century there were enormous opportunities for locating and carrying on business in new countries. If entrepreneurs were mobile, so was their capital. And less well-financed people were also willing to move between continents in search of a better material life. So flows of enterprise, capital and labour could provide a substitute or a complement for trade. Mobility of merchants stimulated trade. On the other hand, mobility of other factors might have reduced differences in comparative advantage. Both trade and factor migration should have tended to equalise the returns – wages, rents and profit rates – between economies.

The cosmopolitan bourgeoisie

International sales and purchases of goods made special demands on credit, information and trust. Because of their overseas connections, foreign merchants were often at an advantage – especially if they had settled long enough to build up domestic contacts in the more commercially open countries. Foreign merchants not only controlled international trade in most low-income states. They formed a world-wide 'cosmopolitan bourgeoisie' by the mid-nineteenth century (Jones, 1987, pp. 27–9, 66–9). A large part of British trade with Europe was carried out by German families with branches in Manchester, Bradford and Liverpool. Odessa's trade was largely conducted by Greeks and German Jews; in 1865 Germany superseded Britain as Russia's premier trading partner. By 1870 the number of Greek firms in Manchester exceeded German primarily because they were able to find markets for British cotton piece goods in parts of the world where British representation was weak – Turkey, Egypt and Africa. They imported grain through the Black Sea ports, especially Odessa, and to a lesser extent the Baltic, helped by the repeal of the Corn Laws. The Ralli brothers, who also operated in India, were pre-eminent, reputedly employing 4,000 clerks and 15,000 workmen in the 1850s (Chapman, 1992, pp. 157–8).

Julius Knoop, a German, began his New York career as a pedlar and sent his son Ludwig to an apprenticeship with De Jersey's, in Manchester. When De Jersey, specialising in the export of yarns and later machinery to Russia, went bankrupt in 1847, Ludwig Knoop effectively took it over, along with the Russian agency for Platt Bros, the textile machine builders. By 1875 Ludwig was probably the largest cotton buyer in the world, with extensive mill interests in Russia and branches in all the cotton markets in Europe and America. Knoop built and managed Russian cotton mills (122 spinning concerns in total by the 1890s), but also was active in the cotton trade with the USA and in banking.

Commodities new to international trade called forth new forms of trading enterprise. In the middle of the nineteenth century, the Siemens brothers, selling electric telegraph equipment, were a family multinational that merged old and new organisations. Werner (1816–92), the eldest, a lieutenant in the Prussian artillery, upset the Prussian postmaster-general with his technological honesty and so lost all Prussian sales. But he gained a massive order for Russia beginning in 1853. Karl was transferred to St Petersburg to conduct the business. He married and engaged in less than successful business diversification in Russia. Wilhelm (1823–83), later inventor of the Siemens open-hearth steel process, went to London, operated a subsidiary of the company, and married the daughter of a Glasgow engineering professor (Siemens, 1957). He was knighted in 1882. When Walter was killed in Tiflis supervising the construction of the Indo-European Telegraph Company's line, Georg took his place in negotiations at Tehran.

In the Far East, British informal and formal control created a vast free-trade area in which Chinese international entrepreneurs flourished (Latham, 1986, 1988). Tan Kim Ching's (1829–92) career earned him political honours and influence from Japan to Siam. He took over the family rice-trading business in Singapore and acquired rice mills in Saigon, Bangkok and elsewhere, as well as two steam-ships and, inland from Singapore, tin-mining concessions whose workers he could supply with rice. For linguistic and political reasons, Europeans were obliged to employ locals, compradores, to conduct their trade with China. The Chinese compradore system could subtly transform a British trading house into an unprofitable formal British business and a highly profitable subsidiary of a Chinese enterprise, headed by the compradore (Wang, 1993). Similar arrangements in Japan stemmed from the obligation on foreign merchants to restrict their commercial activities to an area of 25 miles around the treaty ports. They were therefore dependent on Japanese employees or merchants if they wished to buy raw silk or tea for export from the interior. Jardine Mathieson, a partnership based in Hong Kong and Shanghai, with a Yokohama branch from 1859, lost a great deal of money in the 1860s because of its lack of information about the merchants with whom it dealt, consequent upon the restrictions imposed on them. More generally, western merchants were unable to develop trade as much as they wished, and the view that they 'controlled' Chinese and Japanese foreign trade confuses appearance with reality (Sugiyama, 1988, pp. 52–76).

The fragmented British staple industries created an opportunity for merchant houses, especially in India but also in Latin America and elsewhere. These houses reduced the selling costs abroad of British manufacturers, who were not obliged to establish foreign sales organisations for their exports. Manufacturers could thus remain small and specialised. In the longer term, the merchant house facility may have exercised a harmful influence over exports compared with what sales departments might have achieved in larger firms. Within a unitary organisation, the flow of information back from the customer to the producer is likely to have been greater, and so too would have been the responsiveness to the market. The American style, just beginning to emerge, was to favour unitary organisations because American products were typically complex machinery that could not be distributed in the same way. The Singer Sewing Machine Company, which entered foreign markets in 1854, operated as a multinational company from the first (Davies, 1976).

Trade, factor price equalisation, and income gaps between countries

By creating opportunities for overseas sales and purchases, merchants indirectly transformed economic life at home. Because of the export possibilities open to the wheat-growing areas of the United States, the price of wheat in terms of textiles was higher there than it otherwise would have been. Therefore so too was the value of the wheat-growing land. Conversely, the availability of wheat imports to Britain reduced the market price of British wheat in terms of textiles. As British landlords recognised in their opposition to the repeal of the Corn Laws, imports indirectly also lowered British rents relative to wages. But thanks to transport costs, even in 1870 wheat prices were more than 60 per cent higher in Liverpool than in Chicago, and 20 per cent higher than in New York. The London–Cincinnati meat and animal fats price differential was 93 per cent in 1870, falling to 18 per cent in 1913. Much lower was the cotton textile price gap between Boston and Manchester, at 14 per cent in 1870 and 1 per cent in 1913. Philadelphia's average iron products prices were 80 per cent above London's, falling to 20 per cent in the same years. International trade in goods was reducing differences in relative domestic factor prices that gave rise to trade (Samuelson, 1949). Convergence of factor prices tended to equalise incomes per head. Countervailing forces inhibiting income convergence include transport costs, tariffs and institutional or cultural barriers, any of which prevented the emergence of a single price for internationally traded commodities. In doing so, they also precluded a single price for the factors used to produce these goods, which the factor price equalisation theorem predicts under idealised conditions.[1]

In a later period, between 1870 and 1913, commodity price equalisation served to eliminate one-fifth of the US–UK wage gap, a figure exceeding actual convergence because of stronger US industrial performance (O'Rourke and Williamson, 1992). Wage–rental ratios rose in western Europe and declined in the regions of recent European settlement. Induced factor-saving biases in technology were important as well as price convergence, factor migration, capital deepening and the closing of the US frontier (O'Rourke *et al.*, 1993).

The steam-ship and the railway mainly affected intra-continental trade rather than longer-distance routes in the third quarter of the nineteenth century. Hence regions within Europe were opened up to the international market in these years. If the factor price theorem is relevant for this period, there should have been a greater tendency to equalise wages within Europe, rather than between Europe and the rest of the world. However, even within countries the size of factor price differentials remained substantial. Both between regions and internationally, trade and factor mobility induced changes in the expected directions. In particular, labour flowed towards higher-real-wage and labour-scarce areas. But migration was not sufficiently responsive to eliminate all wage differences between markets, and often needed to fight against countervailing, but temporary, collapses in the demand for labour. According to the evidence available at present, during the 1850s European national income gaps tended to widen (Bairoch, 1976). Britain, the richest country, was the second fastest growing economy, while Russia, with the lowest income, experienced the lowest growth rate. The next decade showed more signs of convergence. France and Germany both attained growth rates greater than those of the small open economies which took the top four places in the income per head league (Crafts, 1983). That these countries – Britain, Belgium, Switzerland and the Netherlands – were clustered together may itself be taken as evidence of the effects of trade, for being small, regional differences within them may have been less persistent and the cost of trading with their neighbours was relatively low.

European income growths diverged primarily because national institutions and political developments limited the extent to which trade and factor migration could take advantage of the new opportunities at home and abroad. The poor performance of the Spanish economy owed much to political instability. Between 1834 and 1868, 74 Spanish finance ministers were responsible for budgets typically arranged for short-term political objectives (Platt, 1984, p. 107). About 20 per cent of the government expenditure was spent on the Army; at the time of the 1868 revolution there were 504 generals on the Army list. Political uncertainty and diversion of resources to non-productive uses reduced overall productivity and lowered tendencies to convergence. However, politics did not entirely shield Spain from the growth-inducing impact of the international economy. The Crimean War boosted the economy into an upswing between 1854 and 1866, and export-oriented mining attracted substantial volumes of foreign capital from 1875 to 1881 (Vives, 1969, pp. 741–2, 744).

Austria-Hungary also remained backward, largely unintegrated with the world economy. Only in 1850 was the customs frontier between Austria and Hungary abolished, and the new tariff of 1852 merely replaced outright prohibitions on the importing of many articles with high duties. Unsuccessful wars with Piedmont and France in 1859 and with Prussia in 1866 similarly did not encourage economic development. On the other hand, from 1853 Austrian trade with the Zollverein was boosted by large tariff reductions. The empire as a whole fell behind the growth of western Europe until 1870 (Ranki, 1983; Good, 1992).

Even after political unification in 1861, Italian economic development was slow. Unlike Germany, Italy's customs union before unification remained at the project stage and economic reforms, with the exception of free trade after independence, were timid. Widespread discontent in the south imposed the additional expense of maintaining 200,000 soldiers there. The relatively small size (and unprofitability) of the Italian railway system reflected the poor prospects for economic growth and extraordinarily high running costs (Toniolo, 1990).

Outside Europe a distinction can be drawn between economies receiving labour and capital from Europe, and the rest.

Worker migration

International movements of merchants were on a small scale and less disruptive of traditional patterns of economic life than migrations of workers. Such mobility can be measured only when the borders are policed and individuals recognise the concept of a nation. National borders were irrelevant to nomad tribes, wintering in the plains and driving their animals up to the highlands in the spring. In Baluchistan on the north-west frontier of India as late as 1911, only 54 per cent of the population lived permanently under a roof and one-third were pure nomads, while the remainder divided their dwellings between tent and house.

Permanent migration was often induced by force of adverse circumstances. Failure of the Irish potato crop for the second consecutive year in 1846 was a disaster. Mortality soared; from a population of over eight million in 1845 nearly one million died in the next six years. Between 1847 and 1854, 1.6 million Irish left the UK mainly for the USA (Harkness, 1929). Table 4.1 shows that the very high Irish propensity to migrate continued through the 1860s. One of the beneficial effects was a rise in the average size of agricultural holding. Those over 15 acres increased and those under declined. Real wages rose strongly from the 1860s, substantially choking off emigration (Hatton and Williamson, 1994; Boyer et al., 1994). The Scots also showed a high tendency to migrate. Andrew Carnegie's family left Dunfermline in 1848 for the United States, after his master-weaver father despaired of finding suitable work. Although the father was no more successful in the USA, his son, Andrew, created US Steel and became a multi-millionaire.

Behind these movements was the 'push' of lack of employment and low wages, but equally important in principle was that there should be somewhere better to go. The same considerations apply to the integration of the labour market as to markets for goods and services, discussed in the previous chapter. If two national

Table 4.1 Intercontinental migration rates 1851–1880

	Annual average rate per 1,000 population, 1914 boundaries		
	1851–60	1861–70	1871–80
Ireland	14.0	14.6	6.6
Scotland	5.0	4.6	4.7
England	2.6	2.8	4.0
Norway	2.4	5.8	4.7
Denmark			2.1
Portugal		1.9	2.9
Germany			1.5
Switzerland			1.3
Italy			1.1
Netherlands			0.5
Sweden	0.5	3.1	2.4
Austria-Hungary			0.3
France			0.2

Source: Baines (1991).

labour markets are connected, shocks in one will spill over to the other. The forces that brought southern and central European labour markets into the world economy included population growth, which had not made its impact felt at the mid-century. Countries that were to rise to prominence as sources of intercontinental migration in the later nineteenth century – Italy, Austria-Hungary and Spain – are hardly represented in this period, only partly because the statistics are poor. Italians heading for South America embarked from Genoa where, since 1846, a regular service via Marseille and Spain sailed to La Plata and Brazil. Spanish arrivals in Argentina during the 1850s were few compared with Italians; in 1860 respectively 930 and 3,349, with respectively 376 and 1,633 departures. Total official emigration from Austria-Hungary in that year was only 2,032. Germans emigrated in large numbers after the 1840s. A Chilean emigration agency recruited Germans from the 1850s. Australian colonial governments took a leading part in encouraging migration by directly offering help with cheap passages and land grants (about 40 per cent of total immigration was assisted). Perhaps this explains why Charles Dickens chose to send the impecunious and fictional Mr Micawber there at the end of *David Copperfield*.

Ending slavery within the British Empire encouraged the demand for cheap labour in the tropics, in particular through indentured labour from India. An indentured labourer signed a contract to work abroad for a specified period, often for specified wages, in return for the passage money, and sometimes the return fare. The contract was a way of overcoming the difficulty that poor people experienced in borrowing even for productive investments. The drawback was that, once abroad, the labourer was often at the mercy of the employer, who tended to control the state apparatus. A fairly steady stream of indentured workers travelled to the British West Indies (including British Guiana) in the third quarter and to Mauritius. Such migration was regulated by the Indian authorities, who stopped emigration to British Guiana in 1848–51 because of the high mortality on the voyages, to Natal in 1866 and 1874 because of unsatisfactory labour conditions, and to Jamaica in 1863 because of economic depression and the unsuitability of the would-be immigrants (Findlay Shirras, 1929).

Japanese labour markets remained unintegrated with the world economy because emigration was forbidden on pain of death, by the 'seclusion law' of 1638. General emigration was not legalised until 1885, but an imperial edict of 1871 urged young men and women to seek *education* in the West. Population pressure could be largely absorbed in the series of Japanese territorial acquisitions beginning in 1872, which by 1910 had increased the land area available to Japan by 75 per cent, while adding only 18.5 million to the population (Ichihashi, 1932).

A great deal of international migration was physically, not just economically, coerced. After 1820 a quarter of a million Indian, French, Spanish, Russian, French and British convicts were shipped across the world's oceans. Russian-registered migrants to Asiatic Russia from the 1850s to the 1880s were predominantly prisoners and exiles, numbering 100,000 in the first decade, rising to 180,000 in the third. Peasant migration which was to rise markedly by the end of the century, averaged 90,000 a decade.[2] The number of non-free British whites sent to Gibraltar and Bermuda was greater than Chinese contract labourers taken to British Guiana and the British Caribbean. More British convicts arrived in Australia after 1833 than before. More European Russians went to Asian Siberia between 1800 and 1918 than the intercontinental flows of Indian contract labour. French whites sent to New Caledonia were more numerous than the African contract labourers brought to the French Caribbean islands. After the mid-century, however, free labour entirely dominated international migration.

The opportunities for the profitable mobility of free labour arose mainly from rural–urban migration and, between countries, migration from the old countries to the new. The movement of indentured labour from India and China to plantations was an exception to the rural–urban pattern. The time and costs spent on long-distance travelling were reduced by the steam-ship even when steam was uneconomical for long-distance freight. The simple Hecksher–Ohlin theory predicts that such migration would normally reduce the volume of trade, but the reverse was true because the natural resources of the United States, Australia and New Zealand were complementary to the mobile factors; without factor mobility there would have been less international trade because the natural resources on which it was based would not have been utilised.

The first major influx of immigrants to the United States was in the decade after 1844 when 2.87 million Europeans joined a population of only 19.5 million. Before the Civil War, immigration preceded railway building and followed coal output. After the war both immigration and coal output lagged behind railway building (B. Thomas, 1973, pp. 92–4). Before the 1870s, when the exploitation of natural resources gave transport developments the principal role, the rate of American expansion was conditioned by new labour which built the railways. The second influx of migrants, from 1863 until 1873, also preceded fixed capital investment. Subsequently, railway building ceased to be the dominant force that it had been, and changes in migration, such as the third expansion of 1878–88, were induced by changes in the general level of investment.

The complementarity of natural resources and capital equipment also suggests why factor prices were not equalised by migration between Europe and the United States. Paul David has maintained that nineteenth-century mechanical technology happened to involve a greater input of 'land' per unit of output when operations were mechanised to save labour (1975, pp. 87–91). The woodworking machinery which was popular in America and neglected in Britain was not only labour saving but also wasteful of wood (Ames and Rosenberg, 1968). American cotton-spinning machinery was not only more capital-intensive than English equipment but also required a greater input of longer-staple cotton (a more costly grade of raw material) per pound of yarn (Sandberg, 1969). The efficient use of mechanical reapers required a level, stone-free farm terrain, arranged in large and regularly shaped enclosures, a specific natural resource input that in the mid-nineteenth century was obtained much more cheaply (relative to the price of grain) in the United States than in the British Isles (David, 1975, ch. 5). Hence in America the capital formation encouraged by the greater possibilities of jointly substituting natural resources and capital for labour may have been responsible for driving up the relative price of labour.

New economies smaller than the United States were more dependent on international factor mobility. The movement of British factors of production to Australia, particularly to supplement local Australian savings, dominated economic development more than in any other growing new country (Butlin, 1964, ch. 1). Australia was a major destination for transportation of convicts. These were youthful workers with high literacy, work skills and physical fitness. In 1840, 71 per cent of the male Australian workforce were convicts or ex-convicts (Nicholas, 1988, p. 53). During the 1860s overseas borrowing from Britain amounted to approximately half of total investment. British funds dominated financing of pastoral assets and were a major part of the finance required for communications development. This transfer, the selling of the greater part of exports to Britain, and the purchase of most commodity imports in Britain, all suggest a dependent relationship usually assumed to be exploitative. In fact, Australian living standards appear to have been considerably above those of Britain (£46 GNP per head in 1861 at constant 1911 prices) and the rate of growth in Australia was higher (1.4 per cent per annum in GDP per head between 1861 and 1877) than in Britain. The flows of British capital and labour increased GDP by 4.9 per cent per annum, much more than labour productivity, from 1861 to 1877. Population grew by 3.5 per cent per annum, considerably above all other countries in the western world, and a substantial portion of the increase came from immigration. Between 1861 and 1890 about two-fifths of the addition to population was due to migration, almost all from Great Britain and Ireland. The Australian economy clearly gained substantially from Britain.

The relevance of Rybczynski's theorem

Suppose we represent the regions of recent European settlement and their trading partners as consisting of two factors, land and labour. Rybczynski's theorem (see, for example, Ethier, 1983, pp. 100–1) implies that, if the conditions of the Hecksher–Ohlin theory are satisfied, then the growth of a country's labour force through immigration will, in the absence of other changes in the endowments of the economy, raise the output and export of labour-intensive goods by proportionately more than the increase in the labour force. The output of land-intensive goods will actually fall. The land-intensive goods are agricultural produce and the labour-intensive goods are manufactures. We do not observe this pattern in the United States, Canada, Australia and Argentina in these years because the supply of land in these economies was being raised by transport improvements at the same time as immigration grew. Once the effective supply of land could no longer be increased – the closing of the United States' frontier is traditionally dated to 1890 – continuing immigration could be expected to bring about the Rybczynski predictions.

Capital movements

When labour migrated, the factor reward, wages, went with it. The migration of capital, by contrast, gave rise to a stream of profits and interest back to the source economy. In the case of labour migration, the total effect is the sum of the factor supply increase in the regions of recent settlement and the corresponding reduction in Europe. The abundant natural resources of Australia and the United States meant that labour productivity was higher there than in Europe, and therefore total world output increased. A similar productivity effect was also generated by foreign investment, but the distribution of the rewards between the source and host economies was likely to differ from the case of labour migration. The economy that bore the 'cost' of raising the capital received a direct reward.

Of the two main capital-exporting countries, Britain tended to invest in the land-abundant regions, thereby reducing the costs of its food and raw material imports, and France tended to invest in Europe. By 1851 Frenchmen had lent about two billion francs to foreign governments, and 0.5 billion francs had been directed to private projects (Cameron, 1961, p. 85). Spanish railways were substantially financed and built by the French. Platt (1984, ch. 5) estimated that, in 1870, 40 per cent of Spanish railways had been financed from abroad.

After 1851 a much increased proportion of French investment, almost one-half of new foreign investment, went into enterprise. Railways alone, on the Iberian and Italian peninsulas, in central Europe and in Russia, accounted for one-third of this investment. Banking, mining and metallurgy were also key sectors. Perhaps the most powerful institution for French foreign investment was the

Crédit Mobilier of the Pereire brothers, Isaac and Emile (Jenks, 1971, pp. 242–5). This was a form of industrial bank intended to assemble resources by the sale of shares and obligations to small investors. At the height of its power it handled about 30 per cent of the new security business arising in Paris. Crédit Mobilier controlled a huge system of state-built Austrian railways and purchased another from the Russian government. It established subsidiaries in Spain and Holland owning mines, gasworks, shipping companies and railways, all financed by the bank. Both Crédit Mobilier and the Rothschilds, who were financing Austrian railways, were interested in persuading the domestic market to hold the shares, so that French capital was not greatly involved after the market potential had been demonstrated. The Dutch were the largest foreign holders of Austrian public debt in 1868 (Platt, 1984, pp. 90–2).

British capital, like French, went mainly to foreign governments (often for expenditures of no benefit to the people), next to the railways, and then to the public utilities. Over the five years before Disraeli bought the Suez Canal shares (without the consent of Parliament) in 1875, British earnings on foreign investment amounted to at least £50 million per annum and capital exports ran to £25 million more than this figure (Imlah, 1958, p. 73). The following year the position was reversed, and Britain collected more than twice as much income from its foreign property as was newly invested abroad.

Direct investment in industry, rather than lending to governments, tended to follow the trading connections which first provided the information about investment opportunities. The Welshman John Hughes founded his New Russian Company in 1871 to take up a concession of coal- and ore-bearing lands in the Donetz basin, after originally becoming interested through supplying iron for the construction of a Russian battleship. A settlement he founded was named Yuzovka after him.[3]

Where the regions of European settlement were concerned, it seems likely that the complementarity of the factor flows from the Old World with the abundant natural resources resulted in persistently higher wages than in Europe, instead of factor price equalisation. In other areas, such as Russia, the informational difficulties and institutional restrictions (Russian serfdom was only abolished in 1861 and the *mir*, village collective, was almost as restrictive) were probably too great to permit the equilibrating tendencies of the international market to operate. The contribution of foreign capital to the development of the United States was probably only about 10 per cent of total capital investment in most decades. In particular sectors, foreign investment could still attain high proportions. Underdevelopment of the domestic banking system explains the high US dependence on borrowing employed for trade and other short-term credit purposes. By the 1860s the United States was already a sizeable economy by European standards, and its rapid growth could hardly have been *largely* financed by other industrialising countries. Their labour contribution was rather greater. In 1870 more than one-fifth of the workforce was foreign-born (Kuznets, 1971).

Income and welfare

Did more international trade make most people genuinely better off? Could associated changes in economic structure have generated offsetting adverse effects? If they did then possibly relative incomes per head are poor guides to relative average levels of welfare. One of the most important structural changes associated with rising money incomes was urbanisation. Britain with the highest European income per head at the mid-century was also the most heavily urbanised. The greater part of the French population in contrast continued to live in the country. The French thereby saved on investment costs of urbanisation, such as sanitation, which may have given an upward bias to British income (O'Brien and Keyder, 1978, p. 188).[4] Urbanisation also replaced unmeasured home production with measured output of market-produced goods, perhaps giving another upward bias to trends in measured British income. Even more fundamental is the possibility that the French land tenure system restrained population growth, while the British pattern of urbanisation and industrialisation encouraged an explosion of population.

These objections to income per head measures are of much wider significance than the comparison of Britain with France. Though Britain apparently had the highest standard of living in Europe as measured by money income, the life expectation of the British male was by no means the highest, despite the likelihood that this would have been regarded as one of the most important elements of living standards. French life expectations were lower than British as the income rankings predict, but the Scandinavians were well above, contrary to the income rankings.[5] In 1840 the Norwegian male at birth could expect to live 43, the English male 40.2 and the French male 38.9 years. Four decades later, the relative expectations, at 48.5, 42.4 and 41.1 years, had not altered.

Another source of bias arises from the use of exchange rates to compare one country's income per head with another (Kravis *et al.*, 1978). The exchange rate is determined by the goods and services that are traded internationally, but these usually amount to only a small proportion of the consumption of the average person. The prices of non-traded goods can vary greatly from country to country precisely because they are not traded internationally. The poorer the country, the lower the wages which usually comprise most of the cost of these non-traded goods, and hence the lower are their prices relative to richer countries. Thus there will be a systematic tendency to understate the real income, the goods and services that can be bought with the available money income, of poor countries relative to rich. Countries with larger non-traded sectors will have a more biased income measure. If the traded sector is increasing – because, for example, productivity tends to increase faster in traded goods – there will be a tendency to overstate the growth of real income using exchange rate conversion factors. On the other hand, if the size of the non-traded sector, such as housing, government administration and domestic service, is expanding in a poor country, the tendency for convergence of factor rewards is understated.

In a comparison of long-term growth rates of real per capita income between India and the United States, Heston and Summers (1980) investigate the order of magnitude of this bias. Between 1870 and 1970 they calculate that real income per head increased in India by 75 per cent and in the United States by 470 per cent, implying a relative decline in Indian income to 31 per cent of its 1870 standing in relation to the USA. On the basis of exchange rate conversions, Indian income was 9.1 per cent of that of the United States in 1870. Using the prices of the two countries to calculate the relative purchasing power of the two currencies, Indian real income per head was perhaps one-quarter of US income in 1870. Exchange rate conversions of national income per head, as this example shows, must therefore be accepted only with reservations.

Economic growth and international economic relations

The comparative advantage theory of trade suggests that income will be increased by opening an economy to trade. This does not, however, constitute economic growth, in the sense of a sustained rise in income per head. Roughly speaking, theories of growth may be classified into those which are demand-led and those in which growth is generated from the supply side. An eighteenth-century conception of trade was that it increased the capacity and the income of the economy by providing 'a vent for surplus' (Myint, 1977). This doctrine closely resembles the later export-led growth theory, in which foreign demand encourages investment in export industries, raising income and stimulating further investment and increased income (Beckerman and associates, 1965; Batchelor et al., 1980, ch. 7).

Supply side growth theories usually emphasise the efficiency of the price system in balancing supply and demand. This equilibration prevents the emergence of unemployed or underutilised resources on any substantial scale unless certain institutions prevent price adjustment. If the price system is working in this ideal fashion, an increase in the demand for exports will tend to divert investment and employment away from those industries mainly supplying the home market, so that there will be no increase in growth. Increased investment can temporarily increase growth rates, but it can take place only at the expense of a reduction in consumption, in current living standards. The only way in a closed economy that growth in income per head can occur is through increased technical progress. The introduction of foreign trade raises the growth rate of the economy temporarily as the gains from trade raise income and savings, and the savings are ploughed back into industry as investment. Diminishing returns to the increased investment set in and the growth rate falls back to that determined by technical progress.

Trade can permanently increase the growth rate when there is a natural input, such as ores, semi-processed metals or agricultural products, for which it is difficult to substitute other factors (Black, 1970). In this instance, the long-term growth rate of the isolated country is set by the slowest-growing natural input. The growth rate can be raised if opening the economy to trade increases the growth of available supplies of this input. This may be considered to be 'import-constrained growth'. France's economic growth may have been raised by the opportunity to increase the supply of coal, and British economic growth certainly owed something to imports of raw cotton. Indian economic growth may have been enhanced by the importing of railway equipment, and Australian growth was boosted by imports of British capital goods.

If trade were important to growth we would expect to observe, as in fact we do for all economies expanding between 1850 and 1873, a rise in the ratio of trade to national income. Demand for poorer economies' products was strongest around the mid-century because of changes in British supply conditions as well as in British demand. The switch to land-economising methods of cultivation in Britain helped create a strong demand for Peruvian guano (see Chapter 1), while the gradual exhaustion of indigenous deposits of copper ore simultaneously with rising demand for copper in the railway and shipbuilding industries fostered the expansion of Chilean copper exports. Export-led growth shifts resources into the export sector. With import-constrained growth, the removal of supply constraints increases imports, which must be paid for by more exports or capital exports. But the same observation is equally consistent with increased trade in a static comparative advantage model, where growth is independent of trade in the long term. However, when 'temporary' increases in growth may last half a century, as seems possible in neo-classical supply-side models, the distinction between these static and temporary increases, and the permanent increases in growth rates, may not matter much in historical interpretation.[6]

In the course of economic growth, comparative advantages will tend to change as industries increase productivity, and the factors of production are accumulated, at different rates. It is even possible that a country need not benefit from growth through the accumulation of its relatively abundant factor: for example, labour in a densely populated poor country ('immiserising growth'). The country will increasingly specialise in the production and export of goods which use a large proportion of the abundant factor, and increase imports of other goods. The price of exports relative to imports (the terms of trade) will fall, and the resulting loss of income may exceed the gain from the increased supply of the factor.

Continental Europe did not benefit from free trade in these years, but was made worse off by such immiserising growth, it has been claimed (Bairoch, 1972). Railways effectively increased the supply of land which was already abundant, and forced down food prices relative to manufactures. Agricultural productivity fell because the industrial sector could not absorb the labour force that was being made redundant by increased imports of cereals, and the labour remained on the farms. National income fell because agriculture was a large proportion of total output. The removal of protection from industry exacerbated the inability to absorb the surplus labour from agriculture. This account assumes a large rise in agricultural underemployment. More food and lower food prices imply higher living standards unless there is a large permanent rise in unemployment or underemployment (as the immiserising doctrine maintains). It is also very pessimistic about the ability of the industrial sector to find comparative advantages.

'Staple theory' is a more optimistic description of trade and growth. As a hypothesis, it proposes that for land-abundant economies the characteristics of the major export influence or determine the pace and pattern of development (Watkins, 1963; Fogarty, 1985). Originally, the theory was formulated as an account of Canadian export-oriented economic development, based first on the beaver fur and subsequently on timber exports to the UK. As logging cleared the land, agriculture began to assume a role as a source of exports from the 1850s. Whereas, in the days of the fur trade, ships returning from European markets carried items for trading with the Indians who supplied the furs, there was no natural return cargo for the timber ships. That created a possibility of cheap passages for immigrants to the United States, and eventually, at the beginning of the twentieth century, to Canada as well. Argentina's cattle hides and wool were comparable staple exports in this period. Australia's 18 million sheep by 1850 provided considerably greater wool exports. Southern slave plantation agriculture in the USA offers a possibility of staple 'path dependence'. The mode of production of one dominant commodity influences the possibilities of future development – a much stronger example of the phenomenon with which Friedrich List was concerned. But for large countries it accounts only for regional development, for few specialise so much as to be dominated by one commodity. To explain the commodity specialisation it is necessary to revert back to the ratios of natural endowments central to Hecksher–Ohlin theory.

Summary and conclusion

Trade and factor mobility tended to equalise national incomes per head among the small open economies of Europe – Britain, Holland, Belgium and Switzerland – which were joined by France and Germany during the 1860s. In other European countries there were political and institutional barriers to this process. Massive endowments of natural resources in the regions of recent settlement, the United States and Australia in particular, were complementary to capital equipment. This allowed the persistence of higher returns to labour than were achieved in the countries from which so much of the labour migrated. Within national economies there were also similar effects which allowed the survival of wide interregional income differences.

The doctrine that there are gains from trade rests on a fundamental definition: being able to buy more of what one wants, for the individual or family, and for society as a whole, is an improvement in welfare. Goods and services that are consumed but not priced directly can create theoretical and practical dificulties for assessing the gains from trade. Increasing international trade required many workers to uproot themselves and find new jobs in towns and in different industries. Typically, an increasing proportion of consumption was supplied by the market and was therefore easily measurable. The loss of unpriced goods and services is not, and remains a potential source of bias in estimating the improvement in real incomes. But however the higher average incomes are evaluated, the extension of international economic relations allowed Europe and the regions of recent European settlement to support larger populations with longer life expectations. This is compelling evidence of the gains from economic development during the third quarter of the nineteenth century, in which the world economy played a key role.

Notes

1. Those conditions include identical technologies among the trading nations. Even with that condition not fulfilled, a *tendency* towards factor price equalisation can be expected.
2. When the trans-Siberian railway was completed in the 1890s, peasant migration soared to well over two million in the first decade of the twentieth century.
3. The name was later changed to Stalino and then to Donetsk (Milward and Saul, 1978, p. 407).
4. Because public health expenditures in Britain were minimal at the mid-century, income then was biased downwards in relation to welfare, compared with French income.
5. Even recent revisions of Swedish national income per head do not place Swedish (monetary) living standards above Britain's (Krantz, 1988).
6. Sato (1966) suggests that 90 per cent adjustment in between 25 and 37.5 years is a plausible result from such models.

5 The world monetary system, 1850–75

An expanding international division of labour was only possible because of the corresponding development of the international monetary system. Specialisation required exchanges between exporters and importers, often separated by great distances, so that goods spent a considerable time in transit and the contracting parties could have little direct knowledge of each other. As well as acceptable commodities or currencies for exchange, trust and finance were essential to foreign trade, since exchanges could not usually be conducted simultaneously. A promise that a payment would be made in the future was an almost inevitable component of transactions. The international monetary system in the third quarter of the nineteenth century evolved to provide these needs. In many respects, international arrangements were merely the domestic monetary system writ large, and accordingly, though providing similar advantages, they suffered similar problems. Periodic financial crises convulsed the international economy; bankruptcies and defaults caused workers to be laid off, unemployment rose, and national incomes fell. One of the tasks of this chapter is to examine the extent to which crises were domestically, rather than internationally, generated, and whether they were exacerbated or alleviated by monetary institutions or policy. But first we describe the organisation of monetary relations and the way they worked.

Merchant bankers and bills of exchange

Among the international monetary institutions of the mid-nineteenth century, their legendary private wealth made the Rothschild family the most glamorous. The Rothschilds were reputedly the best-informed men in Europe, using their knowledge to increase the mobility of European capital and to become even richer. The family fortunes can be traced to Meyer Amschel Rothschild, a dealer in coins, medals and antiques in the Jewish ghetto of Frankfurt. Meyer died in 1812 leaving a vast fortune to five able sons. The eldest remained to manage the ancestral house in Frankfurt; the others separately established banks in Vienna,

Paris, London and Naples, paying particular attention to arranging state loans. James in Paris and Solomon in Vienna were responsible for introducing the railway to their adopted countries. Lionel became the first Jew to take up a seat in the British House of Commons in 1858. In 1875 he helped out his friend, albeit on the other side of the House, the Prime Minister Benjamin Disraeli, by putting up the money to buy the shares in the Suez Canal. Because they lacked an office in Berlin, the Rothschilds adopted Bleichroder as their Berlin agent, thus contributing to the financial strength of Gerson Bleichroder, which permitted him to become Bismarck's banker (Corti, 1928; Stern, 1977; Wilson, 1992).

Almost as powerful as the Rothschilds was the House of Baring. Baring Brothers also originated in Germany; the House had been founded by an emigrant from Bremen to Exeter in 1717. They could claim two British Chancellors of the Exchequer and an Indian Viceroy in the family before the nineteenth century was out. Baring Brothers concentrated their London banking business to a much greater extent than the Rothschilds on extra-European transactions. The bulk of their business in the 1850s arose from trade between the United States and Britain. They bought and sold merchandise and securities on commission as well as for themselves, they operated their own ships, kept the accounts of selected depositors, and acted as financial agents for business houses and governments all over the world, especially in Latin America and the British Empire. But the mainstay of Baring was the 'acceptance' business, and here they had developed a system for assigning credit ratings to potential customers. In 1850 Baring's acceptances amounted to £1.9 million compared with Rothschild's £0.54 million.[1]

The acceptance business was the arrangement of short-term finance by granting 'acceptance credit'. Under these credits, bills of exchange were drawn out and accepted by Baring or another acceptance house, for a commission. The bill of exchange can be seen as a promise to pay a certain sum of money on a particular date – most usually in three months' time. A supplier would issue a bill for the value of goods it was shipping, and for which it expected to be paid at some definite future date; the supplier agreed to 'draw a bill' on the buyer, which acknowledged responsibility for eventual payment by writing on the bill its 'acceptance'. This 'acceptance' by a financial institution signified that the buyer was a good risk for a lender, because the acceptance house was liable in the event of default.

Just as the working of the banking system depended upon public confidence in the safety of their money, so did the bill of exchange system. Reputation and the availability of information about the risk of lending were essential. The London market did not regard Norwegian and Swedish bills during the 1860s as first-class risks because the exports of these countries were almost entirely timber or shipping freight services, supplied by small firms without international reputations. In contrast, the trade of the East Indies and China consisted of items of great value, and consequently was managed by large, wealthy houses, with established reputations, whose bills were first-class risks.

After acceptance the bill was sold to a financier; a lender would then 'discount' the bill (buy it for less than the sum payable in the future) and the supplier would thereby borrow. The difference between the purchase price of the bill and the value of the promise to pay was the interest charged on the loan. When the goods were sold, the supplier was able to pay the debt and withdraw the bill. The bill could change ownership (be rediscounted) during its currency should the original lender suddenly need cash. Bills of exchange were therefore a valuable means of facilitating both national and international trade at a time when transport was slow and communication difficult. By 1875 the total value of acceptances in London was perhaps £50–60 million.

London financial institutions accepted bills even for trade that did not touch British shores. In 1858 it was said that 'a man in Boston cannot buy a cargo of tea in Canton without getting a credit from Messrs Mathieson or Messrs Baring' (BPP, 1858). This type of bill finance by London maintained its importance for trade between ports of small volume in the third quarter of the nineteenth century, but once the volume of trade increased and, along with it, information and confidence, the intermediation of London in trade not otherwise connected with Britain was no longer necessary. By the 1860s, the New York–Bremen trade was no longer mediated by London, but the Bombay–Bremen trade, being small, continued to use London bills.[2]

Not all types of bill were well regarded. The finance bill, issued to provide finance not for a particular transaction but generally for working capital, on occasion was discounted without adequate attention to the collateral. Consequently, the over-issue of such bills, particularly international accommodation bills, was blamed for the financial panic of 1857 (Clapham, 1952, pp. 370–1). The distinction between types of bill was, however, often misleading. Foreign bills were sometimes used to provide short-term capital, and inland bills were often drawn to finance foreign trade. The difference between types of bill was a legal one, not necessarily indicating the type of transaction financed. When the mail from Australia and New Zealand arrived only every few weeks, a London merchant who failed to receive in one mail the payment for his goods would sometimes raise new credit on bill finance until the next mail (Nishimura, 1971).

As might be expected from this example, improved communications reduced the need for bills by cutting down the time spent by goods in transit. By introducing greater certainty into international transactions, the telegraph cut stock levels and allowed the money markets to finance the inventories previously held by great merchants. The level of raw cotton stocks held in British ports and mills was much lower after the spread of the telegraph than in the 1840s, despite the far greater volume of businesss (Newmarch, 1878; Chapman, 1984, p. 138). The two decades from the mid-century saw the international telegraph laid down, with the exception of a transpacific cable. A major advance in speedy international communication came in November 1852 with the first permanent direct communication between London and Paris by electric telegraph, through the lines of the Submarine Telegraph Company and the European and American Telegraph Company (Kieve, 1973, pp. 106–15). Land communication by telegraph was impeded by wet weather in the early 1850s, because of inadequate insulation. The first Atlantic cable of 1858 also stopped working very quickly because of an insulation fault. New York was in telegraph communication with San Francisco by 1864, before a permanent link with London was established and five years before the railway joined them. Not until 1866 did the *Great Eastern* manage to lay a cable that permitted permanent transatlantic communication. Fourteen years later there were nine cables across the Atlantic.

Telegraphic communications between Europe and the Far East and Australia were open from the beginning of the 1870s. The first lines to India were available in 1864–5 using the Turkish state line to Fao via Baghdad, and from Fao through an Indian government coastal cable to Karachi. An alternative route between Tehran and Moscow in 1866 joined with another Indian government line from Tehran to Bushire. Neither line turned out reliable. Even though the Britain–India link was expensive (£5 for twenty words) and erratic, it was a milestone for Australian communications. In 1865 for ordinary people to send a message to Britain and receive a reply would take five months.

Werner von Siemens' Anglo-German Indo-European Telegraph Company completed an improved line to India in 1869, which was operational in 1870 (Figure 5.1). In the same year the British Indian Telegraph Company connected Alexandria and Bombay by a line through the Red Sea and Aden. Singapore was linked via Madras and Penang by the end of 1870 (MacKenzie, 1954, p. 40).

Another communications route between the Far East and Europe was through Siberia (Figure 5.2). The Siberian telegraph was a Russian military project originally, reaching Irkutsk on Lake Baikal by 1863.[3] Use of this line was at first limited by the Chinese refusal to countenance a link on their soil. In 1869 a British businessman constructed a line across the country from Shanghai to a beacon on the bank of the Yangtze-Kiang River. A superstitious mob ripped it up, fearing for the countryside's good fortune. Six years later a Foochow–Amoy line was destroyed by locals and abandoned. It was therefore without permission and under cover of darkness that in 1870 the Danish *Great Northern* landed a cable connecting with Vladivostok in Shanghai. The following year the cable at Shanghai was cut into pieces and disappeared. But none the less Hong Kong was connected in 1871 and Japan was linked at Nagasaki. The Russian administration at last completed the entire Siberian line by the end of 1871, and the northern route was opened to the public on 1 January 1872. A twenty-word message from Europe to the East cost 100 francs. European news that had previously taken 15–30 days to reach the Chinese press now arrived in two days (Ahvenainen, 1981). At Shanghai the system linked up with the cables of the Eastern Extension Company which connected India with China, Singapore and Australia. The Indian exporter could now sell cotton by contract even before it was shipped, and therefore had no need of bill finance. Such were the difficulties of traversing the Australian continent that the Darwin–London link was operational, in November 1871, before the overland route was completed.[4]

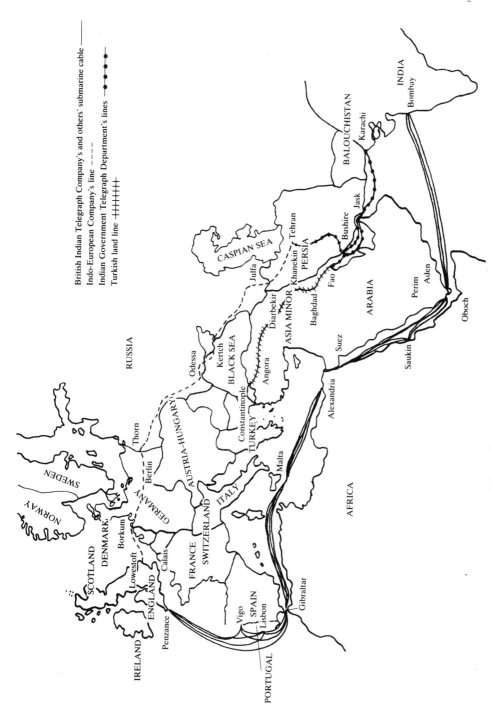

Figure 5.1 Telegraph routes to India in 1870

70 *A History of the World Economy*

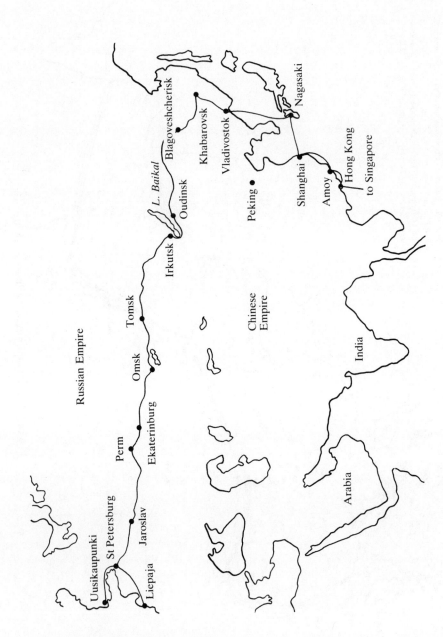

Figure 5.2 Siberian telegraph route to China and Japan in 1871

Exchange rates

The telegraph eventually affected international exchange and interest rates by improving communications. The value of one currency in terms of another, the market exchange rate, was determined primarily by the buyers and sellers of bills of exchange on the foreign exchange. Here were traded debts and claims that originated from exporting or importing or from other international transactions, such as investment or foreign travel. If England was heavily in debt to France, but lacked claims on France falling due to an equivalent value, those few English merchants who held credits in their Paris accounts could obtain rather more for them than when debts and credits were more equally balanced; there was a tendency for the franc to rise against sterling.

A nation with a precious metal monetary standard undertook, at least nominally, to buy or sell gold or silver (or in the case of a bimetallic standard, both) at a fixed and predetermined rate against the national currency. Furthermore, the precious metal was allowed to circulate freely within the national economy exchangeable against the domestic note issue. For every pair of nations that undertook this commitment against the same precious metal, there was a 'mint par' exchange rate determined by the metal content of the two coinages. The franc–sterling mint par, for instance, was determined as follows (Clare, 1891, p.74):

1. The Bank of England coined 480 oz Troy of gold 11/12th fine into 1869 sovereigns.
2. The Bank of France coined 1,000 g of gold 9/10th fine into 155 Napoleons of 20 francs each.
3. 1 oz Troy = 31.1035 g.
4. The gold equivalent of the franc and pound was therefore

such that $£1 = \dfrac{480 \times 11 \times 31.1035 \times 3{,}100}{1{,}869 \times 12 \times 900} = 25.2215$ francs.

It was unlikely that the mint par exchange rate would usually correspond exactly with the market exchange rate. But if the divergence became anything more than slight, there were opportunities for profit from 'arbitrage' in precious metals, from buying bullion in one country and selling it in another. If the sterling rate against the franc fell below 25.10 francs, for example, gold would flow out of England in considerable quantities because a given quantity of gold had become more valuable in francs than it was in sterling at the prevailing rate of exchange. The 'specie points' were the exchange rates on either side of the mint par at which the import or export of bullion (or specie) became profitable. Because of different assaying and melting charges, the specie points depended on the form of the metal, as well as on the transport and insurance costs, as shown in Table 5.1.

Although the United States was also effectively on a gold standard in the 1850s,

Table 5.1 Specie points for the sterling–franc exchange rate

Above 25.35f	Bank bullion sent from France to England	Below 25.20f	Refinable bullion sent to France
Above 25.32f	Shipments of market bullion from France	25.17½–25.15f	Market bullion leaves England
25.20–25.30f	Gold bullion from California or Australia bought by the Bank of England at 77/92d per standard ounce	25.12½–25.10f	Bank of England reserves drawn down
		25.07½–25.05f	Considerable withdrawals of reserves.

Source: Seyd (1870).

the great distance between America and Europe limited most bullion arbitrage to European countries. Between the time of shipment and arrival of the bullion from across the Atlantic, the conditions that had appeared to make the operation profitable could well have altered. Changes in national holdings of American railway bonds were a common way of settling the foreign exchange balance, and later the securities of various governments were widely employed in arbitrage operations (Clare, 1891, p. 74).

In the absence of exchange risk, bond arbitrage, like specie arbitrage, increasingly harmonised European interest rates as telegraph and railway networks spread. In 1869 discount rates in London were only about 0.5 per cent higher than in Paris, Frankfurt, Hamburg and Brussels. At Berlin and Amsterdam, rates were between 0.25 and 0.5 per cent higher than in London. Neither the Austro-Hungarian nor the Russian currency was linked to precious metals. Interest rates in those states therefore diverged from the rest of Europe because of risks of changes in exchange rates. Discount rates in Vienna were nearly 1.5 per cent higher than in London, and the St Petersburg rate was more than double the London rate. The money markets in Turin and Madrid were too small to exert much influence on European conditions (*The Economist*, 1869).

What was true of bonds was also true to a lesser extent of goods. The international movement of goods, or of precious metals, ensured that there were usually no great divergences in the (wholesale) price levels of countries that maintained precious metal standards, and therefore that also maintained fixed exchange rates with each other (Figure 5.3).

The new gold

Opportunities for the profitable international movement of precious metals, goods and bonds were expanded by the discoveries in California and Australia, which massively increased the annual quantity of gold supplied to world markets. As Table 5.2 shows, the annual average supply from 1851 to 1855 was more than ten times that of 1801 to 1810, and over the half-century before 1849 supplies

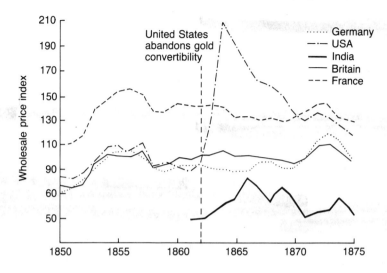

Figure 5.3 National price levels, 1850–75

Table 5.2 The relative value of production and market price of gold and silver, 1801–80

Period	Value of production (annual average)		Proportion of silver to gold production	Average price of bar silver (d per oz)	Ratio of market value
	Silver (£m)	Gold (£m)			
1801–10	8.002	2.480	3.226:1	60 7/16	15.61:1
1851–5	8.019	27.815	0.228:1	61 3/16	15.14:1
1856–60	8.235	28.145	0.292:1	61 5/8	15.30:1
1861–5	9.965	25.816	0.386:1	61 1/4	15.40:1
1866–70	11.984	27.207	0.440:1	60 5/8	15.55:1
1871–5	17.232	24.260	0.710:1	59 1/1	15.97:1
1876–80	19.103	24.052	0.794:1	52 16/15	17.81:1

Source: *Final Report of the Royal Commission on Gold and Silver* (1886, p. 6).

averaged only one-seventh of those available in the 1850s, barely sufficient to compensate for wear and tear of the coinage. At the time many argued that prices in gold standard countries would rise greatly, from the simple notion that prices are merely a quantity of money against which goods are exchanged (for example, Jevons, 1909). At least two other views of the price level in this period have been advanced, however.

The contemporary economists Thomas Tooke and William Newmarch believed that the new gold stimulated economic activity (already encouraged by transport and communication improvements and the freeing of trade). Specialisation and the division of labour were extended by the greater availability of means of exchange. So output rose, rather than prices (Tooke and Newmarch, 1928). The real danger, Newmarch thought, was that new gold supplies would fall. The evidence of the 1870s confirmed his view, he believed: wholesale prices in London in the period 1831/45 to 1870/7 rose by only 10 per cent, which, considering the wars of the 1860s and 1870s (the American Civil War, the Franco-Prussian War and the Russo-Turkish War) and their disruption of production, was remarkably little. Newmarch's hypothesis that the increased spending permitted by the new gold primarily increased output, rather than prices, is a logical possibility. But in developed economies it is likely to be only a short-term effect, as capacity utilisation is increased by greater demand.

A third causal pattern is that during the period 1850 to 1870 it was investment in the new opportunities appearing in coal, iron, railways, steam-ships, textiles and clothing that expanded production, which in turn caused a rise in prices (Schumpeter, 1939).[5] Money, in this account, was passive. Output growth was independent of monetary expansion, although it may have encouraged the development of bank credit.

Consistent with this last view, the period of the fastest expansion of the British money stock, from the late 1840s to the late 1850s, did not correspond with the most rapid growth of British GNP at current or constant prices. Nor did it match the fastest decadal rise in the price index, which was from the late 1850s (Table 5.3). It may therefore have been purely fortuitous that the period of rapid output growth (1846–73) in the premier gold standard nation coincided with the rapid growth of gold supplies.

When the total money stock grew faster than in any other period of the nineteenth century, bank and credit money was the principal driving force. The share of bank and credit money in the money stocks of the United States, Britain and France increased from 37 per cent in 1848 to 59 per cent in 1872 (Triffin, 1968, table 1.2). Banking innovations and extensions were probably a response to the rise in economic activity. The decline in British monetary growth in the 1870s occurred primarily because of a fall in bank deposit growth, rather than because of a fall in the growth of the coin circulation (Collins, 1981). Similarly, an estimate for the whole world shows bank deposits and capital growing at 5.6 per cent per annum from 1840 to 1870, but at only 3.5 per cent per annum from 1870 to 1890 (Mulhall, 1892). Had the new gold been the main cause of monetary growth, bank deposits would not have increased their share of the money stock. If

Table 5.3 The growth of money, national income and prices in Britain, 1846–80 (compound annual growth rates, per cent)

	Gold and silver coin in Britain	Total money stock	GNP at factor cost, constant 1900 prices	Implicit price deflator
1845/50–1856/60	5.2	4.7	2.3	0.49
1856/60–1866/70	3.1	3.9	2.5	0.82
1866/70–1877/80	2.8	2.8	2.0	−0.42

Source: Collins (1981).

the new gold did not cause the monetary growth, it is unlikely to have boosted total spending and output substantially. Hence the decline in monetary growth in the 1870s most likely stemmed from a retardation of world economic growth, not vice versa.

The international transmission of price increases

Why then did prices rise so little in response to the new gold, if Tooke and Newmarch were mistaken? As long as a large part of the growing world economy was linked by fixed exchange rates, the new gold could not drive up prices of internationally traded goods in one country faster than those of that country's partners on metallic standards. Therefore price increases depended mainly on the growth of the fixed exchange rate world output and money stock, rather than on the output and money of any one country.

The initial impact of the new gold was to raise the prices of all domestic commodities in the mining economies of California and Australia, as resources switched from agricultural to mining production. Australia quickly became a net importer of food instead of, as formerly, a net exporter (just as South Africa did with the gold discoveries later in the century). Australia exported gold to Britain (with which Australia effectively shared a common currency) in exchange for manufactures, as Chapter 1 showed, and to other countries in exchange for agricultural produce. Because the new discoveries made gold cheaper by about a half in terms of commodities, Australian prices, being gold prices of commodities, rose. Foreign goods therefore sold more easily in Australia than elsewhere, and their prices were accordingly bid up in gold terms. But many of these prices indirectly constituted a cost of gold mining, for miners had to buy the goods with their earnings. The costs of gold mining therefore rose, stimulating a diversion of resources back to agriculture. By 1870 gold production in Victoria had fallen to half the level of 1856, and exports other than gold had almost doubled, but at a now higher level of prices in Australia and in the rest of the world (Bordo, 1975).

Prices responded less to the new gold than might have been expected also because of the bimetallism of France and, until 1853, of the United States. Bimetallism created considerable difficulties for the monetary systems of the two countries. In the United States the divergence between the legal and market ratios of silver to gold had become sufficiently great for 'bimetallist arbitrage' to be profitable by 1849. The divergence between the ratios continued to widen until by 1853 the market ratio was 15.4:1 and the legal ratio 16:1 (Laughlin, 1897, pp. 75–82). Bimetallist arbitrage worked as follows: 1,600 ounces of silver could be taken to the bullion market and exchanged for 100 ounces of gold with 60 (= 1,600 − 1,540) ounces of silver left over. The payment of debts in gold rather than silver, as bimetallist legislation allowed, was profitable as the above example shows, and this tended to increase the speed at which gold drove out silver. Equally profitable was the sale of silver for gold, and the sale of the newly acquired gold to the Mint. Congress reduced the mint ratio to less than 15:1 in 1853 to stop these arbitrage operations. Had the Mint been willing to take an unlimited amount of silver for coinage at this rate, silver would have tended to replace gold at the Mint, because the market gold–silver ratio was still above 15:1. However, the law provided that the Mint need only take the quantity of silver necessary for making small change. Since it no longer paid to melt down these devalued coins, a supply of them quickly came into use, but gold remained the main coin currency. In the ten years after the Gold Rush of 1849, $400 million of gold coins were struck. Legislation in 1873 dropped the silver dollar from the coinage list, but these pieces had disappeared from circulation many years before. The real abandonment of the bimetallic standard took place in 1853 not 1873, as was to be widely alleged subsequently by those who favoured monetary expansion (see Chapter 8). At the end of 1861 the link to gold and silver was in any case abandoned (except on the West Coast) and it was not resumed until 1879.

As in the United States during the early 1850s, gold was also driving silver from the French currency circulation, and the silver was being shipped to the East. By the 1860s France was virtually on a *de facto* gold standard (Parker Willis, 1968, p. 9). It was against this background that Chevalier's 'Parachute' thesis was formulated. The 'Parachute' limited the rise in the relative price of silver to the increase sufficient to displace silver by gold in the countries whose currencies were fixed in terms of both precious metals. When the displacement was complete, Chevalier predicted that gold prices would fall substantially and silver prices rise (Sayers, 1933).[6] In the meantime, however, prices and exchange rates of the major European gold, bimetallist and silver standard countries were all tightly linked, and the gold price of silver rose only a little (see Table 5.4).[7]

The German price level moved in parallel with those of France and Britain (Figure 5.3), even though Germany held to a silver standard until 1871, France remained with bimetallism until 1874 and Britain maintained a gold standard. The silver that France lost from circulation went mainly to India, a silver standard economy. Because India did not produce silver in any great quantities, perhaps 30 million rupees in silver were needed each year merely to provide for wear, losses and population growth, if prices were to be maintained at the same level (Barbour, 1986, ch. 2). The large imports of silver from 1855 to 1866 were due mainly to borrowing to meet expenditure incurred because of the Mutiny, to the foreign investment in building the railway system, and to the increased value of cotton exports as a result of the blockade of the Southern cotton ports during the American Civil War. Unlike Germany, however, as far as can be judged, the parallelism of Indian prices with European prices was much weaker: between 1864 and 1869 the two sets of prices diverged probably because of the effects of foreign investment, famines and the foreign demand for cotton in India, as well as the greater distances involved in trade. The rupee showed a rise against sterling in the period 1845 to 1855 proportionately greater than the rise in the gold prices of silver. So too did the silver standard Dutch currency. As long as exchange rates moved, price levels had some freedom to diverge.

Table 5.4 Average annual exchange rates against sterling, 1845–59

Year	Paris (franc = £1)	Hamburg (mark = £1)	Amsterdam (guilders = £1)	Calcutta (rupee)	London silver prices
1845	25.92	13.135	12.725	22¼d	59½d per oz
1850	25.40	13.11	12.15	24¾d	60d per oz
1855	25.50	13.8	11.19	25⅛d	61¼d per oz
1859	25.35	13.55	11.16	24¾d	61½d per oz

Source: Newmarch (1860, p109).

Floating exchange rate regimes

The exchange rate of a country with an inconvertible currency gave the greatest freedom to national price movements. An inconvertible currency could not be changed into a precious metal at a legal rate. The value was determined solely by the goods and services that could be bought with it, both in the present and in the future. The external value of the currency, the exchange rate, was similarly determined by the relative strengths of foreign demand for the currency and domestic demand for foreign currency. A government therefore possessed the power to alter the domestic and external value of inconvertible money by changing the quantity in circulation. To European liberalism this power was an 'intolerable evil', because of the disturbance to contracts and expectations caused by the consequent changes in the price level (Mill, 1929, ch. 13, esp. pp. 544–6). This liberal doctrine had been 'tolerable effectually drummed into the public mind' of industrialising countries, but even there, the exigencies of war finance could require the abandonment of the precious metal standard.

In Latin America and in central and eastern Europe, liberal strictures against currencies and exchange rates influenced by state financial needs seemed to have been borne out. The failure to reform the money market may have contributed to the long-term retardation of Brazil (Pelaez, 1976). The Bank of Brazil received extraordinary privileges which allowed it to supply one-third of the banking services in Brazil as well as being responsible for financing the government's budget deficit and the resulting inflation. The growth in the national money stock was absorbed by rising prices, rather than by increases in productivity. Even after reforming its currency in 1843, Russia continued to increase its issue of paper roubles unbacked by gold, and the foreign exchange value of the rouble declined on trend against the metal-backed currencies. The rouble exchanged for 35 British pence in 1844 but for only 24 in 1880. Austria-Hungary was in a similar but less extreme position. During the 1870s, with an uncovered paper currency issue of over 600 million florins, the exchange rate of the florin fluctuated around 20 pence, only five-sixths of the value of the gold florin. Both Austria-Hungary and Russia suffered from erratic development during these years (Mulhall, 1892).

The experience of the United States with floating exchange rates, on the other hand, suggests that the liberal argument perhaps confused cause with effect. Convertibility of the currency into precious metals was not a means of disciplining governments so much as a manifestation of a governmental desire to conform to the liberal ideas of good behaviour. The costs of financing the Civil War meant that from 1862 to 1879 there was no official link between the US dollar and gold, and therefore no fixed parity between the US currency and the pound sterling. The average monthly dollar ('greenback') price of gold, and hence of sterling, varied widely, reaching a peak in 1864 corresponding to a price of more than $12 per pound, or 2.5 times the previous and subsequent fixed exchange rate (Friedman and Schwartz, 1963, p. 85). Prices in the United States needed to fall if the exchange rate was eventually to return to the previous parity, and the massive expansion of national output, especially agricultural production, combined with monetary restraint, achieved this goal. The decline in prices hurt farmers already loaded with debt, and fuelled the Greenback Party, which agitated for a plentiful supply of money to raise prices. This unsuccessful movement was surprisingly the only major political reaction under the floating exchange rate regime; by comparison with the twentieth century, the pressures for government spending in excess of tax receipts, which would have been possible, were small.

A second charge made against floating exchange rate regimes is that they are prone to destabilising speculation. The forces which determined the actual exchange rates of the gold and silver standard countries within the specie points could not be counted on to stabilise the exchange rate in the absence of these points. The American crisis of 1869 has been instanced as a classic example of the instability of floating exchange rates and the detrimental consequences for trade (Wimmer, 1975). During August and September 1869, Jay Gould led a determined attempt to increase the greenback price of gold in the New York gold market. This was equivalent to attempting a depreciation of the exchange. The gold dollar had been $1.31 in greenbacks. By 23 September the price had reached $1.41. The following day the greenback price fell from $1.62 to $1.34 in half an hour when the President authorised the sale of Treasury gold dollars. Gould claimed that he was trying to ease the massive US harvest exports by forcing down the exchange rate. Gold inflows into New York in September were a major impediment to Gould's speculation. The success that he achieved was due to the market's belief that the government supported his activities. Had the government stated its neutrality, the movement would have been broken earlier. Under a fixed exchange rate regime, the government would have been forced to act earlier by the gold outflow, which would have threatened the convertibility of the currency. To this extent the 'discipline' argument of European liberalism is supported.

Much of the public distrust of floating exchange rates was based not on the objections already discussed, but on a mistaken theory of value. A study of the Austro-Hungarian floating exchange rate shows that the contemporary debate on restoring the gold standard centred on a belief that it was precious metals that conferred value rather than the goods and services that could be bought with the currency (Yeager, 1969). In turn this might be accounted for by the comparatively late development and use of general price indices which measured the value of money (Fisher, 1922, appendix 4).[8]

Monetary unions

If all countries in a potential union already linked their currencies to the same precious metal, the general economic policies they could pursue would be very similar. A monetary union would involve little effective loss of sovereignty. Even so, the desire to link political objectives with the formation of economic institutions was paramount in two of the three unions of the period, and greatly reduced their effectiveness. The economic benefits of a monetary union were similar to those arising from trade in general: the integration of national monetary units was merely one facet of the integration of national economies by factor mobility, trade and the establishment of common institutions so that resources were put to their best possible use.

Changing relative prices of gold and silver were instrumental in the formation of the largest monetary union of bimetallist states. Better communications and increasing trade were also major forces behind the establishment of monetary unions. Perhaps the single most important cause, though, was the shared political values of the ruling groups in Europe, a belief in the benefits of trade and industry, and in progress. Mill, for example, asserted that 'political improvement' would eventually culminate in one world currency, as nationalistic irrationalities disappeared (Mill, 1929). Events in the German states underwrote his confidence. The Zollverein treaty required the standardisation of coin systems. At the Dresden convention of 1838, participating states adopted either the Thaler or the Gulden coins, both defined by their silver content. The convenience of the Prussian one-thaler piece, and the absence of legal tender status for bank notes, meant that the Prussian coinage spread throughout Germany, even in the Gulden states (Holtferich, 1993).

With the German example in mind, a consensus emerged at the Great Exhibition of 1851 that not only was an international system of weights and measures necessary, but so too was a common currency. Subsequent International Statistical Congresses reinforced that opinion.

In 1857 Austria and the Zollverein founded the Austro-German Union with a common unit of account, the silver mark. Three distinct units with a fixed rate of exchange provided the medium of exchange (Bartel, 1974). No real integration of money and banking systems took place, however, and the Union was dissolved in 1866 with the outbreak of the Austro-Prussian War. Austria-Hungary failed to restore convertibility of its currency into silver and so could find no place in the Union, which Prussia dominated.

The Latin Monetary Union proved more durable. The falling price of gold relative to silver during the 1850s produced difficulties for the French bimetallic monetary system. In response to the export of silver, the French adjusted the silver content of their coinage, thereby creating problems for economic relations with Belgium, Switzerland and Italy (Parker Willis, 1968, ch. 4). In 1860, 87 per cent of the circulation of subsidiary coin in Belgium was French, the Swiss had adopted the French standard of 1850, and the Italians had done the same in 1862. The Monetary Convention of November 1865 met to resolve these difficulties. According to Willis, France wanted to maintain the status quo because the Bank of France and French financial leaders found bimetallist arbitrage profitable. The other countries wanted to standardise on gold, as did a good part of the French delegation. The political domination of the French ensured that the Latin Monetary Union of 23 December 1865 codified the status quo, although it also militated against the success of the Union, because it inhibited voluntary co-ordination within the Union.

A more ambitious union was soon proposed instead. Thirty-three delegates from twenty countries attended the 1867 Paris International Monetary Conference. They recommended an international gold standard with coins of 90 per cent fineness. The international coin was to be a 25-franc gold piece. With small adjustment, the pound sterling and five dollars would be equivalent to this coin. A British Royal Commission was persuaded, but did not recommend any change in British monetary arrangements (Cottrell, 1992). Austria and Switzerland began minting the 25-franc 'international', but 'nationalistic irrationalities' proved too strong for Britain, and the upshot of the conference was a widening of France's bimetallist Latin Monetary Union.

When Austria joined the Latin Union in 1867, the terms of the treaty specified gold as the medium of exchange. Greece became a member in 1868. Germany moved from silver to gold between 1871 and 1873, and sold substantial quantities of silver. India could not absorb it all because of a growing indebtedness to Britain, and because the introduction of a note issue in 1862 and improved transport reduced the need for silver for transactions. The value of silver relative to gold began to fall, but not, judging by the timing, as a response to German silver sales, or to the demonetisation of silver by the Latin Union in 1878.

Once the value of silver in terms of gold began declining, any Latin Union member state could gain a financial advantage by issuing silver coins. Italy was particularly prone to indulge itself in this respect. The silver coins were then exported to fellow member states whose central banks would be asked to exchange the coins for gold at the agreed rate. From 1860 to 1865 Italian government spending, including expenditure on railways and the 'pacification policy' in the newly conquered south, was twice that of tax revenues (Parker Willis, 1968, ch. 7). In 1866 the Italian government declared its currency inconvertible.

Italy could not be persuaded to cease minting silver coins. As a compromise, the Treaty of 1874 shared the gains from silver minting with a rule of proportionate increases among member states. These were small enough not to be inflationary. France's agreed silver issue amounted to 5 per cent of metallic circulation over three years (Flandreau, 1993). Italy, however, did not acquiesce in the 1874 decision to move to a *de facto* gold standard, and at the 1878 conference announced its intention to continue coining silver. The Union in fact formally abandoned silver in 1878, restricting the amount Italy coined and extracting a pledge that this would be the last silver issue.

Abandoning gold and choosing silver instead as the basis for their currency was an option for the Latin Union. It was bimetallism that was unsupportable when sustained changes in the relative market prices of gold and silver occurred. The Union chose to standardise on gold because that was the currency remaining and because businesses preferred gold; they did so because of the pre-eminence of gold standard Britain in international trade and because large-scale commerce favoured the metal with the greatest value for the smallest bulk (Laughlin, 1897, pp. 167–9).

France's continued adherence to bimetallism and Germany's move to gold encouraged the Scandinavians to form their own monetary union towards the end of the third quarter of the nineteenth century, rather than join the larger Latin Union. The Scandinavian Monetary Union of Sweden and Denmark in 1873, joined by Norway in 1875, turned out to be the longest lasting of the three unions. It ended only when currency inconvertibility with gold, the central banks' response to the First World War, allowed national monetary policies and price levels to diverge (Bergman *et al.*, 1993; Bartel, 1974). Success was due to the essentially similar character of the equally open economies of Norway, Sweden and Denmark, based upon agriculture, engineering, forestry, fishing and trade. The gains from specialisation did not therefore require very great reallocation of resources. Identical gold coins circulated freely within the Union, but by 1885 gold was essentially a reserve currency, with more than half of Sweden's money stock being notes not backed by an equivalent value of gold in the Central Bank. A rudimentary clearing scheme for notes established by the central banks further helped monetary integration.

International fluctuations in economic activity

So far, this chapter has been principally concerned with the structure of the international monetary system and its normal working. We now consider the pathology of monetary relations. Ultimately, the ills of the system can be traced to the structure of bank and trade credit built on the cash base – precious metal coinage for those countries on metallic standards. Monetary policy was directed to ensuring that this structure did not become too large or too small for the needs of the economy, but the main focus was on the bullion reserves of the central bank, for those countries that had such banks. When reserves began to be depleted either because of foreign demands (an 'external drain') or domestic demands (an 'internal drain'), this was a signal for a rise in the price (discount rate) at which the central bank would lend funds to selected borrowers caught short of liquidity.

A higher cost of credit reduced the demand for funds and reduced the credit pyramid relative to the cash base. Thereby the central bank ensured convertibility of claims into precious metal, while at the same time preventing widespread bankruptcy and loss of confidence in the monetary system, by acting as a lender of the last resort. Lacking a central bank in these years, the United States was a source of instability for the international monetary system and was dependent on Europe for funds during crises.[9] The Banks of England and France were the chief regulators of the system, although in 1870 France ceased temporarily to maintain convertibility of its currency and so lost influence.

A shock to one economy maintaining the same metallic standard as another may well be passed on to the second country more strongly than if the two allowed their exchange rate to float. A crisis and depression in one gold standard country lowered the demand for imports and attracted bullion and foreign assets. Fewer imports allowed more room for expansion for domestic production and employment but reduced the exports and employment of the gold standard trading partners. The balance of trade surplus tended to lower interest rates and expand the money supply with similar effects. By contrast, under floating rates a crisis and depression caused a rise in that country's exchange rate, when import demand fell relative to foreign demand for exports, and the domestic currency therefore became scarcer in terms of foreign currency in the exchange markets. This higher exchange rate reduced the marketability of exports and prevented monetary expansion; both forces tended to ensure that the depression continued. The second economy remained largely insulated from the shock by the exchange rate movement.

The United States in 1857 offers an example of the first case, and in 1873 of the second. From the viewpoint of the rest of the world, the United States with a floating exchange rate in 1873 was less harmful than with a fixed rate in 1857, for in 1857 it could export its unemployment to the rest of the world by reducing its imports, whereas from 1873 until 1879 it could not. The greenback appreciated a small amount between 1873 and 1876, and in the following year appreciated almost 7 per cent more. The forces that caused appreciation under a paper standard would have been channelled into stimulating the domestic economy or arresting the fall of prices under gold standard conditions. In 1877 there were signs of an upturn in railway investment, building, manufacturing and mining. This indicates that under a gold standard deflation might well have ended by 1877 instead of 1879, as it actually did.

The classification of exchange rate regimes assumes implicitly the absence of a countervailing central bank policy, which is appropriate for the United States but not for western Europe. Though central banks could prevent widespread financial collapses, so long as they maintained precious metal standards with small reserves, they could not prevent the international transmission of crises. In many instances, the crises occurred simultaneously from similar causes in different countries, and it is therefore difficult to distinguish the extent to which a depression originated at home or abroad. However, the political crisis of 1861 in the United States supplies a clear-cut instance of unemployment being forced upon Britain from abroad, and the 1866 collapse in Britain had a similar impact on Australia. Probably the German crisis of 1873, a direct consequence of the receipt of French reparations, was responsible for the rise in British unemployment then. The French remained largely immune because they had not then resumed convertibility.

The widespread employment of British finance in world trade and the consequent impact of British interest rates on the rest of the world has sometimes led to the inference that the rest of the world had to suffer unemployment forced on it by Britain. Such an interpretation mistakenly assumes that British interest rates were uniquely controlled by British conditions and policy. The majority of the crises that required adjustment of national economies originated outside Britain in these years, that of 1866 being the exception.

The 1857 crisis

The state of the national harvest was an important determinant of national fluctuations in this period, and sometimes good harvests coincided in the major exporting countries. In 1857 three factors contributed to a flooding of the world grain markets: abundant harvests, the re-establishment of normal commercial relations between Russia and western Europe after the Crimean War, and the construction of new railways in the fertile Hungarian plain (Henderson, 1939; Hughes, 1956). The price of wheat fell and the whole US market was affected. The failure of the Ohio Life Insurance and Trust Company in August 1857 was the signal for a general panic. The fall of agricultural prices left farmers unable to service their debts and institutions that had lent them the money were as a result endangered. Panic withdrawals of bank deposits in the United States required cash from abroad, primarily from Britain. Sterling fell against the dollar, and gold was exported to the United States. To protect its reserves, so that Britain could remain on the gold standard, the Bank of England raised bank rate (the rate at which it was prepared to act as lender of the last resort) to 10 per cent by November 1857. The banking panic spread to British and Irish depositors, who demanded their money back; from the Bank's view there was now an 'internal drain' on its reserves as well as an 'external drain' (Hawtrey, 1938, pp. 25–7). In Britain two Scottish banks, one Liverpool and two leading London bill-brokers failed between 12 October and 11 November 1857. These and other failures, with the increase in bank rate, dragged down businesses in Hamburg in Germany: 150 firms with total liabilities of £15 million failed. The 1857 crisis then clearly originated in the United States and the wheat-exporting economies, but was exacerbated by a contraction of demand associated with bank failures, other bankruptcies and a failure to renew bill finance. Under the bill system, traders were liable to postpone purchases that would commit them to finding cash in three months' time, if there was a risk of other traders being insolvent. Traders' ability to meet their commitments depended on the bills they held for the sale of their goods being met on maturity. During the 1857 crises there was a general reluctance to buy goods by bill acceptance; orders for manufacturers therefore fell, unemployment rose, and consumers' incomes fell. High discount rates reinforced the contraction, reducing traders' demands and spreading the recession to South America, South Africa and the Far East (Kindleberger, 1989, p. 142).

The crises of the 1860s

Lincoln's election as President of the United States in November 1860 precipitated the secession of the slave-owning Southern states. New York had financed the greater part of Southern exports. The approach of secession, by removing this finance, caused a banking crisis in the South, and the North began to absorb gold. This switch in the USA from being a gold exporter to a gold importer disrupted the European monetary system at a time when the Bank of France was in any case short of gold (Hawtrey, 1938, pp. 78–81). The Bank of England took over 50 million francs of silver from the Bank of France in exchange for gold. American gold demand subsided only when the greenback issue in the summer of 1861 got under way. A high bank rate and other interest rates in England were deflationary, pushing up unemployment in 1861 and 1862. Recovery began in 1863. Continental Europe suffered a setback in 1864 (bank rate reached 9 per cent in November), but British activity continued to expand.

With the ending of the US Civil War and the blockade of Southern cotton ports, the world price of cotton collapsed, as had the price of wheat in 1857. In London the prestigious bill-brokers Overend & Gurney collapsed in 1866. The Bank of England showed that they were insolvent; their liabilities exceeded their assets when valued at prices normally prevailing, a negative net worth (Batchelor, 1986). The British Foreign Secretary unprecedentedly sent circular telegrams to the embassies stating that the national finances were in no danger (Jenks, 1971, p. 261). Many other firms, mainly those associated with railway contracting (including Peto's), went bankrupt at the same time. The Bank of England's 10 per cent discount rate remained in force for three months, and gold did not flow in even though the Bank of France discount rate was 3.5–4 per cent.

The subsequent British recession was mild, however, and the United States also experienced only a mild decline in economic activity from April 1865 to December 1867. Railway construction continued to increase. In Austria-Hungary and Prussia the prospect of war pulled down stock market prices, and business was reported to be paralysed in Italy and Russia as well. The Austro-Prussian War of 1866 finally ended the career of the Pereire brothers. Crédit Mobilier held large quantities of Austrian securities which depreciated with the Austrian defeat. In 1866 the bank lost 8 million francs and French financial confidence was shaken (Corti, 1928, p. 406).

More distant countries, linked to the British economy by fixed rates, in particular India and Australia, suffered a similar recession with a lag. Not only financial stringency but the collapse of raw cotton prices transmitted the contraction. While the American Civil War lasted, the blockade of the South encouraged a land boom on the other side of the world, in Bombay, as Indian cotton exports expanded to replace those lost to Lancashire. With the ending of the war, the boom collapsed, taking with it a large number of banks that had invested in cotton. Appointed liquidator of the Asiatic Banking Corporation in London, young Jamsetji Tata used his spare time to visit the Lancashire cotton mills. Within three years he had bought and converted a mill in India (Piramal and Herdeck, 1986, pp. 304–5). In Australia the shock was amplified by the failure of the London bank through which the Queensland government had been raising capital to build railways (Pressnell, 1982). When news of the crisis reached Queensland in July 1866, there was immediate panic. The government was expected to be unable to finance railway construction, unemployment would rise and social disorder would quickly follow. Local politicians favoured inconvertibility, which would allow them to pursue an independent, offsetting monetary policy, with exchange rate depreciation. Money was nevertheless raised without adopting an inconvertible paper currency.

The crisis of 1873

The next slump was precipitated on 9 May 1873 by the inflation of German credit with the payment of the French indemnity of five billion francs. This colossal sum amounted to one-quarter of French GNP in 1871 and one-third of Germany's. Yet within two and a half years of the signing of the peace treaty at Frankfurt in May 1871, it had been paid in full. French savings rose and German savings fell. German investment was far heavier than usual between 1872 and 1877 (Gavin, 1992; Hawtrey, 1938, pp. 92–3).

Germany's financial crisis of 1873 quickly spread to Vienna. The British bank rate reached 7.5 per cent in June (in 1873 it changed twenty-four times – a record) and gold imports occurred. On 19 September the American crisis broke. Jay Cooke was unable to borrow in Frankfurt. German investors in United States railways had first supported him, but then cut off their supply of speculative funds. Italy, Holland and Belgium were pulled in.

The effect on Great Britain was less severe than in 1857, but there was still a panic and bank rate reached 9 per cent by November 1873 to staunch the outflow of gold. In Paris and Brussels the discount rate reached 7 per cent. In the most important South American economy, Brazil, the banks almost ceased lending. In Buenos Aires the private banks actually did so, and property values fell by between 30 and 40 per cent. By 1875 stock market prices were lower everywhere. Three years later there were said to be 1.2 million unemployed able-bodied men roaming the USA. One-fifth of US railways were foreclosed or under proceedings (Newmarch, 1878; Giffen, 1880a, 1880b; Kindleberger, 1989, p. 146).

As with the 1857 depression but to a greater extent, transport improvements caused marked and unexpected price declines. The Suez Canal had been opened in 1869 and the railway system of the world had been rapidly extended. In Austria the system had increased from 2,200 miles in 1865 to 6,000 in 1873, almost all the Russian system had been built after 1868, and the railway system of the United States had doubled in the seven years before 1873. Bad harvests in Britain in 1873 and 1876, together with cattle disease, forced food prices up there and reduced the demand for manufactures. The depression in Britain was, however, less deep than in many countries, at least until after the failure of the City of Glasgow bank in 1878. Despite the decline in export demand (of more than 50 per cent to the USA between 1872 and 1876), the British economy experienced no increase in pauperism or decline in state revenues as had happened in 1857 and 1866.

The raw material producing countries suffered more from the recession than did economies specialising principally in manufactures. Transport improvements more immediately affected their earnings. When agricultural prices began falling, the foreign investment that often financed the new railways and port facilities ceased to be profitable. New investment ceased to flow into these economies, which then suffered shortages of liquidity. This was the pattern in Australia and the United States as well as in South America and Russia. In Britain the crisis was largely endogenous, stemming from the realisation of the unsound nature of so many of the foreign loans recently advanced. Lending to Turkey paid interest on previous loans until Turkey defaulted in 1875. Spain was also unable to service its national debt. The entire international financial market was becoming increasingly pessimistic about the likelihood of repayment by primary producers. Egyptian loans which could have been floated earlier became unsaleable and, without the continuing injection of foreign funds, the Egyptian state defaulted on existing debt service in 1876 (Feder and Just, 1984).

Both in depth and in duration, the depression that began in 1873 was the second severest in the history of the international economy. Only the downturn that began in 1929 was longer and deeper.[10] The earliest investigators into these periodic slumps and booms, and modern monetarists, believe they were caused by fluctuations in the supply of money or credit, precipitated by the financial crisis or shocks on which the preceding account has focused (Huffman and Lothian, 1984). It is possible, however, that the great financial crises were responses to downturns in economic activity, and resulted from changes in expectations generated by the collapse (Rostow, 1978, ch. 22; Lewis, 1978b; Kindleberger, 1989). The ensuing panic then accelerated the decline, but did not initiate it. Instead, changes in the determinants and directions of investment, especially in transport improvements, were the fundamental causes. In either case, exchange rate regimes and monetary policy (or its absence) influenced the severity and transmission of these fluctuations.

Summary and conclusion

The international monetary system throughout the third quarter of the nineteenth century depended to a considerable extent on the money and credit of the world's largest trader, Britain. Even trade which did not touch British shores was often financed by British institutions. The acceptability of British money and finance owed much to its apparently unbreakable link with gold for a number of reasons: a belief that gold itself, not social acceptability, conferred value on money; the guarantee that the British government could not reduce the value of the money by increasing the stock, as long as the commitment to gold was maintained; and a strong mercantile preference for gold over other precious metals because its higher value in relation to weight made gold more suitable as an ultimate means of payment for the increasing value of international trade.

The system of commercial banking and bills of exchange, helped by the international spread of the electric telegraph, in fact ensured that very little gold was needed for international exchange. As long as confidence in the financial institutions was maintained, the pyramid of credit built upon the gold base served as a means of exchange. The gold discoveries of the late 1840s were less responsible for the monetary growth of the period than was the development of commercial banking. There was only a gentle rise in prices from this monetary growth because of the relatively rapid accompanying expansion of international and domestic trade. The greater availability of gold provided the opportunity for the major industrial powers to move to gold standards whether *de jure* or *de facto*. This increased demand for monetary gold provided another reason why prices in the industrial world rose very little. France and the United States, whose currencies were linked to both gold and silver, found that the price of gold fell in response to the increased gold supplies sufficiently to drive silver coins out of circulation. But bimetallism seems to have prevented a great rise in the price of silver relative to gold, and therefore limited the appreciations of the silver standard exchange rates, such as those of India and Germany.

The formation of monetary unions in Europe supported the tendency of trade to unify international markets. In turn, monetary unions were a manifestation of the prevailing liberal internationalist beliefs. However, nationalism was strong enough to keep Britain out of an international currency agreement. Where the Austro-German and Latin Unions were concerned, nationalism was as much the driving force of the initiators as liberalism. The success of the Scandinavian Monetary Union derived from the lack of nationalist antagonism, which allowed a genuine integration of financial institutions between the countries.

Monetary unions were facilitated by adherence to a common metallic standard. This created virtually fixed exchange rates between the different currencies within the limits set by the specie points. Although monetary unions and fixed exchange rate regimes created a stability of international relative prices that facilitated trade, they also potentially allowed the international transmission of recessions and booms from which an economy could in principle be insulated under a floating exchange rate regime. In fact, fluctuations in international economic activity seemed to have been caused by discontinuities in the profitability of investment, largely in transport facilities, in agricultural and raw material producing areas. When the investments began to come to fruition and/or wartime restrictions ended, the prices of agricultural products collapsed, creating difficulties for financial institutions which had lent to primary producers and which now found their assets illiquid or reduced in value.

It is doubtful that the exchange rate regime made a great difference to the transmission of these crises, because domestic monetary policies were so important. The depression in the United States after 1873 was probably exacerbated by the floating exchange rate and the contractionary monetary policy designed to allow a return to the pre-Civil War fixed exchange rate. Economic recovery tended to raise the exchange rate at the expense of expanding domestic capacity utilisation. However, in this instance the forces making for depression originated in the United States to a much greater extent than they were transmitted from abroad. Britain suffered relatively little initially, despite maintaining a fixed exchange rate with countries such as Germany which were hard hit, because Britain had, by the 1870s, a relatively small commitment to primary production. Probably of more importance than the exchange rate regime was the willingness of the Bank of England to act as lender of last resort and to control domestic monetary conditions. This facility provided some contribution to preventing marked contractions of the money stock by controlling and reducing internal and external 'cash drains' when residents or foreigners panicked and demanded gold, rather than bank deposits or bills.

The severity and duration of the depression of 1873 began the undermining of the liberal consensus among industrialising nations about the benefits of international trade and investment. A belief that the international market would not raise or preserve living standards adequately became more widespread. Instead, governments or other corporate forms of organisation such as trade unions and cartels were increasingly expected to take remedial or positive action in the face of market forces. Even in Britain the verities of free trade seemed less than eternal. Giffen remarked on the 'continual references to the increase in manufactures abroad'. By the 1870s Britain no longer occupied the same dominant position in the international economy that it had in the 1850s. In the new era that was beginning, the industrial capacities of the leading nations were more equal, trade was less free and the international economy was increasingly used to achieve objectives of foreign policy.

Notes

1. Rothschild's led in loans, by contrast, contracting London issues to a nominal value of £896 million between 1860 and 1890, whereas Baring's managed £560.7 million over the same period (Hidy, 1970; Chapman, 1984, pp. 16–17).
2. Goschen (1866) is a useful contemporary account.
3. The Russian embassy in Peking became a form of European post office, distributing messages sent, once the end of the Siberian line had been reached, by horseback or caravan to the East.
4. The British Australian Telegraph Company charged £8 9s for twenty words to Darwin (Harcourt, 1987).

5. Rostow (1948) adopts a similar line of argument. The three hypotheses can be represented by the Cambridge quantity theory of money. This assumes that the demand for money equals the supply of money (M). Demand is determined by the product of the price level (P), the level of income (Y) and K, the fraction of money income averagely held as money balances. The fraction varies with the rate of interest and the state of monetary institutions among other things. Using lower-case letters to indicate proportionate rates of growth of the upper-case letter variables, the Cambridge relation can be written $M = K.P.Y$ or $m = k + p + y$.

Where '$\rightarrow$' indicates causal direction, the rate of growth equation can represent the first hypothesis of the text as $m \rightarrow p$, the second (Newmarch) hypothesis as $m \rightarrow y$, and the third (Schumpeter) hypothesis as $y \rightarrow p$ and perhaps m.

6. Sayers shows that Chevalier was wrong. When displacement is complete, prices may not fall much because of the size of the new gold area relative to increments in the gold stock.
7. This was the view of the British Royal Commission on the Precious Metals (BPP, 1886, part 1, p. 64).
8. Fisher states that Jevons' work of 1863 entitles him to be called 'the father of index numbers'.
9. Latham (1978b, ch. 5) has suggested another major source of business fluctuation, namely the failure or success of the Asian rice harvests. The high prices of rice when the harvest failed increased the income available to the rice farmers and merchants, who typically demanded British manufactures and semi-manufactures. An increase in the demand for exports tended to send the whole British economy into a boom. The view would have more credibility for the third quarter of the nineteenth century if Asia, and particularly India, was more important in British, European or American trade, but as the figures given in previous chapters have indicated, the poor countries probably accounted for too small a proportion of world trade to be a major cause of fluctuations.
10. In the United States, output expanded rapidly despite the marked decline in prices. This probably reduced the severity of the depression relative to that of the 1890s.

6 International trade and European domination, 1875–1914

European political and economic power reached its zenith in the last quarter of the nineteenth century and the years before the First World War. Those Europeans who had settled in temperate zone lands outside Europe, displacing where necessary the indigenous population with superior military organisation and technology, often maintained a high living standard by participation in international trade. Europeans also assumed political control over most tropical regions with which they traded, without settling there. Over the preceding quarter century, transport and communications networks were put in place. Now, in the four decades before the First World War, they allowed the integration of commodities from the most distant regions, the Canadian prairies, the Argentine pampas, the Siberian tundra, even central Africa, into the world economy.

Additional forces drawing together the component parts of the international economy were population growth and industrialisation, both of which strengthened the demand for imported goods. Radical changes in international specialisation created immediate losers as well as winners. Especially in periodic trade depressions, losers who were sufficiently powerful pressed for policies to defend their position. Within Europe the changing economic and political balance created potentially explosive tensions. The rise of Germany, challenging Britain's former economic supremacy and displacing France from the political leadership of continental Europe, was the most fundamental power shift. The United States' now massive economic strength impacted less on the world economy than it might have because of regional specialisation within the country. Expanding westwards across the prairies and mountains of North America, the economy fed grain and ore to the industrial areas of the north-eastern United States as well as to economies beyond America's shores. China now imported rice, from Burma, Indo-China and Siam, particularly in times of scarcity. India usually exported more than Siam and Indo-China put together. Wheat and rice markets linked eastern and western food grain prices, since some were prepared to eat either if the times required it (Latham and Neal, 1983). Chinese silk supplied the French by the end of the century and, thanks to the Transcontinental railway, Japanese silk went to the Americans (Sugiyama, 1988). The tea trade slipped away from Japan and to a lesser extent from China towards India and Ceylon.

No less striking than the reconfigurations of trade was the spread of economic growth, at a rate unprecedented for most countries, through the international economy. For the sixteen wealthiest economies that accounted for one-sixth of the world's population, growth in real output per head averaged 1.5 per cent per annum betwen 1870 and 1913. The rate of growth of world trade averaged 3.3 per cent per annum (Lewis, 1981; Maddison, 1982). Measured by what had gone before, this sustained performance was remarkable. That it was dwarfed by experience after the Second World War reflects on the historical uniqueness of that period rather than detracting from the achievements between 1875 and 1914.[1]

The following sections discuss some of the international causes and consequences of this growth. First, the changes in manufactures trade are analysed; second, temperate zone primary product trade is considered; and third, tropical primary exports. Then movements in the terms of trade between primary and secondary goods are described and explained. Colonisation, the political consequence of much tropical trade in these years, forms the next topic. Finally, the political responses in the industrial countries to declining prices after 1873 are described.

The changing pattern of comparative advantage in manufactures

The mid-century pattern of comparative advantage in manufactures, described in Chapter 3, shifted as Germany, the United States and other countries industrialised. International exchanges became much more complex. No longer was approximately balanced trade between nations the rule. Figure 6.1 (derived from Saul, 1960) shows the United States earning a surplus of about £50 million in transactions with Britain in 1910 and another £24 million with Canada. Its deficits with continental Europe, India and Japan, were much smaller. British balance of payments surpluses, unlike those of the United States, were earned much more from low-income economies – India, Japan and Turkey – although wealthy Australia was an exception to this rule. Latin America probably falls into this group as far as settlements are concerned. Not distinguished in Figure 6.1, Russia generally ran a surplus with western Europe after 1880. By 1910, however, receipts of interest from investment in Russia generated a surplus for France. Emigrants' remittances from the United States and Argentina financed Italy's deficits with most of its trading partners.

The pattern of settlements did not mean that any particular country's relationships were essential to their balance of payments. Given sufficient time a new configuration could be found, probably with little loss compared with the actual position. It did mean that disruptions of the payment pattern, as occurred in the First World War years and the aftermath, could have considerable repercussions on the international economy.

For no commodity was the shifting later nineteenth-century comparative advantage more apparent than iron and steel. By 1913 German iron and steel exports exceeded British, with American exports not far behind, and Britain had become a major importer of steel (Allen, 1979). New technologies combined with international differences in endowments of natural resources and of specialised human capital to render such changes inevitable. German iron and steel producers had been handicapped by shortages of indigenous ores until 1879 when the Gilchrist–Thomas process was patented (Milward and Saul, 1978, p. 27). This process permitted the use of phosphoric ores with which Germany was richly endowed.

Like phosphoric ores, a scientific chemical education was cheaply available in Germany, whereas in Britain and France it was expensive, difficult to find, and almost non-existent within the universities. Very quickly the German chemical firms established a monopoly of knowledge, which was maintained by the availability of appropriate education. Based on this knowledge, the German chemical industry was quite different from those of France and Britain. It depended much more heavily on applications of organic chemistry, a branch of the industry insignificant elsewhere except in Switzerland. In addition, the much more rapid accumulation of capital in Germany and the United States than in Britain was bound to affect international competitiveness.

The extent of these changes was sufficient to disturb some contemporaries, and subsequent analysts concluded that there was some failure of British industry. On one estimate, the British iron and steel industry was about 15 per cent less efficient than the German and American industries in 1901–4 and was further handicapped by higher raw material prices (Allen, 1979).

Growing competition from Germany attracted attention in Britain to the publication of E. E. Williams' *Made in Germany* in 1896. The impact of the German 'commercial invasion' encouraged a xenophobic scare out of all proportion to the size of the challenge. Five years later, W. T. Stead's *The Americanisation of the World* similarly painted a picture of Britain's commercial decline amidst a flood of American imports.

One cause of shifting comparative advantage from the period of Britain's industrial supremacy was changing coal prices. By 1900 the price advantage that British coal had possessed in the 1860s had disappeared. Pennsylvanian coal was cheaper, and German coal was no more expensive (see Table 6.1). The advantages that British industry had gained from power and heat cheaper than elsewhere in the world had disappeared. Britain's production possibilities resembled more those of its European partners, diminishing the scope for trade of the mid-century's pattern.

The concentration of British exports on coal and cotton textiles offered little opportunity for productivity increases based upon the available science and technology (Levine, 1967, pp. 133–5). By contrast, German exports of chemicals and machinery were proportionately much more important, and textiles and coal less so. British economic growth, which reached its nadir in the decade after 1900, suffered accordingly. But a structural transformation was occurring. British exports did show some of the changes that are more usually attributed to the German economy. Chemicals quadrupled their export share over the twenty-year

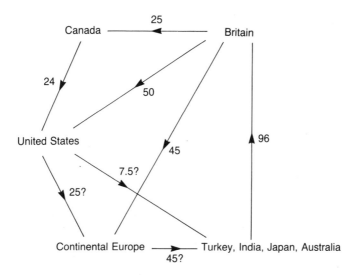

Figure 6.1 Simplified world pattern of settlements in 1910 (£m, arrow points to surplus partner). Modified from Saul (1960).

period, and machinery exports also increased at a rate faster than total exports. When ships are included in the iron and steel manufactures category for 1900 there is a slight increase in the share of that class. Textiles, on the other hand, especially cotton, showed a marked decline in export value share, though not in total value. Thus, although British exports were still less diversified than German in 1900, they were reducing the difference. In 1900, 78 per cent of British merchandise exports were manufactures, whereas the proportion for Germany had risen from only 26.4 per cent in 1880 to 39.2 per cent in 1900.

Germany to a much greater extent than Britain remained an agricultural country, with 35 per cent of the employed population, over 10 million people, officially still working in agriculture in 1913. Because of protective tariffs, there was virtually no reduction in the area of arable land, in marked contrast to Britain, although there was a decline in the land area devoted to pastoral farming. Food exports fell in absolute value as well as in market share because of the rising population and capital accumulation on a relatively fixed agricultural area. Sugar exports are an exception to this generalisation because they received a subsidy.

Table 6.1 The pit price of coal in various locations, 1861–1901

	c. 1861	1901 (average)
France	6–14s	12s 7½d
Germany	7–10s	9s 4¼d
England	6–10s	9s 4¼d
Pennsylvania	n.a.	8s 4¾d
Pennsylvania (anthracite)	8–9s	2–4s

Source: Jevons (1906, p.343).

Berlin and north-eastern Germany imported British coal. The low costs of water transport meant that Ruhr coal could often be sold more cheaply in western Europe than in those regions (Milward and Saul, 1978, p. 30).

Within the total of German manufactured goods exports, cotton textiles greatly increased their share, and the proportion of woollen textiles declined. But cotton and woollen textiles together slightly raised their percentage between 1880 and 1900. Machinery, iron and steel manufactures, dyestuffs and books all increased dramatically. In machinery both Germany and Britain exported their own specialities to each other and the trade was almost balanced (Saul, 1960, p. 32). The steel industries of the two countries also developed complementarily, with each country exporting semi-manufactures to the other.

In the United States changing specialisation resulted primarily in increased inter-regional, rather than international, trade. Although the USA was a great world exporter of wheat and meat, as well as of the older staples, cotton and tobacco, the greater part of the farm output of the newly settled mid-West and western states was sold in the north-east of the country. Hence US participation in international trade was proportionately lower. Rapid growth and sheer size (a population of 76 million by 1900) nevertheless gave the United States a considerable impact on the international economy. The American tendency to engage in import substitution, increasingly to produce domestically manufactured goods formerly imported, slowed the growth of world trade. Throughout the years 1880–1914, the proportion of manufactures and semi-manufactures in US exports rose, as the American economy experienced a process of industrialisation, capital accumulation and population growth similar to the German pattern. The United States' factor endowments shifted so that its comparative advantage was ceasing to be based on abundant natural resources. The natural resource content of US exports, both renewable and non-renewable, declined markedly. Imports of resource products, mainly industrial raw materials such as wool, silk, rubber, hides and skins, increased by more than the decline in imports of crude foodstuffs (Vanek, 1963). Variations in the price of resource products relative to the general price levels of exports and imports did not explain this tendency. Around the time of the First World War, the United States became a net importer of natural resources, as far as direct requirements were concerned. Nonetheless United States technology, and therefore American economic growth, continued to depend on abundant natural resources well into the twentieth century (Wright, 1990).

The transformation of world comparative advantages in manufactures in the last part of the period under consideration is shown in Table 6.2. Changes in exports of manufactures, measured at constant prices, are divided into three parts:

1. The market growth factor: the growth consequent upon the change in the size of the world market for manufactured goods as a whole (this measures how exports from each country would have moved if the area and commodity

Table 6.2 Changes in the volume of export of manufactures attributable to changes in the world market, in the pattern of trade and in market shares, 1899–1913

	Britain	France	Germany	Other W. Europe	Canada	USA	India	Japan	Total
1899 exports	1.33	0.50	0.78	0.46	0.02	0.42	0.11	0.06	3.68
Change due to (a) market growth	+1.02	+0.39	+0.60	+0.35	+0.01	+0.32	+0.08	+0.04	+2.82
(b) diversification	−0.02	−0.07	+0.02	−0.05	–	+0.17	−0.03	−0.01	–
(c) competitiveness	−0.36	−0.03	+0.33	+0.07	+0.01	−0.07	−0.01	+0.06	–
Total	+0.63	+0.29	+0.95	0.37	+0.02	+0.43	+0.04	+0.09	2.82
1913 exports	1.96	0.79	1.73	0.83	0.04	0.85	0.15	0.15	6.50

Source: Maizels (1963, ch. 8).

pattern of trade, and the share of individual markets, had remained unchanged, assuming equal commodity supply elasticities).
2. The diversification factor: the change in exports due to changes in the area and commodity pattern of trade. (This measures how exports would have moved if the world total, and each country's share of each market, had remained unchanged).
3. The competitiveness factor: the change in exports resulting from the change in the share of each exporting country in imports of each commodity group into each market. (This measures how exports would have moved if the world total and the area and commodity pattern of trade had remained unchanged).

One of the more surprising results in Table 6.2 is the small decline in the United States' 'revealed comparative advantage' in manufactures, the competitiveness factor.[2] US exports expanded much more proportionately and absolutely than those of other economies. They did so because they were commodities increasingly in the fast-growing categories and sold to the more rapidly growing markets. The growth of the adjacent Canadian market provided a strong reason for diversification.

Japan increased its comparative advantage, as measured by market share, proportionately by the greatest amount, because it was industrialising most rapidly. Japan's industries traced 'Wild Geese-Flying' patterns of successive waves of rising and falling imports, domestic production and export, for cotton yarn, spinning machinery, cotton fabrics and machine tools. By 1885 domestic production of cotton cloth exceeded imports, and exports were greater than imports in 1910 (Baba and Tatemoto, 1968).

Precocious German economic development has often been suggested as a cause of Britain's deteriorating international competitiveness in manufactures and slow economic growth. One explanation draws analogies between the slowdown in advanced economies during the 1970s and the relative decline of the British economy in the nineteenth century (Beenstock, 1983). In the later period, middle-income countries industrialised rapidly and energy prices were driven higher. Table 6.2 shows that (middle-income) Germany gained the most trade from an improved competitive position at the beginning of the twentieth century. Britain lost almost exactly the amount Germany won, because of a deteriorating comparative advantage in manufactures. But, as already mentioned, there was a good deal of complementarity between the late nineteenth-century British economy and the newly industrialising Germany. The arithmetic relation does not necessarily indicate a causal link.

Could British declining competitiveness in manufactures be the 'normal' pattern of development of a mature economy in which an increasing proportion of resources is allocated to services because of their high income elasticities of demand? French national income per head was close to Britain's, and therefore France might reasonably be expected to experience a similar shift in comparative advantage away from manufactures, but the French decline in competitiveness

Table 6.3 European revealed comparative advantage, 1899 and 1913

Britain			France			Germany		
1899 Rank		1913 Rank	1899 Rank		1913 Rank	1899 Rank		1913 Rank
1	Rails and ships	1	1	Alcohol and tobacco	1	1	Books and films	2
2	Iron and steel	2	2	Apparel	3	2	Fancy goods	7
3	Textiles	3	3	Books and films	4	3	Metals manufactures	6
4	Industrial equipment	5	4	Cars and aircraft	2	4	Chemicals	5
	Alcohol and tobacco	4					Electricals	1
							Wool and textiles	3
							Industrial equipment	4

Source: Crafts (1989).

was smaller even than that of the United States. Thus the 'normal' process of development of the economy is less satisfactory as an explanation for Britain's declining manufactures trade than is low productivity growth in manufacturing, shifting comparative advantage towards natural resources, in particular towards coal.

That leaves unexplained why British productivity growth should have been comparatively slow. Inflexibility of businesses and education is consistent with a comparison of British, French and German 'revealed comparative advantage' (Table 6.3) in 1899 and 1913. The fastest-growing country, Germany, changes industry rankings the most (Crafts, 1989). Britain loses out on growth sectors: cars and aircraft in France and electricals in Germany. Underlying this pattern may well have been Britain's distinctive emphasis upon 'on the job' industrial training. British informal training created a large workforce skilled in particular trades, but one that was perhaps unsuited, or resistant, to potential new industries. By the end of the nineteenth century, industrial development abroad combined with this inflexibility to give Britain a comparative disadvantage in high-wage, high-skill, research and development-intensive manufacturing industries (Crafts and Thomas, 1986).

In shipbuilding and many financial services, Britain still predominated. Germany held an advantage in dyestuffs and organic chemical products that the United States could not match, and the French led the world in internal combustion-engined motor vehicles before 1905. No industrial economy could expect to maintain an advantage across all modern products, and any position was vulnerable to technological advances.

The temperate zone primary product exporters

In the third quarter of the century, shorter-haul (intra-continental and national) freight rates fell. The following generation saw radical cuts in long-distance

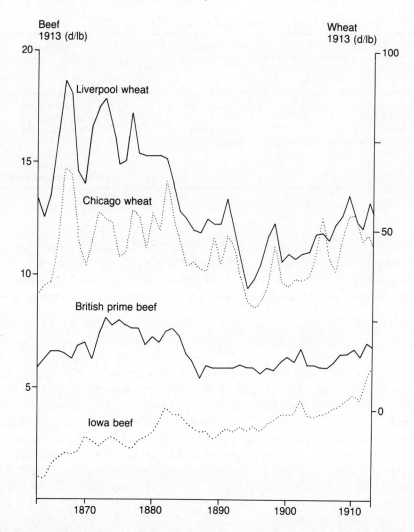

Figure 6.2 The convergence of British and US wheat and beef prices 1865–1913 (Harley, 1992b)

(intercontinental) rates. They contributed to a historically unique characteristic of the late nineteenth-century international economy: the availability of vast amounts of temperate zone agricultural land that could supply food to the world. These freight rate reductions in particular benefited the long-distance wheat trade. The British wheat price fell and the Chicago wheat price rose, the gap between them becoming almost negligible by 1880 (Harley, 1992b). Refrigerator ships of the 1880s allowed Australia and Latin America to supply Europe with meat products, although beef price convergence was less marked than with wheat (see Figure 6.2). Despite the higher European tariffs, the British market remained open, and transport improvements presented the temperate zones possessing abundant land with opportunities for expansion. The low-density indigenous population could be displaced by the threat or use of superior western military technology and organisation. The wiping out of Custer's cavalry by the Sioux Indians at the Little Big Horn in 1876 was a small blip in an otherwise inexorable trend.

All the temperate regions of recent European settlement – Argentina, Australia, New Zealand, Canada, South Africa and the western and mid-Western United States – received labour and capital from the industrialising or developed world. In return they exported a variety of primary products that allowed the migrants to maintain high standards of living. Australia appears to have been the richest country in the world by 1900, but the United States was virtually at the same level of GDP per head. Canada lagged behind Britain, and Argentina was about on a par with Italy, behind Germany (Maddison, 1989, table 1.3).

Despite their different living standards, the Argentine and Australian economies showed similar export development. Both exported large quantities of wool at the beginning of the last quarter of the nineteenth century, when land was by far the most abundant factor of production (Diaz Alejandro, 1970, ch. 1). Australia possessed the largest national flock of sheep in the world, until drought halved the sheep population. Not until 1931 did numbers recover their 1892 peak. Then, with the development of infrastructure, grain exports became the dynamic sector at the expense of pastoral farming. At the same time, chilled and frozen beef exports grew rapidly in both countries. The differences in living standards between the two economies persisted in part because, unlike Argentina, Australia had a valuable mining sector. Gold mining began to revive again in 1886, first in Queensland, then in Kimberly in Western Australia and finally in Kalgoorlie. Broken Hill proved a valuable source of silver in 1887–8. The Australian economy was depressed and facing a severe balance of payments crisis at the time. The contemporary underemployed labour and capital may have actually triggered the revival.

Much more dependent on non-renewable resource exports was the similar South African economy. Small in population (about 0.5 million whites and 1.5 million blacks), but large in area (475,000 square miles in 1890), this vast agricultural and pastoral territory imported nearly all its own food: wheat from Europe, tinned and frozen meat from Australia and preserved fruit (Goodfellow, 1931, pp. 7, 96–7). The paradox is a measure of the magnitude of the demands made upon South Africa's resources by the diamond industry in the early 1870s and by the gold industry in the late 1880s. The chief consuming areas were connected with the ports of Cape Colony and not with South Africa's producing areas: Johannesburg was economically nearer to Australia by railway and steam-ship than it was to many parts of the Transvaal.

The influx of foreign capital and labour to exploit South Africa's natural resources caused greater political disturbance than in any other area of recent

European settlement. Boer farmers resisted conceding political rights to the immigrant English-speaking miners. The Boers defeated the British at Majuba Hill in 1881 and inflicted many reverses on them during the Boer War of 1899–1902, before losing it.

Just as Australia provided the food for the South African miners in the 1890s, so New Zealand supplied the Australian gold fields in the 1850s with grain and potatoes. In the next decade New Zealand exported gold from its own fields and then followed the pattern described for Argentina and Australia, with exports of wool, wheat and refrigerated foods (Gould, 1972).

The pull of the colossal United States economy delayed Canadian economic development relative to other regions of recent European settlement. During the last three decades of the nineteenth century, Canada lost population to the United States as emigration exceeded immigration (see Figure 8.1). But the closing of the American frontier gave Canada its economic opportunity, and in the first decade of the twentieth century the remarkable Canadian wheat export boom pulled over 700,000 immigrants (net of emigration) mainly into the Canadian prairie provinces.

With the exception of the western and mid-Western regions of the United States, the temperate zone exporters clearly depended for their high living standards on their relations with the international economy. The nature of this dependence is less clear. One view, very similar to the 'staple theory' of Chapter 3, is that international trade was the engine of growth of these recently settled areas, including the United States (Nurkse, 1961). Rising populations, incomes and industrial growth in the countries of the 'industrial core' then expanded demand for primary products. This demand raised the returns to labour and capital in the 'countries of the periphery', attracting inflows from overseas to take advantage of the abundant land. On this account the influx of labour and capital boosted economic growth, even in America after the Civil War. For the smaller economies, trade was almost certainly the engine of growth. In New Zealand, life at a European standard of living would have been impossible without international trade. For South Africa, growth was clearly based on exports of gold and diamonds from the 1880s. Some 48 per cent of the variation in the growth of Argentine national income in the period 1906 to 1940 could be explained by variations in export growth (Diaz Alejandro, 1970, ch. 2). The Canadian experience is more difficult to interpret, despite the wheat export boom. Chambers and Gordon (1966) estimated that the spectacular expansion of Canadian agriculture between 1900 and 1910 contributed only 5.2 to 8.4 per cent of the increase in living standards. Canadian income per head would have been between 1.2 and 1.94 per cent lower at the end of the decade without the agricultural expansion, even though agriculture was more important to Canada than to other high-income economies. On the other hand, the marked rise between 1890–1900 and 1910–1920 in international commodity flows as a percentage of commodity output suggests that export-led growth had a role in Canadian development, but a decade later (Green and Urquhart, 1976).

The larger temperate regions were much less dependent on the international economy. Australia was unusual among primary product exporters in the very high proportion of the population that was urbanised; two-thirds of the population were town-dwellers in 1890, whereas this proportion was not attained by the United States until 1920. Exports of rural products, with the possible exception of minerals, could not have provided the dynamic element to an economy so highly urbanised. As indicated in the previous section, the much larger American economy grew mainly on the basis of inter-regional, rather than international, trade. The European economies tended to expand more slowly than the temperate 'peripheral' economies. European imports similarly grew less rapidly than the national products of the peripheral countries, and therefore the European 'core' is unlikely to have been the 'engine of growth' (Kravis, 1970).

Europe had its own 'periphery'. The less dynamic, but small, economies of the European periphery – Bulgaria, Greece, Portugal, Serbia and Spain – attained only relatively low export/product ratios (Hanson, 1986). The experience of other small European economies, especially the Scandinavians, suggests that more outward-looking policies may have accelerated their economic development. For some of the southern and eastern periphery, trade might have been an engine of growth, but was not.

If temperate zone trade had grown purely because of foreign demand, traditional suppliers of primary produce to the core should have benefited in the same proportion as new producers and maintained their market shares. Yet, in the case of wheat especially, the share of Britain's imports supplied by the new producers increased considerably. The extensively farmed prairies could market wheat at much lower prices than the British farmer. Between 1873 and 1894 British wheat production fell by 60 per cent and the United States accounted for 80 per cent of the increase in exports. Traditional suppliers, such as Russia, lost much of their market share. Supply-side changes in the regions of recent European settlement were therefore providing much of the impetus of international growth.

Although Britain was the world's largest wheat importer – by 1914 importing 80 per cent of its wheat supplies – it was not British conditions that were the most important determinants of world wheat demand. British wheat consumption was only about 10 per cent of total consumption in 1896. The British increase in wheat imports amounted to only 6.7 per cent of the total increase in world production and demand in the period 1885–9 to 1909–14, because Britain's GNP and population were not rising as fast as those of most other comparable nations (Olson, 1974). In the same period, Germany increased its demand for wheat sixfold, meeting it, unlike Britain, mainly from expanding home production. The exporting countries raised their own wheat consumption greatly during these years: the USA by 167.7 million bushels and Russia by 359 million bushels. The four decades before 1914 were therefore not as favourable to economies that sought to grow on the basis of wheat exports as the 'trade as an engine of growth' hypothesis suggests.

Tropical trade and the less developed countries (LDCs)

Both the temperate zone primary product exporters and the manufacturing economies maintained high living standards and positive rates of economic growth. The key difference was that tropical countries failed to match the income levels of the western world. This was so despite tropical trade growing at the same rate that industrial production expanded in the main industrial countries from the 1880s to 1914 (Lewis, 1978b). For many subsequent observers, the failure demonstrated that the effects of trade on less developed countries differ from the effects on higher-income economies. They have tried to explain why this should be by introducing a number of concepts. Enterprises producing for export constituted enclaves of modernity. Either they lacked linkages with other sectors or they threw 'backwash effects' on the rest of the economy. They over-rapidly depleted non-renewable resources. They also encouraged foreign consumption patterns which increased the demand for imports, and tended to lower the propensity to save, thereby reducing capital accumulation and growth. Indigenous industries were allegedly destroyed by competition from imports, and the resources they formerly used remained unemployed. Because the market forced these countries to specialise in primary products which have low price and income elasticities, the prices of their exports showed a chronic tendency to decline relative to import prices. These primary products were also often subject to uncontrollable fluctuations in supply conditions, because of drought or disease. The ensuing variability of export earnings further constrained economic development by limiting the ability to import capital goods.

The foregoing analysis, as will be shown, is considerably more pessimistic about trade than is warranted by historical experience. Trade almost certainly benefited tropical countries, although there were losses as well. But except for small economies specialising in products that were not closely competitive with the exports of other low-income countries, the impact was small.

The late nineteenth-century Jamaican economy suffered costs of international trade different from those listed above: the cost of obsolete specialisation. The protection of western European beet sugar production meant that Jamaica could no longer profitably provide sugar from its plantations, and the economy became depressed. Only with the switch to banana production was there any recovery (Eisner, 1961; Hall, 1964).

For a large economy there was little chance that exports would raise growth, but in India there was the additional question of whether exports should be allowed at all when there was a dearth. The (British) Indian government resisted bans on exports of food grains during the severe famines of the late nineteenth and early twentieth centuries on the grounds that, if food was scarce in a free market, it would flow into the country. Exports during famines can be due to the income effects of crop failures on demand. Peasants who lose their crop have no resources with which to buy food. This prevents upward pressure on prices during famine and encourages export. In practice, the observed volatility of Indian prices reflected stockholders' expectations about future, rather than current, output. Average prices were sluggish in signalling local scarcities. Free trade did in fact stabilise food consumption as the government maintained; trade acted as a buffer. But the effect was small (Ravallion, 1987, 1989).

The gains from trade (compared with no trade) are analogous to an increase in productivity of an economy. Exports are the inputs into the 'trade process' and imports are the outputs. For tropical countries, as for others, trade is worth while if the resources used to provide the exports are less valuable than the imports obtained in exchange. The more productive the economy, the fewer are the resources needed to supply a given quantity of exports. The problem of the late nineteenth-century tropical economies was ultimately low productivity in their agricultural sectors, which employed the great bulk of their labour forces. The price at which the products of tropical agriculture could be sold on the international markets was held down by perfectly elastic world supplies of Indian and Chinese willing to travel anywhere to work on plantations for a shilling a day, and by the abundance of tropical land (Lewis, 1978b; 1978a, ch. 3). Because the factors were in elastic supply, so were the products. As demand grew, tea production spread from China to new plantations in India and Ceylon. Burma, Thailand and Indo-China multiplied their exports by four or five times, as they sent rice to feed the plantation workers of Ceylon, Java and Malaya. A rice futures market flourished on the open pavements of Mogul Street, Rangoon, by 1900 (Latham and Neal, 1983). The rubber plantations of Malaya expanded from 5,000 acres in 1900 to 1.25 million acres by 1913. The Gold Coast (Ghana), which had supplied virtually no cocoa in 1883, became the world's largest producer by 1913, as northerners migrated to the wetter south.

Increased productivity, as against increased output, in tropical agriculture might have allowed some of the agricultural labour force to move into manufacturing industry, but this did not happen to any significant extent. Instead, the opening of new tropical areas to international trade by falling freight rates increased the supply of tropical produce and lowered prices relative to temperate zone agricultural goods. Temperate zone workers could find alternative employment opportunities in manufacturing and service sectors, and therefore had to be paid accordingly.

The export growth of tropical countries was low, or the share of foreign trade in their national product was small (or both), and foreign trade therefore could not be expected greatly to affect the tropical economies (Hanson, 1976). But foreign trade was generally unimportant because of the low productivity of tropical agriculture. As with the larger temperate zone economies, supply mattered more for development than demand.

Even so, demand from industrial countries made a difference to the growth of tropical trade. The export growth and diversification of less developed countries was most pronounced in the second and third quarters of the nineteenth century, when high-income countries were pursuing liberal trade policies, and the most open high-income country, Britain, was growing fastest (Table 6.4). The last

104 A History of the World Economy

Table 6.4 Growth in the volume of world trade, 1850–1913

	Average growth (% p.a.)	
	1850–73	1873–1913
Britain	4.2	2.6
United States	5.8	4.1
North-west Europe	5.1	5.2
Other Europe	4.7	2.9
Temperate settlements	5.9	4.3
Tropics	3.4	3.1
East Asia	3.7	3.8
Total	4.6	3.3

Source: Calculated from Lewis (1981).

quarter of the ninetenth century brought increasing competition for tropical exports, actual or potential. Sugar prices were forced down by the expansion of subsidised beet sugar exports from continental Europe from the 1870s; new copper deposits were discovered and exploited in the USA and other countries; and the invention and commercialisation of synthetic dyes greatly reduced exports of Indian indigo. The USA became the world's leading producer of cotton and tobacco, Japan became an important exporter of silk and tea, and the United States, continental Europe and Japan all developed advanced textile industries (Hanson, 1977a). Table 6.4 shows a deceleration of the average growth of tropical trade volume after 1873 from 3.4 per cent per annum to 3.1 per cent. In both periods the pace of expansion was below the average for world trade. By 1913 the share of the tropics in world trade at 16 per cent (Figure 6.3) was thus less than in 1850.

In the later decades of the century, the rate of growth of British consumption of many LDC products fell because British income and population growth slowed. Analysis of Indian exports of manufactures in Table 6.2 supports the position that the declining growth rate of the open British economy was harmful for the exports of less developed countries. India showed the greatest proportionate decline in exports due to the country of destination and product composition, of all the manufactured goods exporters considered. Britain was India's most important market, following the decline in the opium trade with China.

Diversification of less developed countries' exports almost ceased when tropical trade growth decelerated in the last quarter of the century. This reflected a stagnation of relative factor endowments. The typical LDC, as judged by the mean, earned 58 per cent of its export proceeds from one product in 1860, and 52 per cent in 1900. By 1913 at least twenty-two LDCs depended on one product for more than 50 per cent of export proceeds; in 1900 this was true for twenty out of forty-nine LDCs (Yates, 1959). The deconcentration of developed country exports was faster than for LDCs, but several important LDCs such as British India, China and the Dutch East Indies (Indonesia) underwent substantial

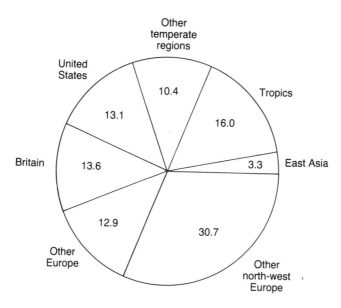

Figure 6.3 World trade in 1913 (%)

declines in export concentration. Others, such as Brazil and Egypt, experienced increasing concentration.

One of the beneficial effects of export diversification is usually a greater stability of export earnings, because uncontrollable supply- or demand-side influences are less likely to affect all export products at the same time. Compared with the LDCs, the group of non-European countries which in the twentieth century can be classified as developed (NEDCs) (Argentina, Australia, Canada, Cape of Good Hope, Japan, New Zealand, the USA and Uruguay) experienced less export instability because they were more diversified (Hanson, 1977b, table 1). On the other hand, the NEDCs suffered more instability than the industrial countries of western Europe in every decade from 1850 to 1906. Nevertheless, the NEDCs generally maintained higher growth rates than industrial countries. Hence, the instability of export earnings is unlikely to have been an insurmountable obstacle to the economic growth of the less developed countries.

Was there something about the nature of the export enterprises or products in tropical countries that prevented economic development? Staple 'theory', among others, suggests that the production function of the primary commodity or commodities that a region first exports governs the region's prospects for future development (Baldwin, 1956).[3] Labour-intensive, plantation commodities, such as cane sugar, restrict economic development, while non-plantation crops, such as wheat, which provide the right technological and institutional factors, encourage development. In many, if not all, primary products there was a choice of production technique. Tea was not a plantation crop in China, but became one in the nineteenth century for India and Ceylon. In Brazil, rubber was not a plantation crop, but it was introduced into Malaya as one. Cotton was grown on plantations only in the United States. Thus if plantations are the root difficulty, it is not so much the product that is fundamental, but the forces that make for the choice of plantation production. In any event, several products representing a large share of LDC foreign exchange earnings in the nineteenth century had good production functions as judged by this criterion, and yet countries expanding sales of these commodities did not obviously experience higher rates of economic development than others.

The tropical countries as a whole in 1913 were participating more fully in the international economy than ever before. Their economies almost certainly showed higher living standards (although redistributed among different groups) and improved infrastructure as a result. As for the larger high-income countries, trade was not a major stimulus to tropical economic growth (nor was it generally harmful). Hard evidence of productivity and living standards for these countries is sparse, but Brazil, Mexico, India, China, Indonesia, the Philippines and Thailand all seem to have increased their real GDPs per head between 1900 and 1913. The best estimate for the average increase of the Asian group was 11 per cent over these thirteen years, while their exports rose by 128 per cent (Maddison, 1989, table 1.3). Their populations grew by only 8 per cent, confirming a very strong rise in the ratios of exports to GDP, even allowing for falling export prices.

The commodity terms of trade

Writing at the end of the period under consideration, J. M. Keynes speculated that the operation of Malthusian pressures would turn the terms of trade permanently against the manufacturing countries and in favour of the primary producers (Keynes, 1913a). The limited availability of land and minerals restricted the scope for increasing the supply of products based mainly on natural resources, whereas technical progress could expand manufactured goods supply without bounds. Admittedly, this did not seem to have happened in the preceding forty years, but eventually the fundamental economic truths would assert themselves. Then economic growth in Britain and industrial Europe would be choked off.

Since Keynes, the most popular predictions about movements in the terms of trade have been exactly the opposite of Keynes' view, and based largely upon the experience of the years 1875–1914. The Singer–Prebisch thesis emphasised the demand side of the relationship between primary and manufactured goods, rather than the supply side, drawing attention to the low income elasticities of demand for food and raw materials (Singer, 1950; Prebisch, 1962). As productivity increases raise real incomes, the demand for manufactured goods will expand considerably faster than the demand for primary products, tending to push up the relative price of manufactures. In addition, the greater monopolisation of the secondary sector, both in the goods and the labour markets, will allow that sector to capture most of the gains from technical progress.

The data originally used to support the Singer–Prebisch thesis were based on the import prices of Britain from 1876 to 1914. Because these were prices of primary products and these prices fell relative to British exports of manufactures, it was inferred that primary exporters must have experienced an adverse movement in their terms of trade, corresponding to the favourable movement of the British terms of trade. Since exports are valued exclusive of transport costs but import values include them, a reduction in transport charges could improve the terms of trade for one country without causing a deterioration for its trading partners. The whole of the apparent deterioration of the terms of trade of primary producers in the period 1871 to 1905 could be accounted for by this valuation bias (Ellsworth, 1956). But thereafter the valuation bias was in fact reversed by rising transport charges, until by 1946 there had been a complete offset (Spraos, 1980). Furthermore, if the unit cost saving from progress in transport technology was equal to the relative fall in price of primary products including transport costs, the importers of the primary goods appropriated the entire gain. This is largely to be expected because importers have price-inelastic demands for primary products. Whether or not it is just, bearing in mind that the manufacturing countries were the ones devoting resources to transport improvements, is open to debate.

The concern with these particular terms of trade arose originally from the belief that they were representative of industrial countries and less developed countries as a whole. Critics of the Singer–Prebisch thesis point out that primary products imported by industrialised countries included commodities predominantly produced in developed countries, and that British data were not necessarily representative of all industrial countries. What evidence there is shows that these objections do not invalidate the thesis. The unit value index for industrial Europe's combined exports and imports of primary products fell by 22 per cent between 1872 and 1938 (Kindleberger, 1956). Over the same period the unit value index for industrial Europe's imports of primary products from poor countries fell by 38 per cent. Poor countries therefore experienced a greater decline in their export prices of primary products than those of industrialised or high-income countries.[4] As to the second objection, reducing the dependence of the terms of trade index on British data after 1900 (and also on oil) does not eliminate the significant deterioration in the terms of trade for primary producers in these years (Spraos, 1980; Sapsford, 1985).

Yet another attempt to invalidate the Singer–Prebisch thesis has drawn attention to the entry of new manufactures into international trade with the passage of time, while the quality of existing products improves. However, the bias introduced into the terms of trade index on this account cannot be ascertained a priori because the quality of primary products is also improved, for instance through processing, and that of manufactures may even decline if durability is reduced, for example.

The likelihood is that movements in the commodity terms of trade of tropical primary producers were relatively unimportant compared with productivity growth, as measured by the factoral terms of trade. These are, for labour, how many exports one person can produce measured by the value of imports that can be bought with them (Lewis, 1978b). If this is so, the policy implication drawn from the Singer–Prebisch thesis may be incorrect. The inference is that specialisation in the production and export of primary products is not beneficial, or at least not optimal. Instead, more resources should be devoted to the development of domestic manufacturing industry and to reducing dependence on international trade. Until the inter-war years, such a policy would not have been justified by the decline of primary product prices, and only then because of the collapse of international trade in general. In part this contraction was a consequence of the pursuit of autarkic policies such as were implied by the Singer–Prebisch view.

Trade and colonisation

Colonisation as a means of trade promotion was no novelty by the last quarter of the nineteenth century, as Chapter 1 has shown, but the speed of the 'scramble for Africa' and other colonial acquisitions in the last two decades of the century certainly was. Figure 6.4 shows the pattern and pace of African colonisation.

The year after H. M. Stanley's discovery of the Congo basin in 1877, Leopold II of Belgium set up a Studies Committee for the Upper Congo to advance his plans for a free-trade colony in central Africa. What thoroughly embroiled Europe in Africa was the German general election, which Chancellor Bismarck wished to manipulate. Foreign adventures were a convenient means of diverting German voters' attention away from contentious domestic issues. As the French attempt to find a new way into the Chinese market had tied them down in Indo-China at the time, and the British concern to secure their communications with India was committing them in Egypt (see Chapter 7), Germany was left with an almost free hand in Africa. Leopold achieved his objective at the Berlin Congo Conference of 1885, with an international treaty guaranteeing free trade and navigation, and Bismarck gained a German empire in Africa, five times the size of the Reich. The new territories were largely unproductive and supported few German economic interests. But they solved Bismarck's electoral problem. By 1886 he could afford virtually to abandon interest in the empire (Fieldhouse, 1973; Stern, 1977, p. 409).

The larger French colonial empire of nearly 1.2 million square miles at the end of the century was, with the exception of Algeria in the 1830s, mainly conquered after 1880. Italy belatedly tried to emulate its northern neighbours, despite a parlous financial condition, until defeated by the Abyssinians at Adowa in 1896.

British colonial activity followed a different pattern, most probably because British economic interests were already well established overseas, and because the liberal tradition of minimal government intervention was more firmly established.

International trade and European domination, 1875–1914

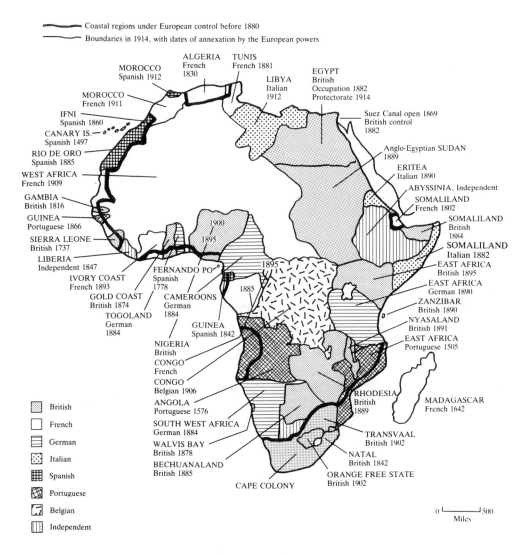

Figure 6.4 The scramble for Africa

British imperial administration was a model of economy. Fewer than 6,000 persons administered the empire during the 1890s, including clerks and cashiers. Only 75,000 British troops were stationed in India and 45,000 served in the colonies (Davies and Huttenback, 1986). When other countries were colonising in the 1880s, British policy was still to hold back on grounds of expense. In 1859 the most powerful chief in Fiji offered sovereignty over the islands to Great Britain, which the British government then declined. Its motivation had not changed in 1883 when the Queensland government's annexation of south-east New Guinea (to prevent further German and Dutch acquisitions) was not ratified by the imperial government until Queensland agreed the following year to guarantee a portion of the administrative expenses (American Economic Association, 1900, pp. 283, 287). Similarly, the British procrastinated in early 1883 when the German government indicated that it would be glad to see British protection extended to German settlers in South-West Africa (Stern, 1977). British policy altered course only in the 1890s when British trading interests were apparently in danger of losing out from other countries' annexations.

All these colonial acquisitions, and most of those of Russia in Asia, differed from most of the temperate zone settlements in their substantial indigenous populations and climates uncongenial to Europeans. The motives for colonisation and the pattern of economic development imposed on them were therefore also quite different. Trade with the new colonies was relatively unimportant for the European powers, as Table 6.5 implies for the 1890s.

Britain was most reliant upon colonial trade, which accounted for one-third of its exports and over one-fifth of its imports. The only other power as closely dependent upon the trade was Spain, with almost one-quarter of its exports going to Spanish colonies. Germany, the most vigorous industrial state, was distinguished by the tiny colonial percentage of its trades. The new colonial empires were in fact even less important to the imperial powers than Table 6.5 suggests, because colonial trade was usually concentrated on one or two possessions.

Table 6.5 Total trade compared with colonial trade (annual average), for selected countries, 1892–6

Imperial power	Colonial/total		Colonial area (000 sq. miles)	Colonial population (m)
	Imports (%)	Exports (%)		
Britain	22.5	33.2	11,090	325.1
France	9.5	9.5	1,195	36.15
Holland	14.5	5.0	785	34.5
Portugal	15.8	9.2	834	7.9
Spain	9.7	24.0	323	8.5
Denmark	1.1	1.6	41	0.127
Germany	0.05	0.09	1,026	9.8

Source: Flux (1899).

Nearly two-thirds of French trade with French possessions were with Algeria and Tunisia, India dominated British colonial trade, East India the Dutch, Angola was much the most important for the Portuguese colonial trade, and Cuba was for the Spanish (Flux, 1899). All of these colonies (except Tunisia) had been held long before the wave of colonisation of the last two decades of the century. That this new colonisation was economically necessary for the European capitalism of the time, as Lenin and others maintained, is hard to square with the small volume of trade involved (cf. Hobson, 1905; Luxemburg, 1972; Lenin, 1934).

If imports from the colonies provided raw materials for the industrial powers and exports to them provided markets, then market factors dominated British and Spanish relations with their colonies, and raw materials dominated Dutch and Portuguese relations. The relative unimportance of French colonial trade in total French trade, at less than 10 per cent, is at first sight surprising given that the extent of France's empire was second only to Britain's. This can, however, be explained by the different ways in which the two countries came to acquire colonies in the 1880s and 1890s, as well as by earlier mercantile history.

Ceasing to hold technological leadership in this period, Britain inevitably lost markets in advanced countries, but this did not trigger a new British drive to boost sales in the colonies. A larger share of colonial trade in the total was an arithmetic consequence of lost markets elsewhere. British colonisation, unlike French or German, was defensive rather than offensive. Even granted this motivation, the effects of the new colonialism do not suggest that it was justified by trade increases. Table 6.2 shows that the pattern of products and of markets was not responsible for any increased sales of manufactured goods by the main colonial powers, Britain and France, during the period 1899–1913.

Not only was new colonial trade of small importance to the major imperial powers, but there is little evidence of the exclusion of one European country's traders from another's colonies. German traders benefited from British power almost as much as the British in China and South Africa (Chapman, 1992). German exports to other west African colonies from 1889 to 1911 grew more rapidly than did German exports to German West Africa in the same period. British exports to German South-West Africa increased by proportionately more than German exports from 1900 to 1911. French exports to British possessions in Africa rose by 60 per cent from 1889 to 1911, while French exports in total rose by almost the same proportion.[5]

The imperial powers staked their territorial claims where their nationals had already established trading interests, however small these were. This is further borne out by the close, but usually decreasing, dependence of colonies on trade with the imperial power. Some 55 per cent of British colonial imports originated from Britain in the period 1892–6, and 49 per cent of exports were shipped there. These percentages were a slight decline from those of twenty-five years earlier. An even larger percentage of the external trade of British colonies, 65 per cent, was conducted within the British Empire. The dependence of the French colonies on trade with France at 61 per cent for 1896 was of a similar order of magnitude.

Trading patterns of colonial areas in these years were the cause, rather than the consequence, of colonisation.

Although colonisation did not generally greatly enhance the trade of the imperial powers, the administration of the colonies became for most of them a considerable financial drain. French colonial expenditures massively exceeded colonial receipts in 1898. Between 1894 and 1913 Germany's expenditure on its colonies, excluding defence, was greater than the entire value of German colonial trade. Similarly for the most recently arrived imperial power, Italy, spending on colonial government exceeded the value of its entire colonial trade in the years 1893 to 1932. The Dutch obtained a financial surplus from the Netherlands East Indies until 1874, but thereafter it was eliminated by the costs of fighting in northern Sumatra. The other Dutch colonies of Surinam and Curaçao demanded subsidies. Japan, whose acquisitions Formosa (Taiwan) and Korea are excluded from Table 6.5, transferred more funds to these territories than were remitted to Japan. Later there was a net payment from Taiwan (Kimura, 1989). Great Britain, in contrast to most other powers, required its dependencies to be self-supporting. Loans were expected to be repaid eventually. Even Britain, though, could not invariably enforce financial self-sufficiency. In the 1890s the depressed British West Indies were one of the few exceptions (American Economic Association, 1900, pp. 27, 73–104, 168–9, 302; Grover Clark, 1936). For the British Empire as a whole in these years, some have seen the costs to Britain as entirely exceeding the benefits, although that conclusion depends upon the alternative international economy that must be assumed to replace the empire (O'Brien, 1988; Davis and Huttenback, 1986; Offer, 1993). Certainly the new European colonies in the later nineteenth century were usually an expensive hobby that could not be justified by the trade generated. Even the old colonies were not obviously a source of abnormally high returns to investment, broadly interpreted (O'Brien, 1982; Foreman-Peck, 1989).

This finding is not so easy to square with the idea of colonialism as exploitation by a foreign society and its agents, who occupy the dependency only to serve their own interests. Some agents may have gained, but demonstrating even that proposition requires a judgement about what they would have attained in a differently organised world economy. Quite possibly all parties lost out from the late nineteenth-century empires. Almost invariably since 1945 the impact of colonialism on the colonised people has been represented as harmful. Forced labour in Java, revived in 1830, was only suppressed in 1870. In the ironically named Congo Free State, particularly barbarous forced labour continued into the twentieth century. Three-quarters of the Herero were exterminated in south-west Africa between 1904 and 1907. The Amritsar massacre in India in 1919 continues to exemplify colonial oppression for some nationalists, though comparable atrocities were not unknown under preceding and successor regimes. There was, however, another side to colonialism. The eminent Victorian economist A. Marshall maintained that the colonies received very good value in the services of the British administration whose salaries they paid:

we export to India a great number of prime young men. If their value was capitalised as it would be if they were slaves, it would be several thousand pounds a piece. We bring them back afterwards, if they come back at all, more or less shrivelled and worn out. Those are a vast unreckoned export. (1926, pp. 312–13)

Many of the colonial territories probably received more effective administration of law and order at a lower cost than they did before, or have since. The colonial impact, considered in more detail in Chapter 10, showed many faces. Colonialism was not a system of international servitude, for it was not a system at all. In some cases it was an imposed solution to the absence of a framework for market relations; in others it was a theatre for European power politics (Fieldhouse, 1981, pp. 8, 48).[6] In all instances the common denominator was European (and American, Australian and Japanese) technological superiority, which made intervention a low-cost strategy. For the imperial economies in any event, colonialism was not an economic necessity to pull them out of a slump.

The political response to depression

The depression of 1873 and after, described in Chapter 3, nevertheless did provoke a political reaction, but in the form of higher tariffs. As the huge agricultural resources of the regions of recent European settlement were drawn into the world economy, most of western Europe refused to accept the further shifts in economic structure that reduced world food prices required. Industrial interests also reacted to lower prices by demanding tariff protection and by forming cartels to restrict output in order to maintain prices. Higher tariffs provoked retaliation, and the ensuing tariff wars temporarily reduced trade.

Tariffs

Under pressure from industrialists and agrarian interests, Bismarck introduced a new German tariff in 1879, which imposed relatively low duties on a wide range of imported manufactures and heavier duties on agricultural produce. No major change occurred in the 1880s, but in 1890 higher rates were imposed on many articles. In 1893 a tariff war with Russia began, culminating in Russian breadstuffs paying 115 per cent higher taxes in Germany than did American, and with German imports to Russia paying 50 per cent more than those from elsewhere. Both sides eventually concluded that the resulting paralysis of trade was dangerous to the peace of Europe, and reached agreement (Ashley, 1904).

France began to embrace protectionism at the same time as Germany, when the flood of US wheat increased agricultural agitation for a tariff in 1878, and the 'American peril' became a matter of popular debate. The 1881 tariff raised rates on manufactures by an average of 24 per cent, but only as a basis for negotiations with other countries. At the same time, the French introduced state subsidies for shipbuilders and for owners of ships engaged on long voyages. The agricultural interests were disappointed in the tariff of 1881 and managed to raise rates on various agricultural products during the 1880s. The French duty on cattle imports annoyed the Italians, who in any case in their search for revenue proposed to revise their tariffs upwards to an average level of 60 per cent *ad valorem* in 1886. Negotiations between France and Italy over the new tariffs broke down in 1888, and each country successively increased duties on the other's products until 1892.

French exports to Italy approximately halved, although they were recouped in other markets. Italy suffered much more, its exports to France falling by more than half, and the decline was not compensated by increased sales elsewhere. The 1892 French tariff raised duties further – on agricultural produce to an average level of 25 per cent, and on textile manufactures. A two-and-a-half-year tariff war with Switzerland followed this increase. If Switzerland, Italy and Russia had been able to foresee the outcome of their respective trade wars, they would almost certainly have pursued more conciliatory policies. They could have avoided the costs of the war and achieved earlier a settlement no worse than the eventual outcome. Each country seems to have underestimated the ease with which larger economies, France and Germany, could substitute for their purchases from the three states. The French reaction was so strong most probably to demonstrate to other small trading partners, such as Belgium, what they could expect if they were not compliant. For both Russia and Italy, commercial conflict was exacerbated by simultaneous political rivalries (Conybeare, 1987).

Russia and the United States had always been protectionist, but increased the height of their tariffs. Russia did so in 1881/2 and 1890/1, and the United States behaved similarly with the McKinley Tariff of 1890 and the Dingley Tariff of 1897. The average US tariff level rose from 47 per cent in 1869 to 49.5 per cent in 1890 and 57 per cent in 1897. The new justification for the American tariff, replacing infant industry arguments, was the need to protect the highly paid US workers against the products of the poorly paid Europeans.

Britain remained loyal to free trade despite the founding of the Fair Trade League in 1881. When in 1906 the Conservative Party fought an election on a platform of protectionism, it lost. The prospect of higher food prices for the greater part of the electorate employed outside agriculture was sufficient to sway their vote. Those parts of the empire without tariff autonomy therefore had to follow suit. Although India was allowed a 5 per cent import duty in 1893 for revenue purposes, Lancashire cotton manufacturers were able to secure a 5 per cent excise duty on cotton goods made in India to eliminate any protective effect on Indian textiles (Dutte, 1970).

The self-governing dominions were able and anxious to promote manufacturing industry by tariff protection. Canada unilaterally granted a tariff preference to British goods in 1897 (McDiarmid, 1948). Unlike Canada, Australia and South Africa were relatively late even in securing free trade between their component states. Domestic free trade was achieved in Australia only in 1900, and in South

Africa in 1903 (Bastable, 1922, pp. 118–19). Taking the total tariff revenue in relation to the total value of imports as a measure of the height of a national tariff, Australia, Canada, the United States and Argentina in 1913 all imposed tariffs of about the same level – 16.5 to 17.7 per cent (Diaz Alejandro, 1967).

Argentina has often been classified as part of Britain's informal empire, and represented as a country whose policy was distorted in the interests of the agricultural landowners and British exporters of industrial goods. It is true that British exports were disproportionately represented on the free import list, yet the two items which dominated duty-free imports, railway equipment and coal, were intermediate goods in which Britain had a strong comparative advantage. These goods would have been prohibitively expensive to supply domestically, and would have raised the costs of many other Argentinian products (Diaz Alejandro, 1967). As it was, from the 1880s, the tariff was sufficient to create viable factories, such as those making galvanised iron. If the Argentinian tariff had been within the British sphere of influence, textile imports would have been taxed at a lower rate.

Earlier in the nineteenth century, Britain led Europe by repealing the Corn Laws and the Navigation Acts. By the last quarter of the nineteenth century, Britain's example was less compelling. Instead a spontaneous international order began to emerge in European economic relations as free-traders expected. International institutions were established covering communications (International Telegraph Union 1865, Universal Postal Union 1874) and trade. This second group included the 'most favoured nation' clause. Agreements were reached to avoid trade wars (beet sugar subsidy conferences 1887, 1898), and a common monetary base was adopted (Latin Monetary Union 1865, Scandinavian Monetary Union 1875, International Monetary Conferences 1867 and 1878). Trade and economic growth were facilitated by free movement. Russia, Romania and Greece were distinguished among European countries in requiring travellers entering their territories to hold passports. (Turkey also did.) For other European states, passports were unnecessary. A form of liberal international economic order was created without leadership or hegemony.

Increasing protectionism in the later nineteenth century must have reduced world trade relative to free trade, although it is less clear that economic welfare consequently declined. Most duties were in specific form, a certain money sum per unit imported. As prices rose after 1896, the real value of such protection declined unless tariffs were increased (Bairoch, 1989). There is a positive association between the revenue raised from imports as a proportion of import value and national income in these years, which may be taken as an indication of the harmful impact of protection, but causation could run in the opposite direction (Foreman-Peck, 1994). Protection in Germany reduced the relative decline of the agricultural sector and therefore cut migration from the country to the towns. This migration was not necessarily beneficial if it was taking place at a rate faster than could be accommodated by the expansion of social overhead capital – schools, roads and sewers. In the British case, the economy might have benefited from protection by utilising economies of scale or learning in some of the new industries in which Britain was weak. Had tariff protection been given to the infant motor industry early in the twentieth century, the greater capacity permitted by increased home market sales might well have driven down costs. These lower home costs might have offset the higher imported motor vehicle prices, forced up by protection (Foreman-Peck, 1979).

Despite political friction over potential colonial territories, massive conscript armies deployed in expectation of, or to deter, a European war, and tendencies towards higher tariffs, the European and the world economies were quite open to the movement of goods and people by 1910. Low rail and steam-ship transport costs, and economic policies that, in historical perspective, were still liberal, underlay these highly internationalised economic relations. Economic nationalism had displaced liberalism to the extent that large European countries (except Britain) maintained higher tariffs than small countries. Small economies lacked monopoly power to exploit. But by 1913 Europe's trade/gross national product ratio, at a peak not achieved again until the 1960s and 1970s, indicated that the late nineteenth-century nationalist/liberal international order supported markets at least as free as those created by the post-1945 institutions of GATT and the IMF. Britain and Scandinavia returned to the 1913 level of integration with the world economy only around 1970, and Italy did so perhaps five years earlier (Grassman, 1980; Beenstock and Warburton, 1983).

In the half century before the outbreak of the First World War, European economic integration was advanced, although most statesmen had not planned it, and therefore the gains to Europeans from trade and specialisation were also increased. Increasing integration meant a tendency towards single European prices for all goods and services, for labour, capital and land. In practice, transport costs, state policies, adjustment costs and other influences prevented perfect integration between national economies, but the proportion of internationally traded to non-traded goods increased. The second category is much less responsive to world market conditions, including as it does rents, wages, services and products with high weight-to-value ratios. Nonetheless non-traded good prices are ultimately affected by world conditions, most immediately in areas specialising in export products. The rising ratio of traded to non-traded goods meant that the European economies were more prone to fluctuate with each other and with the rest of the world economy. Booms and slumps spread more pervasively around Europe and the world, as was to be demonstrated devastatingly after 1929.

Cartels

Among businesses, slumps encouraged a movement parallel to tariff protection to restrict the play of market forces. The Nobel Dynamite Trust Company, one of the first of the international cartels to regulate prices and output, was formed in 1886 (Reader, 1970). Alfred Nobel established a number of companies in different countries in partnership with local interests, largely to avoid tariff barriers.[7] In a later period, the Trust would have developed as a multinational company. As it was, the Trust was supposed to eliminate competition between companies which Nobel had founded and on whose boards he still sat. The General Pooling Agreement of 1889 between the Trust and the German powder companies was an even wider market-sharing arrangement. The American market proved more difficult to control. The European companies and the American firm Du Pont negotiated over it from 1897 until 1914. Eventually they agreed that Europe was to be reserved for the European firms, North America for the Americans and the British Empire, including Great Britain, for the British. Rights elsewhere, as for instance in South America, were negotiable. The basis of these treaties was a shared belief that the market for explosives could not be increased by any action which the companies themselves could take. This belief was almost certainly correct. Explosives were an intermediate good for other industries, and they accounted for a very small part of mining costs. Reductions in price would not therefore induce the mining companies to use more explosives.

The average annual rate of return for the Nobel Dynamite Trust over the whole of its life from 1887 to 1914, a little over 8 per cent, hardly suggests that market regulation was used to make monopoly profits. The only question then is whether the market-sharing agreements reduced the stimulus to innovate and increase efficiency. A detailed study of the United Alkali Company's choice of technique for producing soda showed that it did indeed forgo profits by failing to replace Leblanc process plant with Solvay process plant early enough. The most likely reason is that it operated a market-sharing arrangement with continental producers, which reduced the incentive (Lindert and Trace, 1971). Perhaps in the absence of the Trust, Nobel would have been forced to try to diversify, and would have developed the chemistry of dyestuffs which German companies were using in the 1870s. On the other hand, improvements in operating efficiency were steadily bringing down the costs of production, and this was possibly the best that could be done.

Anyway the international economy was little affected by possible stultification of innovations by cartels, because few cartels were able to divide up the international market entirely among their members; the zinc and lead syndicates, the European plate-glass agreement, and the international pipes and rails cartels were the most important successful examples (League of Nations, 1927). The British Railmakers' Association was formed in 1883 and soon after combined with German and Belgian manufacturers into the International Rail Syndicate. The stability of published steel prices in the following two years suggests that they were set by a committee rather than the market. This syndicate collapsed in 1886. Ten years later the British Railmakers' Association was officially revived, and in 1904 it joined again in an International Railmakers' Association (Macrosty, 1907).

More usual were cartel agreements concluded between firms in different countries. According to one estimate, there were at least one hundred agreements in which Germany was concerned, most of these applying to the chemical industry

(League of Nations, 1927). Writers on cartels were inclined to include companies owning plants in different countries in the same category, on the grounds that they were aiming to produce the same effect. Thus Coats, the British multinational thread manufacturer, controlled spinning mills and thread factories in many countries and was a cartel in intention. The American cigarette trust almost entirely controlled the English cigarette industry and also established itself firmly in Germany. That mergers and cartels were alternative means of security market power explains the strength of the American merger boom of 1899–1900 in comparison to the British. In the United States, the anti-trust legislation of 1890 inhibited cartel agreements and therefore encouraged mergers (Hannah, 1974). The division of international markets by large firms and the tariff wars of governments brought to the fore the element of conflict in international economic relations that had been largely absent in the twenty-five years or so before 1875. The mid-century vision of the free-trade liberals was receding.

Summary and conclusion

The economic development of the industrialising countries shifted the pattern of comparative advantage as well as the political balance of power. Britain's unique mid-century position in manufactured goods, based upon cheap coal, was eroded and its share of foreign markets began to decline in the face of German and American competition. Lower long-distance freight rates, rather than increased demand by manufactures exporters, encouraged the extension of arable farming for export by European migrants in the American mid-West, Canada, Australia, Argentina and New Zealand. These temperate zone countries were able to provide high standards of living for their agrarian population working for the world market. But the tropical countries were unable to emulate this performance, despite a substantial growth of exports. Ultimately, tropical economies failed to raise living standards markedly because the low productivity of tropical agriculture precluded any substantial development of a modern manufacturing sector. In addition, the lower rate of expansion of the British economy in the last quarter of the nineteenth century reduced the growth rate of tropical exports, further limiting the possibilities for development of the tropical economies. Virtually all were constrained by either a low growth of exports or a low ratio of export value per head of the population. Moreover, the commodity terms of trade of the tropical economies deteriorated slightly during these years, the fall occurring mainly after 1900.

European trade with many tropical areas created a variety of problems, which from the viewpoint of the European powers could often best or most simply be dealt with by the extension of formal political control. For reasons of expense, the industrial powers preferred 'spheres of influence' to protectorates, and protectorates were preferred to formal colonisation. Nevertheless, by the turn of the century, except where there were what Europeans could recognise as responsible governments, such as in Siam and China, most areas of the tropics had been converted into formal colonies. Colonisation in these years did not, however, increase the trade of the imperial powers significantly.

Lower freight rates expanded the effective supply of agricultural land and brought down agricultural prices. This provoked a demand for protection against food imports in continental Europe, where sufficient numbers were still employed in agriculture to be an important political force. With the slowing down of world trade growth, the benefits forgone by industrial protection seemed to diminish. Rising nationalism provoked higher tariffs, and the associated tariff wars further reduced the volume of trade relative to what it would have been. The protected nations grew faster, by and large, than when they had been committed to free trade, although the association was probably not causal. Businesses also tried to ameliorate the impact of competition by agreeing upon the division of international markets, but with a few notable exceptions they were not particularly successful.

Notes

1. The growth of output per head averaged 3.8 per cent per annum and trade grew at 8 per cent per annum between 1953 and 1973.
2. The term 'revealed comparative advantage' for trade shares was coined by Balassa (1965).
3. Schedvin (1990) extends the theory to account for longer-term political influences upon economic development.
4. The difference may not be significant in view of the sensitivity of the calculation to the choice of initial and terminal years.
5. Calculated from *Statistical Abstract for the United Kingdom* and *UK Statistical Abstract for Foreign Countries*. Cain and Hopkins (1993, p. 361) see the rising share of Africa in British trade as more significant than the small proportion of British trade involved.
6. This work contains a valuable bibliography and a chronology of colonialism.
7. A Swede brought up in Russia, Alfred Nobel was prone to relieve his feelings by writing in English. He eventually gave his fortune to establish the Nobel Prizes.

7 Capital movements, 1875–1914

The opening-up of new lands to the world market brought into being vast investment opportunities. By the standards of what had gone before, enormous quantitites of capital crossed national and continental boundaries in search of gain. Most, but not all, capital went to regions of recent European settlement. The old trade of lending to foreign governments expanded, and so did the problem of ensuring that they repaid. European political tensions guaranteed that foreign loans would be enlisted in the pursuit of national political goals. Bankers also tried to obtain state support for their projects, generally with less success.

The first section of this chapter discusses the forces underlying national supplies of, and demands for, capital, and the differences and similarities among source and donor countries. The focus shifts in the next section to the capital exporters Britain and France, and to what determined the size and distribution of their foreign investment. The political impact of foreign investment on receiving countries which did not wholly share western institutions and attitudes forms the subject of the third section. The distribution of the gains from foreign investment between the borrowers and the lenders is considered next. To conclude the discussion of capital movements, there is an analysis of the distinctive nature and problems of foreign direct investment, as distinguished from portfolio investment.

The supply and demand for capital

A world capital market integrated a large number of national markets by the last quarter of the nineteenth century. Although domestic and foreign assets did not become perfect substitutes, for many national economies their risks and returns were fairly similar. Between these economies long-term international capital therefore flowed at the internationally determined, risk-adjusted interest rate. Countries with the largest international borrowing grew rapidly and received a high immigrant inflow relative to the indigenous population. In this group are included Canada between 1900 and 1914, Australia in the 1870s and 1880s, and Argentina throughout the period. The United States, Italy (from the 1890s) and Germany, where a high rate of growth combined with a good productivity performance, were largely self-sufficient in capital by this period (see Figure 7.1).

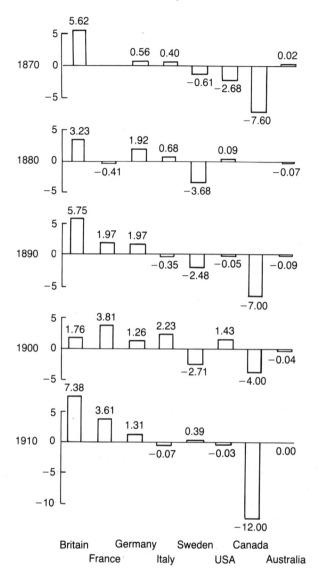

Figure 7.1 International capital flows as a percentage of gross national product, selected countries, 1870–1910

Britain's large and continuous export of capital was associated with an unusually (by international standards) low rate of domestic investment, rather than high savings. As a major net exporter of capital, but not of people, France was anomalous. French savings rates were even greater than the unusually high domestic investment ratios. At the opposite end of the spectrum from France were Sweden and Russia, which imported capital, but sent migrants overseas.

Three possible imbalances or gaps in aggregate spending giving rise to capital movements may be distinguished: between domestic savings and domestic investment, between imports and exports, and between government taxation and spending. These three necessarily sum to zero. Countries with a strong demand for capital (investment) relative to their savings tended to receive capital from abroad. Urbanisation imposed great demands upon capital and therefore those countries with the largest rural-urban migrations, or immigration to the towns, usually received some foreign investment (Lewis, 1978b, pp. 148–50). Inward foreign investment allowed imports to exceed exports by committing the economy to future payments servicing the foreign-owned assets. There was only one other gap which would allow domestic investment to exceed private domestic saving. If a government taxed more than it spent, and lent for investment, or subsidised investment or undertook projects itself, it would, in effect, have been saving on behalf of its population.

Governments sometimes chose to spend more than they taxed. The three-gap approach shows that they must have financed the difference from an excess of domestic saving over domestic investment, or from foreign borrowing, or from some combination of the two. How effectively they used the money relative to what was displaced then determined the desirability of that policy. The Egyptian Khedive of the 1860s and 1870s has been accused of extravagant and unproductive building projects, both 'crowding out' more appropriate investment and saddling the country with enormous future interest payments (see below). His defenders, on the other hand, maintain that his policy 'crowded in' more productive private investment than would otherwise have been undertaken.

Bearing in mind that infrastructure required a great deal of capital but added little to productivity, three principal determinants of the demand for real capital, or for investment, can be distinguished. The first was technical progress, which had to be embodied in physical equipment if productivity was to be raised. The second was the growth of output, which required more capital even with unchanged techniques. The third was the growth or redistribution (urbanisation) of the labour force, which created the need for infrastructure (Green and Urquhart, 1976). Which of these mattered most for international capital flows? Productivity growth did not drive the demand for capital imports, for several of the capital-supplying countries in various decades achieved rates of productivity growth greater than or equal to the rates of destination countries. Output growth is not obviously a candidate for rapidly growing economies, of which the largest was the United States, for they typically invested a high proportion of their gross national product. The major capital-importing countries grew faster than the capital exporters but the proportion of investment that was devoted to construction varied with the national population growths. In the economies of four international suppliers of capital, urban growth typically accounted for up to two-thirds of total population growth. Higher proportions of population increase were absorbed into urban areas in Britain, France, Spain and Germany than in most capital-importing countries. However, total population in each capital-exporting country grew less rapidly than in the United States and Australia. Until 1900 Canada's population growth was only slightly above those of the capital source countries, but rose dramatically after the turn of the century. Population growth is therefore the proximate determinant of an excess demand for investment (investment exceeds domestic saving) in the four decades after 1875.

Rearranging the three gaps and ignoring state activity, the supply of capital can be divided into that portion originating from domestic savings and that arising from international capital flows.[1] Savings rates in the more industrially advanced countries – Britain, the United States and France – were relatively high and steady. The savings ratios of the industrialising countries – Italy, Sweden, Canada and Australia – by contrast rose, and to a lesser extent this was also true of Germany. These observations are consistent with the view that economic growth requires a jump in the savings and investment rates (Rostow, 1972).

Savings may have risen because of changes in the age structure of the population. The life-cycle savings hypothesis asserts that households save or dissave in order to eliminate discrepancies between their consumption plans and their expected income over their entire lives (Ando and Modigliani, 1963). Accordingly, savings rates will be highest for the age group 25–45, the maximum earning age range for most workers. The proportion of people in this group in any country was affected by migration and by natural increase. Because this age group was most prone to migrate, there was a high proportion of 25–45-year-olds in countries of immigration, such as the United States. Savings ratios in immigration economies were therefore high, but typically not high enough to satisfy the demand for investment.

Dependency ratios pulled New World savings in the opposite direction to age structure. Other things being equal, families with high ratios of dependent children to wage earners, saved little. Families in the labour-scarce and land-abundant economies were highly fertile. Their high ratios of children to adults created a US dependency rate in 1850 of 0.415, a Canadian rate in 1851 of 0.560, and an Argentine rate in 1869 of 0.452. These ratios declined until 1900, but they remained above that of the UK. Capital flows therefore in part reflected the operation of an international market where mature life-cycle savers in the Old World were able to lend to young life-cycle borrowers in the New (Taylor and Williamson, 1991).

The capital exporters

The two major capital exporters, Britain and France, grew relatively slowly. Was there a causal link and, if so, in which direction? An absence of high-yielding domestic investment opportunities in these now mature economies may have encouraged investors to turn their attentions overseas. Alternatively, France and Britain may have grown slowly because they exported so much capital. The capital markets of these countries may have been biased, directing too many resources overseas, and not necessarily to the most profitable projects. Whereas in the third quarter of the nineteenth century French economic activity abroad was located mainly in central and western Europe, from the early 1880s the centre of gravity moved away, to eastern Europe, Latin America, Asia and the colonies (Cameron, 1961, p. 485). More than 90 per cent of French foreign investment was in Europe (including the Turkish Empire and Egypt) in 1881, but thereafter almost half the new investment went to other continents. Russia, the Balkans and Scandinavia were major recipients. In part this was merely the shifting locus of opportunities as old areas became self-sufficient in technical knowledge and capital. There was also an important political factor: French governments wanted foreign investment to support national security interests.

French foreign investment was more effective as a political weapon when it came to denying potential borrowers money. The state could forbid the raising of a loan in France, but it was much less successful in persuading investors to pursue projects that they did not anyway regard as sufficiently profitable. In 1887 France's old enemy Bismarck issued a ban on Russian securities, thereby encouraging the Russians to turn to France for money. By 1892, when the Franco-Russian military convention was signed, Russia had received more than 6,000 million francs in French loans. The French General Staff were particularly enthusiastic about strategically important railways in western Russia, which they saw as vital for containing Germany (Born, 1983; Feis, 1930, p. 134). When, in 1887, the Triple Alliance between Italy, Germany and Austria-Hungary was renewed, French banks sold the Italian bonds they held and bought Russian securities. On this occasion German banks took up the bonds, but in 1890 Italian public finances were in such dire straits that Italy was forced to approach the Paris market again. The Paris Rothschilds refused to co-operate, as agreed with the French foreign ministry. Italy was induced to support French policy in Morocco in the first decade of the twentieth century by its need for French loans. Russia also felt French financial pressure in 1904, when requiring money for the war with Japan. French intransigence almost drove the Russians back into German arms, but Germany could not supply Russia's financial needs.

In Russia, which was the largest European borrower, France accounted for by far the greatest proportion of foreign investment. Although Britain was a much larger foreign investor in total, its investments in Russia remained small until after the signing of a treaty in 1906. The greater government involvement with the French capital market also enhanced the opportunities for pressure groups with access to the government to influence the market. Organised employer groups sold their consent to foreign loans for promises that foreign borrowers would give French lenders special consideration in their expenditure, or for tariff concessions. Loans to directly competing industries were sometimes kept off the French market, as in the refusal in 1909 to grant official listing to the shares of the US Steel Corporation.

From the 1870 Franco-Prussian War, Germany was forbidden territory for French investment, despite its proximity, but Germany could manage without

French investment. Germany already invested abroad in a modest way. German finance for the Baghdad and Anatolian railways gave Germany a good deal of influence in the Ottoman Empire. After the 1908 Turkish revolution the new government turned to Britain for financial help. Despite support from the British foreign minister, French machinations defeated British attempts to raise a loan and Turkey turned back to Germany. Similarly, as Austria-Hungary allied itself more closely with Germany, it ceased to be a favourite field for French investment as a result of government restrictions.

These were some of the biases imparted by politics to the French capital market (and to others). Instances can be found of political interference in the British capital market, for there were confidential links between the Treasury and the Foreign Office on one side, and the Stock Exchange on the other, often mediated by the Bank of England. The withdrawal of British government support for the trunk railway line across the Turkish Empire from Constantinople to the Persian Gulf in 1903 led to the withdrawal of the British banks that had arranged to participate in the finance and operation of the line. The signing of the Anglo-Japanese Treaty in 1902 was immediately followed by Japanese loans on the London market. But the British capital market was generally subject to far less overt political control, for whatever reason, and channelled investment very differently from the French. Between 1882 and 1914, 38 per cent of new foreign investment went to eastern Europe from France, and half went to Europe as a whole. In contrast, British foreign investors favoured mainly the regions of recent European settlement, primarily inhabited by people with British traditions, although Argentina was an important exception.

Approximately 10 per cent of Britian's national income in 1913 was earned from foreign investment. Would this investment have been more productive, and perhaps earned higher returns at home, but for the peculiar nature of the London capital market? (See, for example, Liberal Industrial Inquiry, 1928.) This market, the largest in the world, sold primarily safe, well-known securities, many of which happened to be foreign; in order of increasing expected risk these were colonial government securities, foreign government securities, colonial rails, US rails, foreign rails and, finally, foreign company issues. Few domestic British firms raised money on this market. Edelstein (1972, 1982) maintains that this was not a consequence of the rigidity of London's capital market facilities, or of the unwillingness of domestic industrial borrowers to offer debt instruments of the type desired by the new 'trustee' investor, or of a British preference for overseas assets. Rather, provincial capital markets and wealth were quite adequate to satisfy cheaply and flexibly the usually small-scale needs of domestic industry, without recourse to London. As the scale of domestic industry and commerce rose, the London market eventually became involved in long-term funding of industry. American securities on the London market in particular filled a gap in the security market for high returns and medium risk. If there was a failure, Edelstein believes it to be in excessive risk aversion among investors, which may have deterred them from lending to domestic manufacturing industry.

The difference between Britain on the one hand, and the United States and Germany on the other, was that the British capital market, with its excellent facilities for trading first-class securities, allowed small British lenders to indulge their risk aversion. According to Kennedy (1976, 1987), in the United States and Germany banks played a much more important role in allocating funds, and directed them towards the new, advanced technology industries which were also riskier. The published data on firms in Britain before 1914 were so bad that shareholders depended almost entirely on their personal knowledge of the men who ran the business. Such information was harder to come by than information about the reliability of first-class foreign securities. Hence, given that these securities were available, more resources were directed towards foreign investment than if the same information had been provided about domestic firms. One caveat to this argument is the use of the London capital market by foreign lenders as well as British, which allowed American and German savers also to trade off risk against return. But to the extent that their banks provided them with a different range of assets, or supplied information, or intermediated between them and the firms, the German and American savers had less need to do so.

The political impact of foreign investment

Unlike investment within a national economy, international investment must come to terms with the legal, institutional and social environment of the host country, which is not necessarily similar to that of the country of origin. These differences created substantial problems in the forty years before the First World War. In countries with similar social frameworks to those of the western European investor, such as the USA, Canada, Argentina and Australia, the political difficulties were minimal. Countries less committed to economic advancement experienced more adverse effects.

Within a national capitalist economy, investors are prepared to lend for a project if they believe it sufficiently profitable to repay their capital, as well as interest and profits. They will also usually be concerned to have a claim on some asset which will allow them to recover their money in the event of the failure of the project. If a firm defaults on its obligation to pay interest on loans, then it can be declared bankrupt, and a receiver will be appointed to sell off the assets of the firm so as to earn as much as possible to repay the firm's creditors. Where a nation is concerned, matters are rather different. A government may borrow money from foreign countries and then find itself unable to pay the interest or repay the capital, but a receiver cannot be appointed. As long as the economy continues to work, revenue can be raised from taxation. The problem for the foreign creditor is how to get access to these tax receipts. The private citizen who goes bankrupt can be coerced by the state to conform to the bankruptcy laws, which in the nineteenth century often included going to a debtors' prison. The foreign creditor could only appeal to the borrowing country's government, or the creditor's own government, for consideration.

In fact, many defaulting countries in the nineteenth century did submit to treatment much like that of a firm. Foreign commissioners were appointed to administer certain tax receipts, such as the customs duties, and to appropriate some or all of the revenue for the repayment of interest and capital on the debt. Just as under the Bretton Woods system of the post-Second World War period, these countries were obliged to relinquish some sovereignty as a result of their economic mismanagement until the mistakes were rectified. What was more offensive to national feeling was that the commissioners were appointed by the few great powers as national representatives rather than as representatives of an international body in which the recipient country participated, at least nominally. Worst of all was the case where the previous government that incurred the debt was regarded as decadent and corrupt by a new administration that had to bear the cost.

Other defaulting countries were not the victims of direct sanctions, until they wished to raise fresh international loans. As long as they remained in default, bondholders lobbied to ensure that no new funds became available. Since default was often caused by the vagaries of international economic fluctuations, payments were frequently resumed as soon as conditions allowed – a form of unilateral rescheduling of debt. Lenders who felt cheated would often appeal to their governments for support. In general, the use of military or naval superiority by the lending governments was eschewed, unless, as in the case of Egypt, important political objectives were also thought to be threatened. The apparent major exception to this rule in Britain's case would appear to have been governmental support of its investors in the Boer republics in South Africa, which eventually led to war. Here the pursuit of wealth in the diamond and gold fields brought British colonials and nationals into conflict with pioneer farmers of a different nationality, determined that their land should not be dominated by the newcomers. Even here the initiative can be interpreted as a defence of subjects oppressed in a foreign land, with the British government goaded by the threat of German intervention. France dispatched a naval division to Turkey in 1901 to seize the Customs at Mytilene to discourage default. And neither country was averse by the early twentieth century to sending a battleship to the Caribbean to remind foreign governments of their international financial obligations. There was, as we have seen, some difference in attitude between the two major lending powers, Britain and France, in their approach to foreign investment (Feis, 1930, chs. 4 and 5). The French capital market was more closely regulated to achieve government objectives, and because such foreign investment as was allowed served these ends it could usually count on official support where necessary. The competitive effort of Italy in Tunis to procure entry for Italian enterprise was among the influences which decided the French government to assume political control.

Both in the foreign investment position and in the support given to overseas investors, the United States occupied an unusual position. It was heavily indebted to Europe, but investing strongly in the Caribbean and Latin America. The United States was strongly anti-colonial, but concerned with European influence in South America and the closing of the frontier in the USA. Its strategy was to build naval bases and an informal empire of monopolies and concessions to US companies in exchange for state loans in the Caribbean. Those investments were politically risky, which triggered US intervention in Nicaragua in 1909 and Haiti in 1915, and gave rise to the term 'dollar diplomacy'.

The contrast between the successful and unsuccessful use of foreign capital by countries outside the legal and institutional tradition of western Europe is shown most clearly by China and Japan. Although by the outbreak of the First World War foreign investments in China were half as large again as those in Japan, with a population perhaps as much as seven times the size of Japan's, China's foreign investment per head was very much smaller (Allen and Donnithorne, 1954, appendices B and C). China's inability to absorb foreign capital stemmed mainly from the attitude of the government and society. This is well illustrated by the fate of the first railway built in China (with foreign capital). Within a short time of its opening, a Chinese citizen was accidentally killed. In response to the public outcry, the Chinese government bought the line and pulled it up in 1877 (Latham, 1978, p. 20; Feuerwerker, 1980). Compare this with the accidental death of the Trade Minister of Britain, William Huskisson, in 1830 during the opening of the Liverpool–Manchester railway. This tragedy was not allowed to prevent the regular earning of 10 per cent dividend on capital in subsequent years. As a result of ruling Chinese attitudes, China was still poorly served by transport facilities in 1914.

Chinese foreign trade remained small in relation to its population. Imports exceeded exports and remittances of Chinese overseas or incoming foreign investment was needed to earn the foreign exchange necessary to service any foreign debt acquired. The continued fall of silver against gold depreciated the Chinese exchange rate against the currencies of the western powers. Between the 1880s and the first decade of the twentieth century, the rate of exchange of the Haikwan tael halved against sterling. Foreign debts denominated in gold standard currencies therefore became increasingly expensive to service (Remer, 1926).

Before the 1894–5 war with Japan, these problems did not matter because the Chinese government contracted almost no debt, spending as it did very little on administration, economic development or defence. In attempting to finance the post-war indemnity, the Chinese government first sought to borrow from its own citizens. But the wealthy Chinese merchants did not trust their government to repay them any advances they made, and only foreign lenders came forward. The Chinese domestic capital market remained unintegrated with the rest of the world. Chinese merchants and traders faced rates of at least 12 per cent and expected to earn accordingly more (Wang, 1993). Foreign lenders were unable to participate in this market and were restricted to lending to the government. However, they could bargain for security against their loan to the state, and the Chinese Maritime Customs proceeds were pledged. Set up under the control of a

Table 7.1 Foreign investment in China and Chinese trade

Year	Foreign investment in China (£m)		
	Business	Chinese government obligations	Total
1902	116	66	181
1914	250	121	371

Years	Chinese trade (£m annual average)	
	Net imports	Exports
1882-6	21	19
1802-6	54	32
1912-16	76	60

Source: Allen and Donnithorne (1954, appendices A and B).

foreign inspectorate in accordance with the Treaty of Tientsin in 1860, the Chinese Customs had become an efficient internationally manned body, generating a revenue higher than was previously available to the Chinese government, which increased with the growth of China's foreign trade (Allen and Donnithorne, 1954, p. 21).

Because of the Boxer Uprising, and the subsequent occupation of Peking by the western powers, China was required to pay another indemnity for killing European citizens and for the destruction of European property. Once again the Chinese government needed to borrow foreign capital. The revolution of 1911, which overthrew the Manchu dynasty, obliged the republican government to sign yet another loan contract in 1913, primarily to finance the costs of the revolution and for troop disbandment. Thus, to a large extent, China was forced to borrow abroad because of circumstances rather than because of predetermined policy. Nevertheless, direct foreign investment in businesses greatly exceeded government foreign borrowings, as Table 7.1 shows. This investment probably financed most of the balance of trade deficit also shown in the table, the remainder being due to the remittances of Chinese living abroad.

The Japanese employed foreign investment to achieve economic development without losing control over national resources. Direct investment was an extremely small proportion of the total. Railways were nationalised and foreign rights of ownership in land and mining properties were forbidden by law, although foreigners could buy long leases. The government subsidised iron- and steelworks, shipping companies, banks and telephone companies in order to accelerate the diffusion of western technology (Allen and Donnithorne, 1954, ch. 11). To the same end it employed foreign technicians whose salaries in the 1880s amounted to about 20 per cent of government expenditure. This positive approach to development required finance, and in 1867 the Japanese government floated its first loan, on the London market, for the construction of the

Table 7.2 Foreign investment in Japan and Japanese foreign trade (m. gold yen)

Year	Govt loans issued abroad	Total portfolio investment	Direct and short-term (liquid) investments 1904)
1894	2	2	40?
1913	1,525	1,970	70?

	Japanese trade (m. yen. annual averages)	
	Imports	Exports
1883–87	33	42
1904–08	442	377
1914–20	1,300	1,434

Source: Allen and Donnithorne (1954); Allen (1981).

Tokyo–Yokohama railway. In 1872 it launched a pension loan to commute feudal obligations, but thereafter the onset of the depression, and the default of Turkey and Egypt with their associated loss of sovereignty, for Egypt culminating in occupation by a foreign power, inhibited the Japanese from borrowing. As with China, the depreciation of the silver standard yen was also unhelpful to Japanese foreign borrowing. Domestic inflation, a large budget deficit and national debt during the late 1870s made Japan a poor risk for lending countries (Rosovsky, 1968). After the deflation and financial reforms of Count Matsukata, it was possible to raise another foreign loan for building the Nakasendo railway in 1885.

Japan's defeat of China brought Japan into direct rivalry with Russia in northern Asia. This justified a heavy armaments programme. The annexation of Formosa also committed Japan to paying substantial annual development subsidies. Another loan was therefore raised abroad in 1897 and yet another in 1902. The cost of the 1905–6 Russian war, amounting to about 1,500 million yen, was also financed by foreign borrowing (Feis, 1930, ch. 18). Table 7.2 shows Japanese foreign debt held abroad. After a decline from a peak in the mid-1870s, it grew very much faster than Japanese trade from the 1890s, which in turn showed a phenomenal growth. Nevertheless, the excess of exports over imports in the years when Japan was not involved in major hostilities demonstrates that this debt was being serviced other than by the increasing foreign investment. By 1913 more than half of the government debt of Japan was held by foreigners. Table 7.2 indicates that the ratio of Japanese foreign debt to trade surplus was much lower than China's (Table 7.1). Since foreign exchange was needed to pay the interest on the debt, China looked a considerably poorer risk than Japan.

Egyptian, Turkish and Greek defaults on servicing foreign borrowings began the slow emergence of international financial bodies in the last quarter of the nineteenth century, each one set up to deal with a particular problem. In this respect they differed from the institutions established in the inter-war period and as part of the Bretton Woods system. They also differed in that they were not truly international, being managed only or mainly by representatives of the lending powers, or powers who had substantial political status. Comparison of Turkish, Greek and Egyptian experience with foreign loans illustrates the process by which a country could lose its sovereignty. In the case of Egypt, default was associated with massacres of Europeans and a perceived military threat linked with a loss of access by Britain to the Suez Canal, deemed crucial to communications with the Indian empire. In the case of Greece and Turkey, default led only to the imposition of international control over certain tax administrations and the obligation to control the money supply and the government's budget until the creditors were satisfied. There was no breakdown of law and order. The survival of Turkey as a buffer against Russian expansion was of sufficient importance that a war loan could be raised in London during 1877 while Turkey was in default.

The depression of 1873 and after ushered in the defaults on their foreign obligations of a considerable number of states. Honduras and Santo Domingo were in default from 1873, Costa Rica, Liberia and Paraguay from 1874, Bolivia, Guatemala and Paraguay from 1875, and Peru and Turkey from 1876 (Council of the Corporation of Foreign Bondholders, 1878). Italy restricted payment, and Greece needed persuading to resume payment on its long-outstanding debts. Only with these last two countries did the British government take part in negotiations on behalf of bondholders.

With the other major European powers, the British government signed a declaration at the Congress of Berlin in which Turkey acquiesced by creating an international debt administration to manage the tax receipts assigned to the foreign debt. Negotiations at Constantinople (Istanbul) between representatives of British, Dutch, French, Austrian, German and Italian bondholders settled in 1881 the taxes which the Council of Administration was charged with collecting and administering for servicing the debt (Council of the Corporation of Foreign Bondholders, 1928, pp. 419–20). Egypt proved more of a problem.

Under the Khedive, Ismail Pasha, from 1863 the Egyptian national debt more than quadrupled. The money was used to rebuild Cairo in imitation of Paris, to extend Ismail's properties until they included 20 per cent of Egypt's arable land, to build railways, a telegraph network, a road system and irrigation canals, and to operate factories. Loans were raised to pay interest on earlier loans, and the tax system became a system of arbitrary tribute and confiscation (Landes, 1958; Feder and Just, 1984; Feis, 1930, ch. 15). Ismail appointed a receivership in 1876 which put under control of the representatives of creditors the taxes, such as customs duties and the salt and tobacco taxes, which had been assigned as securities for loans. The receivership was composed of four representatives appointed by the Khedive on the separate recommendations of the French, British, Austrian and Italian governments. They were to have the status of Egyptian officials who could not be recalled by their governments without the consent of the Khedive. Debts were unified at a lower interest rate. Another Khedival decree named two European controllers, one French and one English, to watch over the state finances. Interestingly, in the light of later developments, the British government refused to make its nominations, but did not object to them (Feis, 1930, ch. 15).

In view of the parlous state of the nation's finances – debt charges were over £8 million, while total government revenue was only about £9 million – it is not surprising that the system did not work. A Commission of Inquiry in 1878 recommended as a solution the dismissal of army officers, the taxation of the upper classes and further restrictions on the powers of the Khedive, thereby securing an alliance of all three parties against the European creditors. Although Ismail was deposed in 1879, the foreign controllers became locked in dispute over financial matters with the Egyptian Assembly, and the strength of nationalist sentiment had by 1881 virtually made ruler an Egyptian of peasant origin, Colonel Ahmed Arabi. Throughout the country there was a breakdown of order. The British intervened militarily after the murder of Europeans in Alexandria, and when the Egyptians began fortifying the town. They feared their access to the Suez Canal would be blocked.

Under the Constitution of 1883, the British Consul-General became the real governing power. The Egyptian representative bodies could advise and criticise, but not directly oppose. Their consent was necessary to new taxes, but not to the spending of the revenue from existing taxes. Between 1882 and 1913 the British administration reduced taxes per head, but almost doubled net revenue. It regularised tax collection, revised the system of public accounts and enforced Treasury control of expenditure. Out of the limited budget surplus, roads, railways, irrigation and drainage canals were built and the external debt was slightly reduced. Until 1914 the improvements in financial administration were matched by a rising standard of living among the peasantry.[2] In 1923 Egypt regained its formal political independence.

Greece unilaterally reduced interest payments on its external debt in 1893, without causing any international reaction. By that date the national debt was absorbing one-third of state revenue. But in 1898, defeated by Turkey, Greece needed more foreign financial aid. The creditor powers therefore established an International Financial Commission to control the revenues pledged to the defaulted loans (Dakin, 1971, pp. 146–8, 154–5, 201; Cameron, 1961, pp. 497–8; Feis, 1930, pp. 284–92). The six members of the Commission were appointed directly by their governments. Russia, Austria and Italy were represented because of their political sensitivity to Greek affairs; their citizens held only very small amounts of the Greek debt. British, German and French representatives were primarily concerned with protecting their nationals' financial interests. The pledged revenues were similar to the Egyptian. They were directly administered by an organisation with a Greek staff and a director elected by the bondholders subject to the Commission's veto.

The debt settlement also included provisions calculated to improve the nation's finances. Restrictions were placed upon government borrowing, and the money supply was to be controlled. These measures raised the foreign exchange value of the drachma. The yield of the taxes did not rise, except for the Piraeus customs revenue, in contrast to the Egyptian case, either because the administration was less chaotic and thus had less scope for improvement, or because no direct foreign control was placed over the tax systems. In any event, the very existence of the Commission enhanced the foreign borrowing power of the Greek government. This allowed the Greeks to sustain wars that increased their territory by 68 per cent. Between 1905 and 1911 Greece spent 193.7 million drachmas on armed forces, and about 411 million on the Balkan wars themselves (25 drachmas = £1 in 1913). The yield included 72 million from the new provinces.

Viewed from a nationalist perspective, such foreign control was a humiliation. Employing instead the nineteenth-century liberal's welfare judgements, the loss of sovereignty by Egypt as a consequence of foreign lending was, until the First World War, desirable. For Greece, which maintained control over most of its policy, foreign investment increased the ability to pursue policies which did not enhance the material welfare of the great majority of Greeks.

When foreign investment is put to productive uses, it almost certainly increases world output. What makes the rate of return on investment higher in the host country than in the country of the donor is the greater contribution of the marginal unit of capital to real output. Thus the increase in output in the recipient country brought about by foreign investment must be greater than the reduction in output in the investing country. This conclusion does not hold when the foreign investment is directed to unproductive purposes, such as war for Greece, or extravagant building programmes in Egypt.

The gains from foreign investment

Defaults notwithstanding, investors earned returns on the overall portfolio of loans to the ten highest-borrowing governments between 1850 and 1914 similar to those that they would have received from investing in US or UK government securities. The ten highest-borrowing economies, dominated by Russia, are listed in Table 7.3. Chile is the exception for inclusion in the top ten. South Africa and New Zealand borrowed more but, like Canada and Australia, those empire countries were able to borrow at privileged rates and were particularly safe (Lindert and Morton, 1988). This might, however, bias the judgement as to the overall profitability of foreign loans.

Column (3) of Table 7.3 indicates the average premium that borrower states expected to pay for being perceived as less reliable than the US or UK governments. Egyptian and Turkish loans were judged unusually risky at the times they were made, but column (4) shows that, supported by British intervention, foreign investors in Egyptian bonds received a better return than if they had chosen safe Canadian loans. Lenders to Turkey were not so fortunate. Thanks to the Juarez government's repudiation of Maximillian's debts, and to

Table 7.3 Return on state bonds issued 1850–1914 or outstanding in 1850.

	Internal rate of return from issue price and repayment terms	Safe interest rates at home	Expected return over safe interest rate	Realised return over safe interest rate
Argentina	5.07	2.91	2.15	1.71
Brazil	4.86	2.95	1.91	0.88
Chile	5.39	2.98	2.42	1.48
Mexico	5.78	2.91	2.87	−2.27
Australia	4.35	3.01	1.34	1.01
Canada	4.47	3.17	1.30	1.27
Egypt	7.18	3.11	4.07	2.92
Japan	4.36	2.90	1.47	1.25
Russia	4.94	2.92	2.01	−1.63
Turkey	7.39	3.16	4.23	−1.56

Source: Lindert and Morton (1988).

defaults after the 1911 revolution, Mexican investors were even worse rewarded. The Bolshevik Revolution similarly upset investors' expectations of Russian loans. Summing the net present values for all ten loans yields a positive figure, but the realised average return over the safe assets was barely positive. Apparently there was virtually no additional realised investor benefit, or borrower penalty, to offset the greater risk on average from investing overseas.

When such foreign investment was devoted to productive purposes, how was the increased output divided between the lending and the borrowing country? Foreign-owned enterprises are often accused of exploiting labour in the host country, or taking away natural resources at less than market value. Such exploitation can occur only when there is monopoly power, such as conferred by the exclusive rights of the British chartered trading companies, the British South Africa Company (BSA Co.) and the Royal Niger Company (these powers were conferred to compensate the companies for bearing administrative costs normally borne by governments). The BSA Co. did behave exploitatively, expropriating land in southern Rhodesia (Zimbabwe) and selling it to European settlers. The African population was restricted to land ownership in reserves or to wage labour on settler farms (Palmer, 1977). Most foreign investment during these years, however, was the result of competitive bidding for contracts, examples of which are given below, and hence exploitation was unlikely to occur.

The distribution of the benefits may have depended upon the type of investment. Investments complementary to the labour and capital employed in the donor country, such as those calculated to increase the availability of raw materials, or reduce the price of food, especially by improving foreign transport facilities, must offer gains to the investing country, as well as to the private investors, who receive a higher return on their capital than if it was confined to the domestic economy. Because so much of British and French investment was of this nature during these years, especially for the provision of foreign railways, the receiving economies must have gained. This is not to concede, however, that the returns might not have been higher with different investment patterns, strategies and institutions.

On the other hand, when foreign investment reduces labour costs, the savings occur in the employment of factors that are competitive with domestic resources. Such foreign investment enhances the productivity of domestic capital, but lowers the productivity of domestic labour relative to what it would have been if similar investment had been undertaken in the domestic economy (Kreinen, 1975, pp. 394–400). American investment abroad tended to be in manufacturing industry and thus, in contrast to British and French investment, may not have benefited the American investing economy so much.

Tying exports to investment was a widely alleged source of gain from foreign investment to the donor country. The other side of the coin was that the host country was exploited. In continental Europe this policy was not unknown. The French secured orders in return for loans most consistently and effectively in loans to the Balkan states and Turkey. Often in these negotiations a representative of Creusot, France's greatest manufacturer of steel and war materials, or some bank on whose board Creusot was well represented, took a direct part. Such tying was frequently justified by reference to German demands. The Germans similarly explained their own restrictions. In other cases, the relationship between trade and foreign investment was exactly the other way round. In the years before the First World War, the allegation of preferential treatment was most frequently heard about investments in China and Latin America. In fact, for railway contracts in these countries, competitive bids were normal for both the loans and the construction (Winston, 1927). The lender's nationality was irrelevant if better offers were received from other sources, in seventeen important Chinese government railway loan contracts with Americans and Europeans between 1898 and 1914. And these were the majority in which the informal tying of investment with trade by the great industrial powers could occur, amounting to about 2,800 miles out of a total of 6,000 miles of railway. Of the remainder, 1,000 miles were built with Chinese capital, 1,900 miles by the Russians and the Japanese, and 600 miles by French and German companies, their purchases being closed to international competition by the terms of their charter. Table 7.4 shows that the smallest investor in Chinese railway loans, the United States, obtained the second largest exports of rails and rolling stock to China, and the largest investor, Great Britain, obtained the second smallest value of export orders.

In Latin America experience was similar. Three-quarters of Argentine railways were owned by the British, yet British companies secured less than half the imports of railway equipment. The French owned almost half of the Brazilian railway system, yet their share of railway imports into Brazil remained smaller than into Argentina except where rails were concerned, with just over one-quarter of Brazilian imports. The United States, with only a small investment stake in Brazil, secured more than half the imports of railway engines.

Table 7.4 Railway loans and exports to China, 1898–1912.

From	Railways loans to Chinese govt ($m)	Exports of rails and rolling stock to China (value $m)	(2) as % of capital loaned for rail
USA	3.0	11.48	382.8
Germany	29.76	9.47	31.8
Britain	52.39	6.90	13.2
France	27.1	0.65	2.4
Belgium	11.6	12.85	110.8
Total	123.85	41.35	

Source: Winston (1927).

The small size of their investment stake was not the cause of the relatively small trade of France and the United States with Argentina. Rather, their investments were a consequence of their lack of interest in trade with that country. Argentina was a temperate zone region of recent European settlement, and therefore was a potential supplier of beef and wheat. Britain was far more dependent on basic foodstuff imports than France and the United States, and therefore was the Argentine's principal trading partner. As nations of coffee-drinkers, the French and Americans were much more interested in the Brazilian trade, which the tea-drinking British neglected. The citizens of each country stood to gain from lower import prices from these countries. Investment in the improvement or construction of railways, ports and public utilities in Brazil and Argentina was the means of reaping these returns, and trade provided the information about investment opportunities.

Direct investment

Symmetrically, investment was also encouraged in countries to which exports were consigned, although here the Americans led. The bulk of this investment was controlled by the investor; it was direct, rather than portfolio, investment. Two criteria have in the past been used to distinguish these types of investment: the medium through which shares were distributed, and control by the investor in the undertaking in which the capital was embodied. The present-day definition uses only the second criterion. Thus investments floated on host stock exchanges and controlled by foreign investors have been counted as portfolio under the earlier definition. Methods of valuation of private investment have also changed. The earlier approach used only the nominal value of new issues, while the later method employed the book value of total assets, which tends to be much greater. Taking these points into account and with better data on foreign direct investment (FDI), the share of direct investment in all private investment in the poorer non-industrial economies in 1914 was as high as between 44 and 60 per cent (Svedberg, 1978).

American direct investment abroad dominated FDI because the large-scale, multi-unit industrial firm first emerged in the United States. By the 1880s, the telegraph, cable, railways and steam-ships were sufficiently widely diffused to allow the control of these enterprises. FDI was one manifestation of the enhanced technology of communication and control. The large firm often moved overseas by first setting up branch offices and warehouses abroad. Then with the expansion of demand and the appearance of local tariffs, the enterprise built plants abroad which it soon began to supply from nearby sources (Chandler, 1977, 1980). By 1914 at least forty-one American companies, concentrated in machinery and food industries, had built two or more operating facilities abroad. The largest number were in Canada, but by 1914 twenty-three had factories in Britain, and twenty-one in Germany, with a small number scattered in other countries. Transport costs and problems of the scheduling of shipping could be as important as tariffs in the foreign investment decision. The number of plants established in Britain despite the absence of tariff barriers testified to the contribution of the shipping problem.

The large-scale firm in Britain flourished in the consumer sectors of chemicals and food, and in the processing and distribution of perishable goods – meat, dairy products and beer. Not surprisingly then, when these firms first moved branches overseas, it was to areas with tastes and income similar to those of the British market – Australia, Canada, New Zealand and South Africa. In the twentieth century, more ventured into the United States and continental Europe. Occasionally before the First World War, overseas sales grew to a size that warranted building factories abroad, usually prompted by tariffs rather than by transport costs. Two-thirds of a sample of around 400 pre-1939 British multinational enterprises were operating in high-income-country markets, among which the empire was most important before 1914, followed by developed Europe. The USA was less significant as a destination for direct investment than undeveloped Europe, at least in numbers of production plants, although not necessarily in their size. Food/drink and chemical firms appear to have been especially advantaged (Nicholas, 1991).

Manufacturers large enough to engage directly in foreign trade originated from the smaller economies of western Europe far more than their relative GNPs at first sight would suggest. Their domestic markets were too limited to support the overhead costs without exports, and so in that way trade contributed to their higher living standards. Foreign investment displaced trade when tariffs or other barriers threatened export markets. Swiss chemicals companies, electrical companies such as Brown Boveri, and dairy goods manufacturers such as Nestlé, all substantially depended on overseas business. Sweden gave birth to LM Ericsson and other major hi-tech multinational companies, while the Netherlands was home to the electrical firm Philips and shared parentage with Britain of Unilever and Shell (Hertner and Jones, 1986).

Most interest has centred on foreign direct investment from the United States, while that country was already receiving substantial flows from the rest of the world in the last quarter of the century. Some, such as investment in cattle ranches, served to increase US exports of primary produce. Other investment – for example, in Southern iron and steel – was innovative, creating the basis for import substitution. The French tyre maker Michelin, the German chemical and pharmaceutical company Hoechst, the German machinery maker Orenstein & Koppel and many others not only manufactured but also operated national sales organisations in the USA. For a brief while, Edison General Electric was German owned. The largest flour producer, the biggest mining equipment manufacturer, the top four meat packers and the leader of the new radio industry were all foreign owned before 1914. All industrialising countries were represented by multinational enterprise in the USA by 1914. The Japanese soy sauce maker Kikkoman built a small factory in Denver, Colorado, in 1892 (Wilkins, 1989, pp. 336–7).

The two-way international flow between the USA and much of the rest of the world highlights the different determinants of direct and portfolio investment. In any pair of economies between which portfolio capital passed, gross flows were generally similar to the net movement. Portfolio investment was largely concerned with financial questions of risk and return. Direct investment, on the other hand, required that the parent firm possess a specific advantage that allowed a subsidiary to be competitive outside the market in which it originally traded. These advantages were to play an increasing role as the complex technology expanded during the twentieth century, and flows were more likely to occur in both directions between a pair of countries.

Summary and conclusion

The demand for internationally mobile portfolio capital was due largely to the high rates of population growth in land-abundant countries. The supply came mainly from Britain and France. Britain's unusually low demand for domestic investment, rather than a high savings ratio, accounted for its supply. By contrast, the French savings ratio was high and the French demand for domestic investment was relatively strong. French investment was more subject to political influences than British. Once Russia became a French ally against Germany, it also became the major destination for French capital exports. The London capital market dealt in foreign investment to the exclusion of all but the largest domestic manufacturing concerns, probably less because of a structural bias than because for most of the period domestic capital requirements were adequately supplied through the provincial stock exchanges.

In the receiving countries, major problems occurred among those states that did not share western attitudes and institutions. Here borrowing abroad for unproductive purposes could easily result in the loss of national control of taxes, the financial administration, or, in extreme cases, sovereignty. The gains from such foreign investment to the borrower in such instances were negligible, except in so far as the consequent foreign administration increased governmental efficiency once and for all. The means of dealing with national default on foreign financial obligations bore a close similarity to post-1945 methods. However, the groups of national representatives that took steps to rationalise defaulters' financial policies were constituted for the occasion, rather than originating from permanent institutions, such as the IMF and the World Bank. Although lending to foreign governments was risky, the loans to the largest borrowers yielded on average about as much as investment in US or UK government bonds. Foreign investment in which the investor maintained control of the project was also widespread before 1914. What was changing was the number of large manufacturing businesses that were increasingly supplying a world market by investing overseas. Politics continued to trigger these moves. The most multinational American company before 1914, Western Electric, acknowledged the need for a manufacturing plant in each market because only in that way could it hope to win orders for telephone equipment from increasingly nationalistic governments (Wilkins, 1970; Foreman-Peck, 1991b).

Notes

1. Measured by the 'residual' method described in Chapter 2.
2. Despite asserting that Egypt's lack of political autonomy constrained its economic development, Issawi (1961) concedes that living standards did rise, as did population until 1914.

8 International migration, 1875–1914

For the first, and perhaps the last time, really cheap long-distance passenger transport coincided with the availability of unsettled land which could supply world food and commodity markets. When the transport revolution had worked through by about 1890, population growth and industrialisation particularly in Europe continued to expand the margin of profitable cultivation. Millions of ordinary working people, especially from the United Kingdom, southern Europe, India and China, travelled thousands of miles to improve their lot, by either permanent or temporary settlement in a new continent. The surprise is less that they were willing to migrate than that they were allowed to do so. Emma Lazarus' sentiment, engraved on the base of the Statue of Liberty outside New York habour, 'Give me . . . your huddled masses, yearning to breathe free', was not echoed by any government of the last quarter of the twentieth century.

In this chapter, first the pattern and measurement of international migration are described. Next are discussed the forces that were powerful enough to make so many people permanently uproot themselves. An evaluation of the gains and losses of international labour migration is attempted in the following section. Finally an explanation is offered for the, with hindsight, remarkably liberal immigration policies of the period.

Magnitudes and directions

Gross outflows are an overestimate of permanent emigration because many of those who left eventually returned. Ocean transport was so cheap by the end of the nineteenth century that agricultural workers from Spain and Italy could sail to Argentina every year, work on the wheat harvest there, and then return in time for the harvest in their own countries (Diaz Alejandro, 1970, p. 22). In this instance, migration was deliberately transitory; gross migration to Argentina exceeded net migration. However, most intercontinental movements of labour in this period were intended to be permanent, even if some turned out not to be. As well as the seasonal versus permanent categories, distinctions can be drawn between intercontinental and intra-continental migration, and between European and Asian emigration.

Figure 8.1 gives Green and Urquhart's (1976) summary for selected countries.[1] Total emigration from Europe grew in each decade until the outbreak of the war, although the outflow in the 1890s was little higher than in the previous decade. Britain ceased to be the largest source of migrants after the 1880s, being overtaken by Italy and Spain in the 1890s, and also by Austria-Hungary in the first decade of the twentieth century. The United States received far more immigrants than any other economy throughout the period. Despite the almost trebling of Brazilian immigration with the coffee boom of the 1890s, total emigration declined in that decade because of the contemporaneous fall in US immigration. For the same reason, the depressed state of the US economy, Canadian emigration, which normally went to the United States, also fell in the 1890s. Thereafter, the Canadian wheat economy boomed and Canadian immigration became strongly positive. After the Boer War, large numbers emigrated to South Africa as mining expanded. Australian immigration was held at a low level for twenty years by the depression of the 1890s, followed by drought. Apparently immigration to New Zealand was greater (although the New Zealand data are suspect because virtually all arrivals were included in the immigrant category).

As Thomas Malthus, theorist of population growth driving down wages to subsistence, would have predicted, rates of population growth were considerably lower in source, than in destination, countries. This was caused not merely by the impact of immigration, but also by the effects of natural increase. Economies of immigration were land abundant and paid high wages compared with the countries of origin. British population growth exceeded those of most other source countries, despite substantial migration, and population decline in Ireland. Russia's low proportion of migrants in the population, despite the highest rate of population increase of all source countries, may be attributed to the system of collective responsibility for the village land established after the emancipation of the serfs in 1861, and to the difficulties of emigrating legally. The similarly low proportion of migrants from France can also be explained by the land tenure system, but here the family farm arrangement probably constrained fertility to give France the lowest population growth rate in nineteenth-century continental Europe. Farmers wanted neither to divide their land nor to leave children without land.

France, in fact, was an important host country for European migrants. In 1881 a million foreigners lived in France, and in the first decade of the twentieth century a third of a million immigrants entered the country (although this was less than emigration), more than half of them from Belgium and Italy (Bunde, 1931). Germany also hosted a considerable number of foreign workers. The summer census of 1907 showed that 1.3 million German residents were born abroad. Some were alien seasonal migrant agricultural labourers. In 1910 almost two-thirds of a million entered Germany and received identification cards. They came mainly

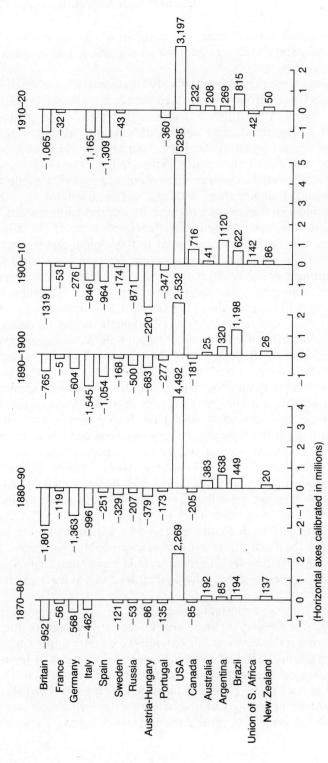

Figure 8.1 International migration, selected countries, 1870–1920 (in thousands)
(Horizontal axes calibrated in millions)

from the East, from Russia and Poland (Burgdorfer, 1931). Seasonal agricultural labourers, perhaps 30,000 a year at the end of the century, also emigrated from Ireland to Great Britain at harvest time (Harkness, 1931).

Between 1871 and 1915, 15.8 million people left India and 11.7 million returned, leaving 4.1 million as net migrants (Lewis, 1978b, pp. 185–7). The much greater proportion returning than was found in Europe is to be explained by the indenture system, which bound a migrant to work for a plantation employer usually for five years in return for the cost of his pasage. By the end of the century 30-day verbal contracts for work in Ceylon and Malaya were increasingly common (Latham, 1978b, ch. 4). Figures for the gross migration of Chinese are not available, but the number of Chinese living abroad increased by perhaps 5 million between 1880 and 1922. As with Indian migration, contract labour was important and therefore the gross flow must have been some multiple of this approximate net flow. Even so, as a proportion of the total Indian and Chinese populations, migration was very small compared with European experiences because the populations were so large – probably over 400 million for China and over 250 million for India. The benefits to the economies were also proportionately small.

Population densities in the destinations of Asian migrants, like those of European emigrants, were lower than in their home countries. Ceylon thereby escaped the famine and disease that regularly checked the nineteenth-century Indian population, and the new tea plantations, among other activities, attracted a net immigration of 1.275 million between 1871 and 1911. Burma increased its population by more than half over the same period, compared with a 20 per cent increase for all British India, suggesting a similar flow of immigrants. Indians also went in considerable numbers to East and South Africa, Mauritius, Fiji and the Caribbean. Chinese net migration was concentrated on southern Asia, especially the Dutch East Indies and, to a lesser extent, Malaya. The Japanese outflow came late, because of emigration restrictions. Beginning in the mid-1880s, emigration reached about one million in total by 1913, mostly travelling to temperate zone countries, to Hawaii and to Brazil.

The causes of international migration

Why did all these people move so far away from the places where they were born? Because of labour market pressures, and in particular because national labour markets were integrated to create world labour markets. In these conditions, a stronger national demand for labour than supply (given the prevailing wage rates, conditions of work and leisure, availability of information, transport costs and legislation) gives rise to immigration. Conversely, emigration stems from a stronger national supply of labour than demand. The relative strengths of national supplies of, and demands for, labour determined the late nineteenth- and early twentieth-century patterns of migration. People moved to 'equalise the net advantages' of the jobs and life-styles. Changes in these conditions, or those determining national labour supplies and demands, 'pushed' or 'pulled', increased or decreased the numbers of migrants between countries.

Workers migrated from higher to lower population density economies, because their labour was more productive in the areas of recent settlement. This high productivity was based on new technology, which lowered transport costs and which could make good use of relatively abundant land. High productivity meant high earnings, but migrants tended to drive down these earnings to the point where the present value of the differential in earnings between the receiving and sending economies was sufficient to compensate for the relative advantage in the sending and receiving countries. Other things being equal, the differential needed to cover the costs of transport and the earnings forgone while travelling. If Europeans regarded the United States in the nineteenth century as a land of religious and political freedom compared with Europe – and they valued those freedoms highly – then a lower long-run equilibrium American wage than the European earnings would have been established by immigration, other things being equal.

The direct costs of passenger transport across the Atlantic showed no pronounced increase or decrease in these years, varying between £3. 10s and £5 steerage class from Liverpool to the United States. What did change was speed of passage and the conditions. In 1867 the average crossing-time by sail was forty-four days and by steam about fourteen days. The White Star Line had reduced this to 9 days 16.5 hours by 1875, and to 7 days 15.5 hours by 1890. Greater speed reduced the time when the migrant would not be earning, and so cut the total costs of moving (Gould, 1979).

One of the biggest alterations during the period was in the information available to potential migrants. The profitability of steam-ship lines and of railway companies came to depend upon a continuous flow of migrants, so they advertised aggressively to ensure that the numbers were maintained. Probably more important, though, was the knowledge of conditions provided by friends and relatives who had already migrated, or who had returned from the prospective host country. As this stock built up overseas, information about opportunities for improvement became more widely spread across Europe. Friends and relatives abroad play a key role in explaining the pattern of intercontinental migration (Hatton and Williamson, 1992; Gould, 1979). Past events, such as famines, created a national or local propensity to migrate by building up the migrant stock abroad.

The overseas stock of migrants has considerable value in explaining the destination of migrant groups. Even in the 1890s Wisconsin and Iowa accounted for 52 per cent of Norwegian-born American residents, because of the pull of kinsfolk, soil and climate, together with the availability of free land on the frontier in the 1850s. The United States remained the main destination for migrants from Norway, Sweden and the Balkans. For British migrants, the attraction of empire countries, which often subsidised fares, reduced the proportion going to the United States as the period wore on, while the opposite was true of the Italians. The Italians, formerly migrating to Latin America, increasingly went to the United States. Perhaps for linguistic reasons, Spanish and Portuguese migrants consistently avoided the United States and travelled to Latin America.

By the first decade of the twentieth century, those European countries with the highest propensities to emigrate as a proportion of population were, in descending order, Italy, Norway, Ireland, Britain, Portugal, Spain, Finland and Austria-Hungary. Some of these were well-established sources of migrants in the preceding quarter century, but there are notable newcomers on the European periphery. The boom in migration from southern and eastern Europe in the years immediately before the First World War cannot be explained by the removal of restrictions on movement, such as were imposed by serfdom. These were removed many years earlier, and income differentials were little if at all wider.

The diffusion model does not necessarily explain changes in the total volume of migration, which can still be understood in terms of the 'push' and 'pull' doctrine: that is, the strength of changes in the excess demand for labour in the receiving country and changes in the excess supply of labour in sending countries (Jerome, 1926). Where figures are available, the most obvious indicator of 'push' and 'pull' is unemployment, and unemployment in the host country has proved a consistently good predictor, in a statistical sense, of immigration (Kelly, 1965; Gallaway and Vedder, 1971; Wilkinson, 1971; Tomaske, 1971; Richardson, 1972; Quigley, 1972). Unemployment in the source country is a less satisfactory variable because, while high unemployment may increase the desire to migrate, it will also reduce the ability to move of those with the greatest need. Changes in the population structure might also be expected to 'push' people out of source countries; domestic industry may be unable to absorb a large number of people arriving at working age without a radical realignment of wages. This shift in labour supply in the country of origin would increase that in the destination country as migration occurred, tending to reduce wages relative to what they would otherwise have been. In the source country, wages would also be reduced, but by less than if there had been no migration.

Immigration into the USA flagged in the last quarter of the nineteenth century because the north-west European 'push' declined by more than the increase in American 'pull' (Williamson, 1974, ch. 11). Industrialisation in north and west Europe raised relative wage incomes and increased jobs well before the First World War. Switzerland became an area of net immigration from 1888, and the western industrialised German states from the 1890s. The Po valley in Italy before 1914 was also receiving migrants from outside the region. If Sweden had industrialised twenty years earlier, there would have been no emigration from that country. The retardation of the British economy, by contrast, explains the rise in migration from Britain in the decade before the war.

Rapid industrialisation in north-western Europe, apart from Britain, required the new immigration to the United States in the first decade of the twentieth century increasingly to come from new sources. The diffusion of information in southern and eastern Europe about employment prospects abroad, and/or population growth there, provided these sources.

An alternative explanation for the new immigration emphasises instead changes in demand in the host country (Piore, 1979, ch. 6). Just before the Civil War, the extension of the division of labour began to reduce skill requirements, as production expanded in the American shoe and textile industries. This expansion created a demand for unskilled and more or less transient immigrants. Towards the end of the nineteenth century, the consolidation of American industry by trusts segmented the labour market into a stable section and an unstable part where the immigrant jobs were to be found. The new immigration consisted of unskilled and often illiterate workers, mainly from southern and eastern Europe, who were deliberately recruited on a temporary basis to fill these jobs at the bottom of the labour market. Of the 1.5 million employed by United States railways in 1910, immigrants amounted to only one-fourteenth of the ticket and station agents, and one-tenth of the engineers, but one-quarter of the track foremen and one-half of the labourers (Taylor, 1971, pp. 197–8). One-quarter of the total United States labour force consisted of immigrants.

Were these unskilled and unstable jobs essential to the economy? Although industrialisation and higher incomes in one country of emigration would have meant that other sources of migrant supply would have been found to fill these jobs, it is not necessarily true that the US economy would have collapsed if the restrictions on immigration in 1921 had been passed earlier. There would have been other adjustments, such as black migration from the Southern states in the nineteenth century, as actually materialised in the twentieth century. But what if racial discrimination had prevented migration to the North? Then the development of labour-displacing technology by industry, for which the United States had been noted from the middle of the nineteenth century, would have proceeded faster. At the same time, prices of services which were difficult to mechanise would have been higher.

More fundamentally, this demand-side explanation founders when it is recognised that there were other countries to which migrants went apart from the United States, with different levels of industrial and agricultural development. If illiteracy was of prime importance, as the demand-side account maintains, Spaniards should have been attracted to the United States as were Italians, but they were not. The increasing transitory component of migration can be explained equally well by the reduction in transport time and costs as by changing industrial structure, especially when the seasonal agricultural migrants to Argentina, and the Asian plantation workers are recalled. Thus it is plausible to regard the unskilled jobs in American industry as more a response to the supply of migrants than a cause of migration.

The gains and losses from migration

Migrants gained the increased wages in the host country. The world as a whole benefited because labour was more productively employed. Those who stayed behind were rewarded with higher wages than if no one had emigrated. Employers in the host economy paid lower wages and workers born there received them. The actual difference between real wages in Britain and the United States in 1913 was 54 per cent. Without the late nineteenth-century migrations, the wage gap would have been much wider, at almost 105 per cent (Hatton and Williamson, 1992).

A short-run benefit to the receiving countries and a corresponding loss to the sending countries arose from the concentraton of migrants in the young-adult age group. The sending countries bore the costs of supporting and training from infancy, but reaped little benefit when they reached working age, except through payments sent back to relatives (which, however, could be very large). An estimate of the cost to Germany of emigration took into account the above costs, and the effects of the age structure of migrants. The study assumed that each worker generated a surplus of production over consumption that would have repaid the present value of early education and maintenance costs by the age of 60 to 65. Emigration then cost Germany over four billion marks in the 1880s, nearly 1.7 billion marks in the 1890s, and over one billion marks in the following decade. In addition, the money that immigrants took with them would probably have raised the loss to the German economy by another 10 per cent. In 1880 and 1913 German net social product has been calculated at about 17 and 52 billion marks, respectively. As a proportion of social product over the decade of the 1880s, the costs of emigration from Germany were a little over 2 per cent (Burgdorfer, 1931; and calculated from Mitchell, 1975).

Conversely, the receiving countries obtained the benefits of this injection of human capital without bearing the costs. Population pressure was temporarily relieved in the sending countries, and wages and the productivity of labour were at least temporarily raised. Whether there was any permanent effect depends upon whether birth rates rose to compensate for the emigration, on Malthusian lines. The evidence of falling birth rates in the last quarter of the nineteenth century suggests that there probably was a lasting impression on European populations.

Australia might have generated a higher income per head if it had received fewer immigrants (see, for example, Gould, 1972, p. 18). The natural resources would have been shared among smaller numbers. There are, however, certain minimum populations necessary to make some services viable. Whether the gains from increasing these types of service offset the losses from sharing the natural resources with more people is an open question.

Even if these various effects can be accurately measured, an estimate of the gains and losses from the great migrations depends upon what standpoint is adopted: that of a person in a sending country, in a receiving country, or in the international economy as a whole; and whether the person earned the bulk of his or her income from work, from land or from capital.

The political economy of international migration

The distribution of the gains and losses from migration explains whether migration is permitted. For international migration to take place, not only must economic conditions be favourable but accommodating policies must be adopted in both source and destination territories. This depends on who controls policy and where they see their interests lying.

When both source and host state are run by landowners, an analogy between immigration policy and slavery or serfdom is clear. The source state will never encourage migration and will generally discourage it, to hold down the cost of labour. Potential source countries must be run either by labour or neutral ruling groups, or the landowning rulers must be less than fully effective if there is to be much international migration (see Table 8.1) (Foreman-Peck, 1992).

Nineteenth-century empires add another constraint and perhaps policy motivation, if not type of state. In at least one respect, a concern with security, the empire is analogous to the neutral or coalition state. But because the empire spans continents or partly self-governing boundaries, migration policy may appear different. Perhaps more important is that the geographical range of empire places a greater policy emphasis on security and sometimes on the related goal of 'nation building' (ensuring cultural homogeneity). In principle, empires could be any of the three types of state described in Table 8.1, neutral/coalition, labour or land/capital dominated. Typically, land/capital empires will be more concerned to control emigration so as to prevent the escape of conscripts and subversives.

In a closed economy, implicitly assumed in the discussion based on slavery and serfdom, and in the some migration analyses, trade does not equalise factor prices (Williamson, 1974; Greenwood and McDowell, 1986). Therefore (intra-marginal) migrants motivated by economic gain will be more productive in the host country than in the source (governed by absolute advantage).

However, most migrations in the nineteenth century were between areas that participated in substantial international trade. In such circumstances, movements of goods and factors are substitutes. With the free international movement of

Table 8.1 International migration regimes: a first approximation.

	Pro-labour	Pro-land/capital	Neutral/coalition
Source country policy	encourage	discourage	encourage
Destination country policy	discourage	encourage	encourage

goods there is a strong tendency for one price to emerge on world markets. If these goods were made with the same technology, factor productivity and rewards would converge in trading countries. With similar technologies available, and incomplete specialisation, more migration is likely to reduce comparative advantage and therefore to reduce exports of the migration host countries' labour-scarce goods. In turn this has consequences for wages and rents in host and source countries. In the limit, when tariffs and transport costs are negligible, factor price equalisation takes place, once allowance has been made for non-pecuniary advantages. Immigration is then unnecessary to enhance incomes, and immigration controls alone are ineffective in raising or protecting labour income. Higher tariffs would have to accompany such measures. Equally, trade restrictions would have to offset falling freight and passenger transport costs merely to maintain a wage differential (and therefore an incentive to migrate).[2]

In an open world economy, the effects of migration on factor prices will be stronger than the impact of movements of goods.[3] Migration can be a once and for all movement, whereas goods must be continually traded for factor prices to be influenced. Even, say, a 5 per cent difference between similar products in two countries, induced by tariff and transport costs, in turn raising wages in the host economy 5 per cent above those of the source, would have justified the spending of perhaps up to 50 per cent of a young migrant's annual wage on the move. Much less than that was necessary to cross the Atlantic by the end of the nineteenth century. The results of the simple (no trade) model are therefore applicable to the trade in goods case, within the limits of wage variation allowed by transport costs and trade restrictions.

Explaining policy

The behaviour of Latin America and the tropical dependencies, land abundant with landowner-dominated governments, corresponds with the model predictions of encouraging immigration. The more representatively governed land-abundant dominions and Boer republics were less enthusiastic, but were constrained, one way or the other, by the international regime. Generally labour-abundant Europe exhibited a spectrum of policies from West to East, varying with the degree of democratisation and distribution of income. European policies ranged from low-cost minimal encouragement of emigration in Britain to bureaucratic obstruction in Russia. Host policy on non-white migration to the land-abundant regions was discriminatory by the end of the century as a means of maintaining host wages.

In Brazil and Argentina, landowning oligarchies recognised that subsidies to labour could permanently expand the labour supply by the once and for all payment of a transport grant. Their subsidy policies of the 1880s were prompted by the opening-up of the interior, which enhanced the effective supply of land. Almost at the same time that the USA was increasingly pressing for restrictions on immigration, between 1885 and 1913, £11 million was being spent on immigration subsidies by the São Paolo and federal governments (Leff, 1982). Only in the 1880s did Argentine military and political organisation become adequate to defeat the Indians and drive them off their land (di Tella, 1985). When that expropriation was releasing new land for commercial exploitation, Argentina even experimented with providing free passages to immigrants during the 1880s (Martinez and Lewandowski, 1911, pp. 118–21). Most Russian emigration was illegal, especially since many young men wished to avoid conscription (British Parliamentary Papers, 1903, p. 464). Collective responsibility through the *mir* for repayment of the emancipation loan was another great deterrent to peasant migration. Authoritarian empires like Russia were also interested in ridding themselves of minorities, such as Jews, to promote national security, or 'national building'.

Half a million indentured migrants, 80 per cent from India, went to the British West Indies between 1834 and 1918. Unable to obtain a legally enforceable claim on the extra income generated by migration, capital markets were unwilling to advance loans for migration, thereby restricting labour movements to workers wealthy enough to finance their own movement. By enforcing an obligation of labour to work for the landowner who had paid the passage money, the law ensured that such advances were forthcoming. More low-income labour could move and earn higher wages when indenture contracts were enforced.[4]

The land-abundant dominions were constrained in their immigration policy by membership of the British Empire. But other influences operated as well. Despite somewhat similar endowments, Canada pursued policies less favourable to immigration than Argentina because of a different income distribution and franchise (Solberg, 1985). Australia, with a fairly representative government, did subsidise white immigration, but aimed to discriminate in the payment of the fare subsidy in favour of immigrants who would obtain rural employment (Condliffe, 1930; Diaz Alejandro, 1970). Rural labour was complementary to the majority of Australians who lived in towns and held secondary or tertiary sector jobs. Moreover, the subsidy varied with the pace of expansion of the Australian economy. Some 50 per cent of arrivals were assisted in the 1870s, but only 10 per cent in the 1890s (Kelley, 1965). In 1878–9 New South Wales labour temporarily lobbied effectively against supported immigration (Malchow, 1979, p. 262).

The British response to Russian and Polish immigration after 1882, the 1905 Act, is consistent with the model predictions for a democratic government in a labour-abundant economy. On three occasions during the 1890s the British Trade Union Congress passed motions objecting to immigration (Holmes, 1982, p. 181).

By contrast, the French willingness to absorb migrants may be related to the more equal distribution of land, thanks to the Revolution, and a low natural rate of population increase, further retarded by the impact of the Franco-Prussian War (Williamson and Lindert, 1980). Despite about fifty legislative proposals between 1883 and 1914 to restrict or tax immigrants, the only substantial French restriction was the limitation in 1899 of the proportion of immigrants employable on public works (Green, 1985). Elsewhere in Europe emigration was a matter of concern by the end of the century, but policy measures were ineffective. Even in Belgium, emigration was strictly not merely a matter of individual decision: an Emigration Commission was the sole judge of whether emigration was permissible. The Swiss went as far as to create the unenforceable offence of urging a man to leave his native land (Johnson, 1913, p. 336).

If pure racism had been the sole reason for anti-Asian immigrant legislation in temperate zones, Asians would have been excluded altogether from the dominions rather than being subject to special taxes or literacy tests. True the United States was unusually exclusive. Following an inquiry in 1876, which falsely represented Chinese immigration to California as contract labour, Chinese workers were largely excluded in 1882 (Cloud and Galenson, 1987; McClain, 1990). But the United States attempted to keep out contract labour from all sources. The Chinese were charged a special poll tax in Australia, New Zealand and British Columbia (British Parliamentary Papers, 1903, p. 828). Literacy tests were also introduced at the end of the nineteenth century by Natal and later by Australia to limit non-European migration. Australian attempts to restrict specifically Indian immigration were vetoed by the London government on the grounds that India was an empire country (Bevan, 1986, p. 67). Both dominion special taxes and literacy tests, and the objections to contract labour in the USA which motivated Chinese exclusion there, are entirely consistent with the labour objective of maintaining wages at levels determined by the productivity of temperate zone agriculture (Lewis, 1978b). Similarly, the South African Mines and Works Act 1911 is more consistent with a policy of 'insider' wage maximisation than with pure racism. African immigration was not at issue, but African employment threatened white wages. The Act therefore restricted employment of African labour in skilled and semi-skilled jobs.

The American anomaly

Some questions remain over the policy of the principal destination of European migration, the United States. Britain reacted politically to a very small volume of immigration, yet the USA maintained a virtual open house to Europeans for the entire period, despite political agitation for controls. Had land abundance been the sole reason, we would expect the closing of the frontier around 1890 to mark the beginning of the restriction period. The slumps of 1893 or 1907 would have precipitated legislation.

Nor does the explanation advanced for French policy seem so promising for the USA. Although the Homestead Act's offer of 160 acres of land to western settlers might have been expected to encourage a more equal distribution of income, industrialising tendencies predominated. By the end of our period, the United States' distribution was considerably more unequal than France's and only a little less unequal than Britain's.

American policy towards migration was as limited as British at the mid-century. Acts requiring literacy tests were passed by Congress, only to be vetoed by President Cleveland and, later, by President Taft. Through the influence of the Knights of Labour, contract labour from any source (workers whose fares were paid out of subsequent wages) was excluded by Acts of 1885 and 1889 (Erickson, 1957, 1984). Between 1907 and 1910, the US Immigration Commission sat to consider complaints that immigration was harming domestic labour, but no general US restrictions were passed until the Quota Act 1921.

The reason for the persistence of the liberal US immigration policy lies in the great variety of immigrant groups that were drawn into the USA and the ease with which they could become enfranchised. Even when they were able to wield political influence, their interests remained sectional. Many would have liked to exclude other immigrants, but to maintain an open door for their relatives and fellow countrymen, a difficult combination to achieve politically.

This account has the merit of also explaining the timing of the US policy change. The First World War cut the USA off from Europe, forcing immigrants to identify more with their host country and less with their country of origin. At the same time, the war reduced the proportion of recent immigrants in the population and in unions. The particular form taken by immigration restrictions when they came, historically determined quotas by national origin, is consistent with the continuing electoral importance of national immigrant groups. The longest-established, most influential, enfranchised immigrant groups received the largest quotas, ones which in fact were not fully taken up. For landowner-influenced states, the strategy of franchise restriction obviously makes sense. The Argentine state gave an incentive for immigrants not to register as voters by allowing aliens exemption from military service (Diaz Alejandro, 1970). The Prussian treatment of Polish immigrants performed a similar function (Hoerder, 1985).

Summary and conclusion

The forty years before 1914 saw the greatest migration of people and capital that the world had ever experienced. Many millions crossed the Atlantic to settle temporarily or permanently in North and South America. Substantial numbers temporarily left China and India to work on plantations around the world. Smaller numbers of Europeans also went to Australasia and South Africa. Malthusian pressure was thereby relieved in Europe, but not in Asia. Much of the pattern of the origin and destination of migrant nationalities is to be explained by wage gaps and by the information flows sent back by friends and relatives already abroad. The timing of migration depended upon economic conditions in the sending and the receiving countries. The temporary decline in migration to the United States towards the end of the nineteenth century is attributable mainly to the reduced pressure to emigrate that came with the industrialisation of western

Europe. The increase in the first decade of the twentieth century was mainly from eastern and southern Europe. Migration policy accounts for distinctive characteristics of the nineteenth-century world economy which cannot be explained merely by economics. Had the United States pursued a franchise policy similar to the Boers, the twentieth century would have been very different.

Notes

1. Unfortunately, only gross migration figures are available for Germany, Spain, Russia and Brazil.
2. The imposition of the McKinley and Dingley US tariffs in 1890 and 1897, nominally intended to protect US workers from low-wage European competition, may have received an impetus from falling transport costs. Such tariffs merely provided a stronger incentive for immigration to lower US wages, yet paradoxically (if the tariff justification is believed), European immigration was left unrestricted.
3. Information flows between source and destination territories were very effective, and migration was often financed by remittances from the host economy. See, for example, Baines (1985, pp. 27–31). By 1875 the Atlantic could be crossed in less than ten days. Consequently, wage arbitrage was quite efficient.
4. With the abolition of slavery in Jamaica, for example, former slaves abandoned plantation work and became self-sufficient peasant proprietors in the formerly unoccupied central highlands (Green, 1984). Other emancipations reinforced labour shortage for landlords throughout the nineteenth century: France 1848, Russia 1861, the Netherlands 1863, the United States 1863–5 and Brazil 1888.

9 The heyday of the international gold standard, 1875–1914

By adopting, or returning to, the gold standard in the four decades before the First World War, the majority of the world's economies established fixed, or 'par', exchange rates with each other. Replacing the varying exchange rates of the bimetallist, silver and paper standards increased the predictability of the returns and costs of international transactions. Investment and trade were encouraged by the reduction in risk. Apparently the international monetary system operated smoothly without the intervention or oversight of any authority, or so many among later generations were convinced, when they looked back on the 'golden age'. A great deal of subsequent analysis first concentrated on how the gold standard worked; how the balance of payments adjusted to various shocks, the distribution of the adjustment costs between countries, and why the system lasted as long as it did. More recently, concern has focused on what the gold standard achieved, its contribution to the stability of prices and output growth, and to economic development.

In assessing the international gold standard, it should be noted that a pure gold standard, with free convertibility between the national currency and gold at a rate guaranteed by the central bank, was rarely chosen. Only Britain, the Netherlands and the United States (New York) allowed free markets in gold (Rosenraad, 1900). For normal purposes this mattered little because international transactions did not typically employ gold. Instead they used a few key currencies, among which sterling was still pre-eminent when the First World War broke out. Many governments, including those of Japan, Austria-Hungary, the Netherlands, Scandinavia Canada, South Africa, Australia and New Zealand, therefore held foreign currency assets to maintain their convertibility into gold.

Despite this saving of monetary gold, the spread of the gold standard and the reduced expansion of gold supply until the 1890s persuaded some contemporaries to attribute the simultaneous downward drift of world prices to monetary causes. An assessment of the relationships between money and prices in these years forms the subject matter of the first section of this chapter. What consequences for the low-income countries flowed from the changing relative scarcities of gold and silver is discussed next.

The fixed exchange rate regime of the gold standard was in some respects similar to the shorter-lived Bretton Woods system after the Second World War (see Chapters 13 and 14). The third section compares and contrasts the different aspects of the two systems: parity changes, liquidity and reserve currencies. The fourth section examines whether the exchange rate commitment was facilitated by contemporary attitudes towards public finance, or whether the link with gold itself constrained government expenditures.

Another difference between the gold standard era and the years after 1945 lay in the greater instability of banking and financial institutions (outside Britain) during the earlier period. Most of the downturns in the international economy in these years were marked by banking collapses. These and other determinants of international economic fluctuations are discussed in the fifth section. Finally, the role and methods of monetary policy in coping with these disturbances are analysed, and an example of the benefits of a policy of joining the gold standard is described.

Money and the international price level

All economies sharing a common metallic base to their money supplies were linked by their convertibility commitment to a world price level. The price level was strongly influenced by a world money supply, the sum of all the integrated economies' money supplies. (Only the precious metal markets needed to be integrated.) Prices may be expected to track the money supply over long periods if there are no radical changes in output growth or in interest rates. With a gold-based money, the supply of that metal will be vital for the quantity of money in circulation. But the growth of the money supply may differ from the growth of the monetary gold stock because of changes in all, or any, of the three proximate determinants of the money supply:

1. The relationship between high-powered money (currency and bank reserves) and the money supply.
2. The deposit–reserve ratio, determined by the banking system.
3. The deposit–currency ratio, which depends upon the preferences of the public.

Because of the spread of branch banking, the British money supply was able to economise on notes and coin. The ratio of currency to deposits halved in the last quarter of the nineteenth century, and the value of gold coins in circulation actually fell (Capie and Weber, 1985; Bordo, 1981). Figure 9.1 shows declining prices for the gold standard countries until about 1896 when the trend was reversed. Short-term price movements in the gold standard economies were fairly closely synchronised. The price levels of France and Germany were particularly highly correlated and, to a lesser extent, so were those of Britain and the United

156 *A History of the World Economy*

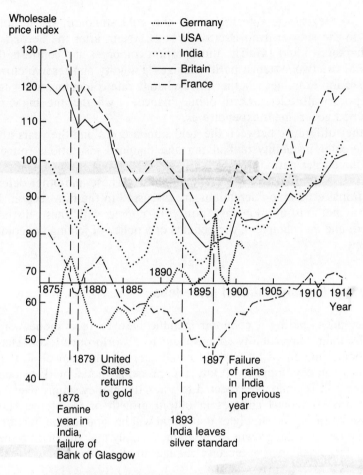

Figure 9.1 National price levels, 1875–1914

States. The rise of the American price level in the early 1880s, not matched by a British increase, may be explained by the excessive price fall in the years immediately before the 1879 return to gold, and a subsequent reaction. Silver standard India exhibited the opposite price pattern to the gold economies until the closing of the Mints in 1893. The Indian price level rose as silver fell in value relative to gold. Famines only partly disguise the tendency; food prices soared in 1878, 1897 and 1899.

The growth rate of the world money supply, as measured by Triffin's calculations for Britain, France and the United States, fell from 4 per cent per annum between 1848 and 1872 to an average of 3.2 per cent per annum between 1872 and 1892 (Triffin, 1968, p. 28). The figures for uncovered credit money (that portion of credit money, bank deposits and notes in excess of total gold and silver reserves) showed a similar fall in growth rate from 6.5 to 4 per cent per annum over the same period. Evidence that this slowing of monetary growth is attributable to the decline of gold production is the fall in monetary gold increase from 6.2 per cent per annum in 1849–72 to 1.4 per cent per annum in 1873–92. The simultaneous depression of prices in the last quarter of the nineteenth century invited an explanation in terms of the quantity of money (see, for example, Marshall, 1925).

Gold discoveries, especially those in Transvaal in 1886, eased the monetary restraint from the mid-1890s. Between 1893 and 1913 Triffin's measure of the world monetary gold stock growth averaged 3.6 per cent per annum, more than double the rate of the preceding twenty years, and from 1896 prices began to rise again. The growth rates of the total money supply and of uncovered credit money also rose to 4.3 and 5.4 per cent per annum respectively.

The proximate cause of the more rapid growth of the British money supply after 1896 was the expansion of high-powered money by the state of the balance of payments and Bank of England policy. The findings for both phases are consistent with a monetary interpretation of price movements: in the first phase, the effects of gold on the money supply were only partly offset by changes in the banking system; and in the second phase, the link between gold expansion and monetary expansion was precise and central. Similarly, virtually all of monetary expansion in the United States between 1879 and 1914 is attributable to the increased output of gold.[1]

An obstacle to accepting a monetary account of price level movements has been the parallel movement in interest rates, Gibson's paradox (Phelps-Brown and Ogza, 1955). When monetary expansion drives prices up by increasing expenditure on goods, it should also increase the supply of loanable funds and thereby push interest rates down. But falling prices in the last quarter of the century were asociated with declining interest rates. A resolution of the paradox is that lenders are interested in what their money will buy when it is returned, and in choosing the interest rate at which they are willing to lend, they therefore make a judgement about future price levels. If they expect prices to fall in the future, they will accept a lower nominal than real interest rate. Similar considerations apply on the demand side of the market. Only if the slower monetary expansion than output from 1873 to 1896 was not anticipated to lower prices (regardless of whether or not it did) would it have raised interest rates. A full resolution of the paradox requires a number of downward revisions of price expectations (Capie, Mills and Wood, 1991).

Alternative views are based on supply-side shocks (Lewis, 1978b, ch. 3; Rostow, 1978). Given the lesser price flexibility of manufactured than agricultural goods, it is quite possible that one or more supply-side shocks could generate the price movements of the period (Mokyr and Savin, 1976; Bordo and Schwartz, 1980). As in the expansion of the 1850s, there are accounts for the years after the mid-1890s stressing that banks increased loans not because gold reserves were more readily available, but because of greater secure trading opportunities that warranted the advances (Laughlin, 1919). Prices in the USA in 1896 were exceptionally low because of an industrial depression. The subsequent rise in American prices was explicable by high tariffs, agricultural readjustment, higher wages and the increasing expenditure of the rich. Two points tell against a purely US supply-side explanation. First, since the price movements were a world phenomenon, the supply shocks and investment opportunities need to be large enough to affect the world economy. Second, the different behaviour of the silver economies, and the movement of the silver–gold price ratio, discussed next, is rather compelling evidence for the monetary story. Money can only be half the explanation because it does not cause the growth of output. To that extent 'real' and monetary accounts may be separate blades of the explanatory scissors.

Gold supplies were also influenced by world prices. When the general price level declined, the rise in the purchasing power of gold first reduced the proportion held for non-monetary purposes and then encouraged more investment in gold mining, production and prospecting. When and if that investment bore fruit in more gold, the price level began to rise and the profitability of investment in producing gold declined. Hence a world economy linked to a gold standard could expect a long cycle in prices, rather than price stability (Barro, 1979; Rockoff, 1984). Although this is the nineteenth-century pattern, the underlying process actually owed more to geological chance.

Gold and silver prices

When the western economies adopted the gold standard, they raised the demand for monetary gold at the same time as increasing the supply of silver for non-monetary uses. Figure 9.2 shows the consequent rise in the ratio of the price of gold to the price of silver after 1870. By 1889 the USA had acquired nearly 20 per cent of the world's monetary gold stock (Friedman, 1990). This must have pulled down prices in gold standard economies. Certainly small farmers, silver miners and great landowners in the United States believed in the connection when they campaigned for the remonetisation of silver to reverse the trend. The United States in 1878 therefore passed the Bland Act, requiring the coining of not less than two million dollars in silver every month at a ratio with gold of 1:16 (Barbour, 1886, p. 67). The Act was a response to the 'Crime of '73', by which America had altered its nominally bimetallic currency to a nominal gold standard, although its currency was not then fully convertible. At the same time as the Bland Bill became law, the United States vainly tried to bring the Latin Union back to its former bimetallism at the International Monetary Conference in Paris, also to prevent the decline of prices. Despite the disruption of Franco-Italian relations over Tunisia, the Union presented a united front against silver at the 1881 International Monetary Conference.

Another American Silver Purchase Bill was passed in 1880, and the gold price of silver began to rise, helped by the formation of a silver cartel. The cartel could

Heyday of the international gold standard, 1875–1914

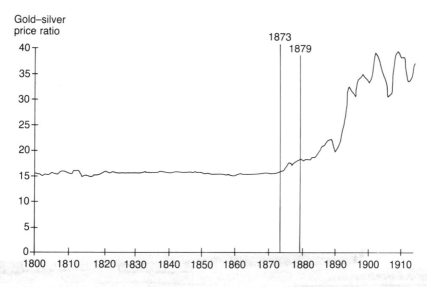

Figure 9.2 Ratio of the price of gold to the price of silver, 1800–1914 (Friedman, 1990)

not, however, hold the market because European governments took the opportunity to rid themselves of their demonetised silver stocks. There were wide oscillations in the price of silver, which convinced Hungarian farmers of the need for the stability of the gold standard even if it did mean lower prices, and in 1892 Austria-Hungary adopted the gold standard. Japan went on to gold in 1886. In 1893 the US Silver Purchase Act was repealed. India closed its Mints to the free coinage of silver in the same year, and in 1896 the United States survived the last serious attempt to return to silver in the presidential campaign of that year. Russia also joined the gold standard in 1895–7.

Thus, the main trading nations of the world by the end of the century had turned to gold, increasing the demand for it even faster than the growth in world trade, tending to depress prices of gold standard countries further and to raise prices in the remaining silver countries. Between 1896 and 1915 world monetary gold demand rose probably by more than $4 billion, whereas silver demand fell by about $1.8 billion (Laughlin, 1919). American resumption under a bimetallic standard would effectively have linked the USA to silver and prevented the massive rise in the ratio of gold price to silver price, except between 1891 and 1904 (Friedman, 1990). The United States would have stabilised the ratio in much the same way as bimetallic France did in the 1850s and 1860s. The US price level would not have fallen by anything like as much as it did, because the money supply would have risen faster. This, and the stabilised gold–silver ratio, would have affected all other currencies linked to the metals; British prices would have fallen by less and India's would have risen less markedly.

The silver standard economies

As it was, the impact on the silver standard countries of the fall in the gold price of silver was dramatic (Latham, 1978b).[2] Most of the less developed economies were linked to silver, and they found their exchange rates continuously depreciating between 1873 and 1894 against gold standard countries, in total by about half. If resources were unemployed or underemployed in these countries, depreciation would have raised national incomes by increasing the demand for exports, and for import substitution. In turn, higher incomes might have induced more investment, enhancing export competitiveness and further raising incomes. Without this induced investment, higher exports paid for in silver must have inflated the money supplies of silver standard countries. This will have raised domestic prices and eliminated the competitive advantage of the initial depreciation.

Silver countries with obligations denominated in gold to other countries were less fortunate. In India a major government concern was the payment of the 'Home Charges', the interest on debt, civil servants' pensions and so on, which were denominated in gold, so here the burden increased as silver depreciated. The Home Charges also contributed to the falling Indian exchange rate (Foreman-Peck, 1989). In general, importers, obliged to pay higher prices, objected to the depreciations, and exporters, whose products became more competitive, did not.

Export growth of silver and gold standard countries in these years under some conditions measures the impact of silver depreciation (Nugent, 1973). The 'average' gold standard economy achieved a growth in the gold value of its exports of barely 1 per cent per annum, although as world prices were falling steadily at about 2 per cent per annum over the period, in real terms the increase in exports was higher. Exports from the silver countries present a very different picture. Even when Korea and Bolivia (the two fastest-growing countries) are excluded, the average silver country's exports grew at more than 4 per cent per annum in gold values, and there is no evidence that any silver country experienced a decline in the gold value of its exports. Depreciation, therefore, seemed to favour the trade of silver standard countries. The impact on foreign investment could well have been harmful, though, for with repayments and interest denominated in a gold currency, the burden would have risen as silver fell.

These results may, of course, have been purely fortuitous. Other major and relevant changes may have occurred at the same time. Regardless of the depreciation of silver exchange rates, productivity and population growth in the gold standard countries from which India imported exceeded those in India. Hence the otherwise unexpected observation that the price of Indian imports fell by more than silver fell in relation to gold, or by the rupee–sterling exchange rate, so that they became cheaper despite depreciation. Some Indian export prices rose, but in general they did so by less than the fall in the exchange rate. Therefore, the competitive advantage conferred by depreciation was maintained.

The Indian price level was a little higher in the mid-1890s than at the beginning of the 1870s. Net imports of silver were especially heavy in the quinquennium ending 1893/4 and probably pushed the general price level upwards (Narain, 1926; Barbour, 1886, ch. 21). If the rupee had not depreciated, the growth of exports from India between 1873 and 1895 would have been even further below the growth of imports than it actually was (respectively 89 and 108 per cent). This would have tended to lower prices as silver flowed out of the country. At least prices would have been lowered as long as the outflow did not push down further the silver price in terms of gold, as gold standard exporters exchanged silver for gold and further depreciated the rupee exchange rate.

As expected from an export stimulus with sticky prices, there was some evidence of rising real incomes. In the interior of India, the money wages of skilled labour tended to rise without a corresponding increase in food grain prices, although unskilled rates showed little improvement. The rapid growth of Indian state revenues since the fall of silver indicated general prosperity, as did the increased imports of luxury items such as clocks and watches, corals and glass (Barbour, 1886, p. 149). But it would be rash to attribute this prosperity solely to the depreciation of the exchange rate.

In China the inflow of silver following the depreciation of the exchange rate similarly does not seem to have raised prices in the country as a whole to offset the depreciation so China also maintained an export advantage (Remer, 1926).

The working of the gold standard

Commitment to an indissoluble link between the national currencies and the same precious metal implied a long-run relationship between the price levels of any pair of countries and their exchange rate (Chapter 5). Their exchange rates were determined by the relative gold contents of the two national currencies. Demand- or supply-side shocks, at home or abroad, disturbed this purchasing power parity relationship. Three types of adjustment occurred over different time scales, and with different magnitudes, depending on national institutions and economic structure. The most immediate response was typically an interest rate change, followed by an alteration in the level of income and economic activity, and finally price level movements (Schwartz, 1984; Eichengreen, 1992b; Ford, 1989). A financial crisis abroad would trigger an outflow of capital, including gold, that would tighten domestic interest rates. The process could be reinforced by central bank policy in an attempt to retain the gold so as to maintain currency convertibility. Higher interest rates would rein back bank lending, and domestic spending and incomes would fall. Lower incomes and employment cut imports, reducing pressure on the balance of payments. As goods, services and labour remained without buyers, their prices might be bid down. One estimate, across a range of countries in the nineteenth century, found that the typical half-life of a shock to the price relations of purchasing power parity was three years (Diebold *et al.*, 1991). 'Periphery' economies, with undeveloped capital markets, could not hope that interest rate adjustment would be as effective as in more advanced economies. Income adjustment was likely to be more prominent. For Argentina, a case usually cited in this context, the problem was less the gold standard than domestic policies: the Argentine public hoarded gold between 1886 and 1896 when confidence in domestic policies was low, a process begun before foreign bondholders began to sell (in 1889) (della Paolera, 1988).

More liquidity, higher gold reserves, might have delayed the raising of interest rates and the reduction of business activity so early in the expansionary phase of the business cycle. As it was, defence of very limited international reserves in Britain required an active interest rate policy. Concern was often voiced about the inadequacy of the Bank of England's gold reserves, the frequent rises in bank rate, the contrasts with the stable, cheap money regime in France, and the constraining effect on economic development (Bloomfield, 1959). Because the Bank was still privately owned and obliged to earn a return for its shareholders, low gold reserves were desirable for the Bank. They earned no money for the Bank, whereas bank rate variability imposed on the Bank no costs.

Anxiety continued to be expressed into the twentieth century after gold production had accelerated. Between January 1904 and December 1914, the Bank of France changed its interest rate eight times, the Bank of England forty-nine times and the Reichsbank thirty-seven times. In 1907, when the Bank of England raised bank rate to 7 per cent, the Bank was sufficiently concerned to institute an inquiry about the effects outside the London money markets (Sayers, 1976, vol. 1, pp. 43–5). The inquiry concluded that only an increase above 5 per cent damaged businesses. Even then, the commercial banks tempered the wind to their customers, charging less than the bank rate. In the north, notably in Yorkshire, the traditional 5 per cent still reigned. The Bank's actions might have impinged on domestic business activity also through the cash reserve of the commercial banks. The Bank tried to influence these reserves only to affect the market rate of discount, not the level of bank lending. But commercial banks' willingness and ability to make loans, if not the price at which they would do so, depended on their own costs of getting funds, which were influenced by bank rate. The volume of credit, and therefore the extent of credit rationing, must have responded to bank rate.

Reserve currencies

For most economies, the problems for international liquidity that the move to gold generated were partly met by the use of the key or reserve currencies. Their easy convertibility made these currencies 'as good as gold' and, unlike gold, some interest could be earned by holding them. Most private international settlements were conducted in sterling, French francs or German marks, and these key currencies composed a share of the world's official reserves as well – about 20 per cent by 1913 (Lindert, 1969).³ Their use economised on the gold that was becoming scarcer in relation to economic activity. Slightly over half of the known official foreign balances were held in Russia, India and Japan. Some countries, most notably Britain and the United States, held no official exchange assets at all.

A consequence of the key currency system was that Britain's liquid liabilities to foreigners were several times greater than the Bank of England's gold stock (for which the liabilities could in principle be exchanged), even when the sterling balances of the colonies were excluded. Germany's liquid foreign liabilities to a lesser extent, though perhaps not those of France, were similarly greater than gold reserves. Before 1914 the international economy was willing to accumulate the reserve currencies, and the central banks of these countries, especially the Bank of England, saw little wrong as long as their own gold reserves were not declining relative to their own liabilities. There was a downside for British industry though. First, combining the Bank's small gold reserves and the key (or reserve) currency role of sterling made frequent rises and falls in bank rate almost inevitable in the face of international shocks. Second, when linked with the much greater earnings from overseas investment, the reserve currency role of sterling allowed the British economy to run an increasing trade balance as a proportion of GDP. Domestic expenditure increasingly favoured imports over home-produced goods and/or switched goods from exports to the domestic market.⁴

The dominant role of sterling in the international economy evolved from Britain's unchallenged superiority in international trade from the mid-nineteenth century. As confidence in the international role of sterling grew, so did foreign holdings of sterling. British industry needed to export less in order to buy a given quantity of imports than if sterling had not been a reserve currency. The adjustments of the British economy, necessary to maintain a balance of payments equilibrium, were reduced. If Britain had been forced to modify the structure of its industry faster, the eventual transformation after the the First World War would have been less wrenching, moreover the rate of industrial growth in the late nineteenth century might have been higher as the demand expanded for the products of the new industries, which had much greater scope for productivity increases than the old staples.⁵

Instead, British trade competitiveness declined (Chapter 6), and the currencies of other countries came into increasing use as reserves. Eventually confidence was lost in sterling convertibility and the British economy was forced to adjust radically in the inter-war years. This is not to deny that the financial and political consequences of the First World War and the rapid economic development of the United States were responsible for the timing and much of the magnitude of the transformation. Even without the war, sterling's international role would have been challenged.

The longevity of the gold standard system as a whole may have owed something to expectations. Since exchange rate changes were not permitted under the gold standard, speculators' beliefs in the viability of the system encouraged short-term capital flows that allowed the maintenance of the chosen rate.⁶ The credibility of the commitment to the gold standard in the core industrial countries enlisted speculators in support of the system.

The gold standard and public finance

Perhaps the limited role of government in the heyday of the gold standard also contributed to the viability of fixed exchange rates. Lower expenditures may have reduced governments' propensity to finance budget deficits by monetary expansion. Differential national rates of inflation were then less likely to threaten exchange rates.

However, there were strong pressures to increase government spending in the late nineteenth century. Budget deficits were financed by debt accumulation, which may have crowded out private investment and certainly increased the tax liability of future generations, with possible inflationary consequences. Causality may have run in the opposite direction: the need to maintain currency convertibility into gold at a fixed price restrained the growth of government expenditure. The continuing deficits of the Austrian Treasury and the privileged position of the government in the money markets, which allowed the payment of up to 8 per cent on long-term issues, almost certainly excluded private investment unable to pay these rates (Cameron, 1972). Austria-Hungary did not join the gold standard, abandoning its floating exchange rate, until 1892. Even so, the money supply had increased only slowly in the previous decade (Yeager, 1969). Neither was the gold standard necessary to impose financial probity on the Russian government in this period, despite an inconvertible paper currency until 1895–7. During the Russo-Turkish War of 1877–8 there were large increases in the paper money circulation, but then, from 1880 to 1885, the money supply was contracted. The money supply similarly grew rapidly in 1905–6 during the war with Japan, despite the existence of nominal gold convertibility (severely limited in these years) (Yeager, 1969; Drummond, 1976). Budget deficits quadrupled the Italian national debt in the first twenty years after unification. But Italian deficit financing preceded suspension of convertibility in 1866 (Fratianni and Spinelli, 1984). Conversely, between 1894 and 1913 tight Italian monetary policy co-existed with inconvertibility. Spain and Serbia showed similar penchants for unproductive government expenditures that slowed their economic development and kept them off the gold standard.

There is little question of the gold standard constraining government budgetary policy in Argentina either (Ford, 1958). Once inflationary finance had caused the collapse of the standard, a depreciating exchange rate moved the distribution of income in favour of exporting and landed interests, and the representatives of these interests saw no reason to return to the standard. Stabilisation in 1900 was engineered by these groups largely to stop the unfavourable redistribution of income they were experiencing from the appreciating exchange rate, once foreign investment began to yield higher exports, and primary product prices rose.

Even in economies whose finances were sounder, there were powerful forces pushing up public expenditure. Germany introduced old-age pensions and unemployment insurance in 1888, and Britain did the same two decades later. Armaments became increasingly important in the German budget as the German navy was expanded to become capable of challenging the British. German expenditure on the army and navy in 1909 was half as high again as it had been in 1905, and the national debt increased by the same proportion (BPP, 1903 and 1914). In the sixteen years after 1893 the German national debt increased at an annual average rate of 6.4 per cent, considerably faster than national income. Between 1904 and 1913, the budget was in surplus for only three years.

Not all gold standard countries financed their increased expenditure by increasing their debt. Italy raised its military expenditures by more than a third between 1905 and 1909, but slightly reduced its outstanding debt, as did France, whose military expenditure rose 14 per cent between 1905 and 1909. Russia reduced its military expenditure by 20 per cent between 1905 and 1909 after defeat in the war with Japan, but debt nevertheless rose 15 per cent.

Those countries which did raise the ratio of debt to national income confronted the prospect of reducing expenditure on other items to pay for the debt charges, finding other sources of taxation, raising existing tax rates, or abandoning the convertibility of currency into gold and printing money. Germany faced particular difficulties in raising more revenue from taxation because of its federal structure, but this does not seem to have constrained debt expansion (Balderston, 1981). The experience of hyperinflation after the First World War suggests that the gold standard would have been abandoned first, and that for Germany at least the gold standard did not inhibit government expenditure, although it did restrain the expansion of the domestic money supply.

Among the benefits of gold standard membership was that financing budget deficits was facilitated. The scope for taking up this option with productive public sector investment clearly differed between countries, as did the willingness to use state finance. With their abundant natural resources, the regions of recent European settlement possessed greater opportunities for profitable investment in infrastructure than the older European countries. The remarkable contrast between the debt per head in Australia and Argentina on the one hand, and the United States and Canada on the other, indicates the varied possibilities for state participation in economic development. There were limits, however, to the extent to which the state could depend on foreign investors underwriting development. The size of the national debt held by foreigners, and the problem of earning the foreign exchange to service the debt, forced Argentina off the gold standard from 1885 to 1900. In the late 1880s Argentina was therefore obliged to pay twice the interest rate on British securities for government borrowing.

The gold standard countries adopted a wide range of financial strategies; taxation, debt and and debt service differed greatly between them. These created varied pressures on the balance of payments, and therefore on the ability to maintain convertibility with gold. The gold standard exercised little restraint over public finance, and the underlying pressures to raise state spending were strong. By comparison with the years after 1945, the doctrine of the balanced budget except in wartime undoubtedly made par value exchange rates more viable. But even in the most prosperous industrial countries, the disintegration of the nineteenth-century liberal consensus on government expenditure was making the maintenance of the gold standard more difficult by 1914.

Fluctuations in economic activity

Gold standard countries, and others, needed to adjust to sizeable fluctuations and severe financial crises in the international economy. World economic disturbances were exacerbated by the absence of an American central bank to act as lender of last resort, and to provide seasonal elasticity to American money and credit. Some have seen the business cycles in these years as distinctive for each country (Lewis, 1978b). But the pervasiveness of relatively fixed exchange rates implies that, wherever the cycles originated, they would be transmitted, perhaps with diminished intensity and some delay, to other economies. Supply shocks abroad (such as the devastation of French vineyards by phylloxera) were demand shocks for the domestic economy.

Accumulation of monetary gold by the United States, Germany and France dragged down economic activity in Britain in 1878. British interest rates were raised to protect the gold reserve and then the City of Glasgow Bank failed (Hawtrey, 1938). Another slump began in 1882. In France the ravaging of the vineyards was at its peak, the Freycinet plan to spend massive sums of public money on transport improvements was cut, and the Union Générale failed. Gold flowed from London to Paris to quench panic demands for cash, and the Bank of England's discount rate reached 6.6 per cent. Germany, Britain and the USA moved into recession. In the spring of 1884, the American banking system suffered from the railways' financial difficulties, although on a smaller scale than in 1873. The recession soon spread to Canada and to South America, but demand in most of the British Empire remained buoyant. A redirection of British foreign investment stimulated activity in India and Australia. After 1886 the United States economy recovered, gold was discovered in the Transvaal, and Indian demand stagnated.

The collapse of an attempt to rig the copper market and the bankruptcy of the Suez Canal enterprise in 1889 in France was a prelude to the Baring crisis in London the following year. The French crises also provided a blueprint solution when the Bank of France collaborated with leading finance houses to support the Comptoir d'Escompte (Pressnell, 1968, p. 205). Massive British investment in Argentina initially generated rising incomes and imports there, only later supplying the exports from the pampas (Ford, 1956). Much borrowing was at fixed interest, denominated in gold or sterling and payable or guaranteed by public authorities. Domestic inflation was stimulated by irresponsible banking and weak, corrupt, financial administration. In 1889 a bad harvest signalled problems in paying for the overseas debt, and the following year there was a coup and a temporary default. The Argentinian default found Baring in a position of having lent for long periods and borrowed for short periods. To save such a prestigious and well-connected firm from bankruptcy, the Bank of England did what it had not been prepared to do for the 'Norwich upstarts' Overend and Gurney in 1866, organise a consortium of financiers to set up a rescue operation. A key difference was that Overend and Gurney were insolvent, but Baring was merely illiquid. The consortium cut out a more powerful group of upstarts, the joint-stock banks.

The problems of the 'external drain' of gold prompted by the panic could not be dealt with by the consortium. The Bank of France disliked the disruption to the loan operations in Paris that high interest rates would cause, and was easily persuaded to allow the (Paris) Rothschilds to ship £3 million in gold across the Channel (Pressnell, 1968, p. 199).

The Bank of England itself purchased some £1.5 million in German gold coins from Russia, which offered more. This the Bank declined, but Russia supported the reserve position by agreeing not to withdraw substantial deposits from Baring, as had been its intention. The Governor of the Bank of England assured the Russian ambassador of the safety of these deposits.

The Bank of England and international co-operation prevented the Baring crisis from becoming a national and international monetary crisis. Prosperity therefore continued until two years later. The American depression began in 1893 when the Australian land boom also collapsed. For the first time since 1853, annual gold output exceeded £30 million in 1892, continuing to rise rapidly. Low discount rates in London from 1893 to 1896 were made possible by the new gold easing the Bank of England's reserve position, and the British slump was therefore not severe.

By now multilateral trade pervaded the world economy: no longer was Britain at the centre of the majority of international transactions. Hence countries of the British Empire could expect to share more in the booms and slumps of the United States (Saul, 1960, ch. 5). Intra-European trade dominated European trade with the United States, so that if conditions were right elsewhere, an American recession, such as that of 1903/4, could pass unnoticed in Europe. In 1907 conditions were not right. Financial panics occurred in Egypt in April, in Japan in May and in Germany in October. The Knickerbocker Trust failed in New York, and in October and November the US banking system suspended general cash payments. This was the climax of an increasing demand for gold imports which the Bank of England met by raising the bank rate through 1906 and 1907 (Ford, 1964). The drain of gold was met by temporarily reshuffling the destinations of newly mined gold to such an extent that Britain remained a net importer. Temporary accommodation came from Europe, in particular from the Bank of France and the Reichsbank, not merely because of a change in short-term interest rate differentials, or because of the German financial collapse, but because of central bank co-operation. The Bank of France placed resources at the disposal of the Bank of England for fear that, if England needed further rises in the discount rate, France too would have to increase its rates exorbitantly and injure its own trade and industry. The Societa Bancaria Italia was less fortunate. In the spring of 1907, Paris and London cut off credit to Italy and the United States, and doomed the Bank (Kindleberger, 1989). Short-term interest rates dropped in 1908 and British exports recovered in 1909.

In 1914 fear of war caused heavy sales of internationally traded securities in the European stock exchange, which, in turn, triggered a partial breakdown of long-term capital markets (Sayers, 1976, vol. 1, pp. 43–5). Short-term international credit also collapsed. The London bill market and the London Stock Exchange no longer provided the London clearing banks with liquid assets. The Bank of England could not supply enough cash to the bill market to prevent new business there drying up on Thursday, 30 July. On Friday morning the Stock Exchange did not open and gold moved from London to Paris. Imports from New York would take some days to arrive. Despite their substantial gold reserves, the clearing banks refused to pay out sovereigns, passing the burden on to the Bank of England. Keynes accused the commercial banks of short-termism; de Cecco maintained that they used the opportunity to attack rivals. The Treasury and the Bank of England resisted commercial bank proposals to pool gold holdings and issue emergency notes. Central bank policy consisted of direct assistance to the money market, closures and moratoriums to prevent bankruptcy, and issue of Treasury notes. The Bank Holiday of Monday, 3 August was extended for three more days, by which time war had been declared.

The financial crises of 1914 did not break the gold standard even before war was declared (de Cecco, 1974). The crisis was caused by the war scare, and if war had not broken out, it would have been remedied by the arrival of gold from New York. In any event, British gold convertibility was not abandoned in 1914, and the joint-stock banks' refusal to pay out gold sovereigns was part of their struggle for power over monetary policy with the Bank of England, and did not represent an intrinsic inability of the gold standard institution to cope with the crisis. Britain asked for repayment of outstanding debts, and the rest of the world responded that they needed British advances first. The crisis in fact gave an indication of Britain's continuing financial strength (Seabourne, 1986).

War scares threatened international monetary relations by undermining the confidence that shored up the pyramid of credit. Prospective illiquidity of financial institutions was equally dangerous. Suitable monetary institutions and policies were therefore fundamental to stability. The contrast between the Baring crisis and the rescue operation of the Comptoir d'Escompte on the one hand, and the failure of the Knickerbocker Trust on the other, demonstrates that proper monetary management and regulation could alleviate downturns in the business cycle that originated domestically. Whatever determined the cycle, instability of financial institutions almost certainly increased the severity and duration of slumps. The British slumps of the late 1870s, the early 1880s and 1907 owed much to the transmission of foreign disturbances through the gold outflows and high interest rates. But low Bank of England reserves left little discretion to British monetary policy to offset these shocks.

Monetary policy under the gold standard

National monetary policy under the international gold standard was primarily concerned to maintain the convertibility of national currencies into gold in the face of international or domestic disturbances. The 'rules of the game' to achieve this end were that a drain on gold reserves was to be countered by an increased discount rate and that a rise in the reserves was met by a reduced rate (Clare, 1891; Neuberger and Stokes, 1979). The Bank of England, however, followed rather than led the market down in the latter case. These interest rate changes attracted or repelled short-term assets that would restore equilibrium in the balance of payments and eliminate the accommodating gold flows. Actual practice differed somewhat from the 'rules' because central banks pursued other objectives, as well as maintaining their reserves. They were also concerned to insulate the domestic economy from foreign disturbances as far as possible.

The Bank of France regarded its gold reserves more as a national war chest and therefore tended to prevent an outflow by paying a premium on gold. Large withdrawals for export required special permission. The French did not like to raise the discount rate to protect gold reserves because of the impact on domestic industry. From 1885 to 1888 the Bank of France did not alter the rate from 3 per cent. After 1900 at least, when gold did flow out of the country, reserves were so high that a higher discount rate was unnecessary. The premium policy to prevent a gold outflow was of doubtful effectiveness because, by contrast with Britain, in France the cheque system was almost unknown and trade was conducted with gold or notes. Gold could therefore be withdrawn from circulation and exported, if exchange rate depreciation made it profitable, without having to go to the central bank. When the Bank of France did choose to raise the discount rate, it could count on attracting gold from Belgium and Switzerland.

Berlin in the late 1880s operated a discount policy similar to the British. Gold could be drawn into Germany from Denmark and Scandinavia, but the Reichsbank was loath to let gold leave the country. Although it had never refused gold (and had no power to do so), there was a general impression that, when the Reichsbank agreed to gold exports, it generally took action to indicate disapproval when inconvenienced by requests for gold. In the struggle to accumulate gold in the 1880s, the Reichsbank was more ingenious than the Bank of England, allowing importers of gold free finance. Thereby Germany's gold reserves rose by one-half between 1884 and 1890. Nevertheless during the 1907 crisis, when there was almost a flight from money in Germany, none of the Reichsbank's devices, including a higher discount rate than in London, prevented a massive outflow of gold.

The United States handicapped itself and created problems for the rest of the world by the refusal until 1913 to establish a central bank. Since the USA lacked a 'lender of last resort', American domestic banking crises were more intense, depressions were therefore deeper, and seasonal and panic demands for gold had to be met by other countries. But interest rates in the United States did follow the rules of the game, rising rapidly when the reserves of the principal banks, especially in New York, fell to the legal one-quarter of deposits. Government policy exacerbated the difficulties of monetary control through the national budget. Whereas in Europe tax revenues were paid into the central banks and recirculated, in the United States the receipts were drained from the economy into the Treasury vaults. The unwillingness of the Republicans to reduce the customs revenue and objections to increased government expenditure combined to produce budget surpluses of over $100 million in the late 1880s. This surplus was eventually used to retire the national debt, but as the debt diminished, the policy became increasingly difficult to pursue.

A wider study of central bank policy showed an ambivalence towards the 'rules' (Bloomfield, 1959). For only six out of the eleven central banks considered was there a close inverse correlation of discount rates and reserve ratios (on the basis of annual average monthly statistics) between 1880 and 1914. The rules prescribed that a fall in reserves should be countered by higher discount rates. For five of these six banks, reserve ratios tended to move inversely with domestic business cycle fluctuations. During a slump, reserves were high. Discount rates therefore moved positively with the cycle, consistent with a policy of domestic stabilisation. Discount rates were low during periods of low economic activity, tending to encourage borrowing and spending.

Bank of England policy

The Bank of England's behaviour conforms with this pattern (Goodhart, 1972, chs. 14 and 15; see also Beach, 1935). During an upswing in income, the banking system accommodated the increased demand for money. Consequently, there was an 'internal' drain on the Bank's reserves. A larger trade deficit also accompanied the rise in economic activity, but did not cause an 'external' drain of gold. Instead the increases in interest rates largely brought about by the Bank of England to protect its reserves led to such inflows of short-term capital that gold actually flowed into the Bank from abroad during periods of increased domestic activity, and thus partly, but not entirely, offset the internal drain of reserves from the Bank.

To make the Bank of England's discount rate effective, the Bank needed to ensure sufficient monetary scarcity that the bill market would be forced to borrow from it. It had to engage in open market operations. The Bank did not sell bills to make bank rate effective because it carried no suitable portfolio (Sayers, 1976, vol. 1). The supply of Treasury bills was insignificant. Instead it sold securities, thereby affecting the bill market indirectly through the Stock Exchange. Another method of making bank rate effective was by the direct control of funds that would otherwise have been lent directly to the bill market, such as those of the Council of India. In the 1890s, the Bank of Japan's funds also became subject to the Bank of England's direct borrowing. The ability of the Bank of England to control the British and the international monetary system by 1914 was reduced by the rise of ten large British joint-stock banks. The Bank's power to build up its own gold stock depended to some extent on their goodwill. To make the bank rate effective, the Bank tended to borrow from the clearing banks from 1905. Larger gold reserves, which everybody agreed were necessary, required that the clearing banks place bigger reserves with the Bank of England, but the Bank was unwilling to concede any power over the gold reserves that the banks wanted in exchange for the deposits.

Sometimes the Bank experimented with methods more direct than discount rate policy to influence international gold flows. The Bank was left a free hand in fixing its buying and selling prices for foreign gold coin – the most readily available gold – subject always to the limits implicit in the possibility of melting these coins into bars. To check an export of gold to the USA, the Bank would raise the selling price of American gold coins or refuse to sell them at all, forcing a diversion of demand to bar gold. Interest-free advances on gold shipments were another option, as was paying over the odds for bar gold, especially at the weekly South African gold sales from 1903.

The Reichsbank also borrowed in the market to make its discount rate effective, by selling Treasury bills in 1901, 1903, 1905 and 1906. Like the Bank of England, it did not engage in open market operations to ease the money market (Bloomfield, 1959).

Could central bank policy influence events at all? Or was each individual economy too small to change the world money supply significantly by altering its own monetary conditions? If prices were internationally, not nationally, determined, monetary policy that required the central bank to influence the price level was bound to be ineffective (McCloskey and Zecher, 1975; Keynes, 1930, pp. 306–7). David Hume's 'price–specie flow' theory of the gold standard did require price changes in response to gold movements in order to maintain the gold link.

According to the simplest of monetary theories of the balance of payments, the Bank could only contract (or expand) money supply growth so that the excess supply did not (or did) spill over into the balance of payments and drain (add) gold from (to) the reserves. This is an implication of the notion that the balance of payments is merely the difference between the national demand for money and the national supply of money. The level of national income influences the demand for money, but is not itself influenced by the balance of payments adjustment process. During an upswing, the tendency for excessive credit expansion was checked by a gold outflow, as national monetary growth exceeded the world average. The contraction of the monetary base tended to cause financial stringency, putting up interest rates, unless rationing of now scarcer loans was preferred, and the rise in bank rate merely reflected and signalled this stringency.

The great limitation of this model is the assumption that there is effectively only one type of asset apart from money. In fact, domestic investment was not a perfect substitute for foreign investment. Short- and long-term assets and the assets of different countries were equally imperfect substitutes. Hence different market interest rates on various types of asset were established, and the Bank of England influenced rates on sterling assets that foreigners held (Dornbusch, 1980, ch. 10; Eichengreen, 1987). It is certainly true that interest rates in different financial centres did diverge during crises, such as in 1907 (Ford, 1964, p. 34).

The benefits of the gold standard

One of the choices of national monetary policy was whether to join – and stay on – the gold standard. The benefits of maintaining gold convertibility are shown by a comparison of Spanish and Russian experience. Spain briefly joined the gold standard between 1876 and 1883. During these years, foreign investment flowed into the country. Thereafter investment fell markedly and growth was accordingly lower (Martin-Acena, 1993). Later than Spain but more determinedly, Russia committed itself to gold. The main cost of Russian convertibility was the two-thirds of official borrowing abroad between 1885 and 1897 which was used to acquire gold reserves, rather than being diverted to productive investment. The benefit was the increased inflow of foreign capital, attracted by the reduction in exchange risk (Gregory, 1979). Foreign capital flowed into Russia after convertibility at a rate at least 50 per cent greater than in the earlier period. If this increase is attributed to the adoption of the gold standard, then the consequent Russian net national product (NNP) growth can be found from a Harrod–Domar-type calculation. Foreign investment as a proportion of NNP rose from 0.5 to 1.5 per cent between the two periods. The marginal Russian capital–output ratio was about 2.5:1. Hence the percentage increase in national income from the rise in foreign investment, being the product of the increased investment share of income and the marginal product of capital (1 per cent × 1/2.5), was about 0.5 per cent per annum. If Russia had not joined the gold standard, the growth rate of the economy would have been 3.5 per cent per annum instead of 4 per cent per annum, and by 1913 national income would have been approximately 7 per cent lower. Using the same method to assess the maximum costs in terms of growth of the official borrowings from 1885 to 1897, growth would probably have been 0.2 per cent per annum higher if the gold reserves had been devoted to productive investment. Since 0.2 per cent is less than 0.5 per cent, there was a net benefit to the Russian economy from the gold standard.

Sweden attracted foreign capital after gold standard membership as well (Jonung, 1984). But if there was a causal connection, foreign capital may have allowed Sweden to retain gold convertibility, rather than the gold standard 'causing' the inflow (Lindert, 1984). One possible piece of evidence is the London market prices of government debt. Every gold standard economy in 1889 was quoted at a lower interest rate than every non-gold standard country (Table 9.1). Exchange rate risk, and perhaps country risk, was clearly lower for gold standard

Table 9.1 Bondholders' interest on loans to various nations at market prices in 1889

Britain	2.7	France	3.6	Austria	4.3	Egypt	5.1
United States	3.1	Victoria	3.6	Chile	4.3	China	5.4
Belgium	3.2	Cape Colony	3.6	Russia	4.6	Buenos Aires	5.9
Holland	3.3	Prussia	3.7	Brazil	4.8	Hawaii	6.0
New South Wales	3.4	Norway	3.7	Spain	4.9	Santa Fe	6.0
Canada	3.5	Sweden	3.8	Portugal	5.0	Japan	6.3
Switzerland	3.5	New Zealand	3.9	Argentina	5.1	Uruguay	6.6

Source: Mulhall (1892, p.260).

economies, so they must have been able to borrow more easily than those not on the gold standard; some capital inflow must have been induced by membership of the gold club.

Summary and conclusion

The thirty years before the First World War saw an international fixed exchange rate regime, unprecedented in history. Linking national currencies to gold left the international price level vulnerable to changes in world gold supplies. This is most probably why prices fell between 1873 and 1896 and thereafter rose to 1914. On the other hand, in the most advanced economies, the greater part of the money supply consisted of bank deposits. These were linked to gold in a manner that changed with the emergence of branch banking. Reserve ratios otherwise were not sufficiently flexible for bank money merely to respond passively to changes in world output.

Silver demonetisation cut the demand for silver and hence reduced the gold price of silver in the last quarter of the century. The demand for gold was correspondingly increased, so that countries remaining on a silver standard experienced a depreciation of their exchange rates with gold standard countries. On average the growth of exports from silver standard countries exceeded those from gold standard countries. Strong export growth created the possibility of more pervasive economic development in the silver economies.

Under a gold standard, an overvalued exchange rate could only be remedied by price and wage deflation, or by persistent unemployment. In this respect, the standard differed from the international monetary regime established after 1945. The same feature may account for its relative longevity, by encouraging stabilising speculative capital flows, in contrast to the Bretton Woods system. Like that system, the gold standard saw the emergence of reserve or key currencies that economised on the use of gold, and reduced the problem of international liquidity.

During the late nineteenth century, as after 1945, there were strong pressures to increase government expenditures, which threatened the viability of fixed exchange rates for some countries. There is little evidence that the link to gold constrained public finances. The operation of the gold standard was not dependent on a tranquil international economy. World trade and finance were convulsed by periodic crises, exacerbated by inadequate financial institutions, especially in the United States. Harvests, capital movements and income effects all played a part in synchronising fluctuations in national economies.

Central banks pursued monetary policies calculated both to maintain currency convertibility into gold and to avoid as far as possible adverse repercussions of international fluctuations on their domestic economies. Discount rates, raised to attract gold from abroad when reserves fell, were the main instrument of policy. Open market operations to 'make the rate effective', gold devices and direct borrowing from the market were subsidiary instruments. During crises, central banks co-operated to some extent because they recognised their interdependence.

Notes

1. For a somewhat different interpretation, see B. B. Aghlevi (1975).
2. Latham (1978b, ch. 2) provides a useful summary of the experience of LDC silver standard economies.
3. See also Keynes (1913b) on the emergence of the key currency system.
4. This 'absorption' effect was not complemented by a 'Dutch disease': a fall in the price of exports relative to domestic prices and a rise in export prices compared with those of competitors (Rowthorne and Solomou, 1991).
5. Kirby (1981, pp. 14–16) maintains that industrial structure was deformed by overseas investment.
6. By contrast, since rate changes were permitted under the Bretton Woods system, speculative capital flows tended to precipitate them. Williamson (1977) believes this to have been the main reason for the failure of the Bretton Woods 'adjustable peg'.

10 International trade in the twilight of liberal capitalism

1914 altered the course of the international economy. Inevitably, the unprecedented destruction of life and physical resources in Europe during the ensuing four or five years spilled over into the whole world, for European political and economic weight was then so great. Even so, for the new economic giant, the United States, for its trading partners especially in Latin America, and for most of the huge populations of Asia, the end of the nineteenth-century liberal, or liberal/nationalist, pattern of economic relations is more often dated to the Great Depression, beginning in 1929. That trauma was even more pervasive than the First World War. Throughout the world of the 1930s, new domestic and international economic policies were introduced to alleviate the collapse of trade and employment.

Whether 1914 or 1929 is the break point, the period labelled 'the inter-war years' has acquired a poor reputation for achievements in the fields of economics and politics. What were later regarded as the mistakes of the inter-war years fixated policy-makers at the end of the Second World War and subsequently. These were years of metamorphosis of institutions and beliefs. The old international liberal economic order seemed unable to cope with the disruptions of the 'war to end all wars' and then with the Great Depression. Rival doctrines of fascism and communism both offered remedies which emphasised greater state economic control. And all policy-makers chose more corporate ways of organising domestic and international economic relations.

This chapter begins by outlining the disruptions caused by the First World War and then examines the subsequent tendency for international trade to decline. Whether this trend mattered much for well-being is considered in the third section on overall economic performance. Contemporaries believed that, among the failures of the market economies, the persistence of agricultural trade depression was prominent. The chapter therefore goes on to assess the relative importance of deficient demand from industrial countries, technical progress and the expansion of colonial supply as explanations for agricultural distress throughout the world economy. In the same section on primary products, we discuss the impact of oil competition on the coal exporters and some of the political elements of the young oil trade which were to continue into its middle age.

Like the primary product sector, manufactures also experienced substantial technical progress. How different economies took advantage through the world economy of such developments is described next. The 1930s saw international trade fall to one-third of its 1929 gold value as the Depression ushered in a host of controls and alliances intended to raise domestic prosperity, regardless of the rest of the world. These policies and their impact are the central concern of the final section of this chapter.

The First World War and European economic relations

Such a break in the pattern of life was the mayhem of the First World War that many people of the 1920s and later were encouraged to see the years before 1914 as a golden age which the war had destroyed. The war and its aftermath caused a structural maladjustment of the international economy, it was widely thought. Hence economic policy among the victorious powers was largely an attempt to return to pre-war conditions.[1] Others have seen the old order (variously defined) as on the verge of disintegration or transformation quite independently of the war and the associated political changes. Thus de Cecco (1974) argued that the gold standard was no longer viable by 1914. Dangerfield (1966) asserted that the liberal consensus in England disappeared, leaving the country on the verge of civil war over Ireland, while the rapid industrialisation of the United States was eliminating British and French trade supremacy and potentially causing balance of payments problems. German industrialisation and trade expansion was one more potentially destabilising influence on international economic relations before 1914 (Hauser, 1917). The most favoured nation system was already eroded by two-tier tariffs for bargaining (Conybeare, 1987, p. 237).

The survival chances of the old order of international economic relations in the absence of war might be assessed by considering what did actually break it in peace time, the Great Depression. That crisis itself was related to the reconfiguration of the post-war settlements (Chapter 11). Without the war, it is unlikely that the shock would have been so severe and therefore that the system would have collapsed, whatever the difficulties of individual economies. A great deal is required to pull down a regime perceived to confer advantages on all major participants. Avoiding major depressions or wars probably would have been sufficient to ensure the survival of the liberal regime.

If war was the direct outcome of international economic relations, however, there is little sense in posing the question about the impact of war on those relations. Largely on the basis of German experience, Lenin (1934) formulated his hypothesis that the war was an extension of the economic and political turmoil of the years before 1914, and not an exogenous event. Lenin represented war and territorial acquisition as the outcome of monopoly finance and the struggle of capitalists to avert the tendency for the rate of profit to fall. The state, according to Lenin, was necessarily the instrument of the predominant type of economic organisation. In their attempts to secure raw material supplies such as oil and iron ore, the monopolists enlisted the state to acquire territory by force. The German invasion of Belgium in 1914 supposedly revealed this tendency, as did most international conflicts in the previous four decades. In fact, the part played by finance in the international disputes leading up to the war, and in the outbreak of war itself, was almost the reverse of that proposed by Lenin (Robbins, 1939; see also Staley, 1935). Between strong states there was hardly any evidence of finance causing friction or war. Rather finance was continually used as a political pawn, as shown in Chapter 7. National commercial policy employed to support national security goals was more effective in creating an antagonistic climate favourable to war.[2]

Europe's key role in world trade before 1914 ensured that international economic relations were thoroughly disrupted by the war. First the currencies of the European belligerents ceased to be reliable international media of exchange, as panic seized traders and bankers. Then trade became further disturbed by naval blockades and by attacks on merchant shipping. International migration and investment slowed to a trickle, as the former lending nations, Britain and France, became net debtors to pay for the war. Resources were reallocated to provide war materials and expand the armed services, to the benefit of heavy industry – steel, coal, shipbuilding and heavy engineering – while domestic agriculture in industrial Europe was encouraged to replace imported foods (Hardach, 1977, chs. 2–6).

With perhaps 8.5 million European military dead and 5 million civilians killed, the human costs of the fighting probably damaged the European economies more than the destruction of capital. Nevertheless, nature exceeded man's brutality with the influenza pandemic of 1918, estimated throughout the world to have killed as many as 20 million (Aldcroft, 1977, pp. 13–23).[3] Far-reaching political changes reinforced the mortality shock. In 1917 the Russian Empire finally collapsed with the strains of three years of total war, and, on the promise of 'Bread and Peace', Lenin and the Bolsheviks seized power. The newly created major communist state directly or indirectly influenced economic policy in the non-communist world for the next seventy years.

Fear of further communist revolutions encouraged expansionary monetary and fiscal policies in Britain after the war. Boom conditions were maintained during demobilisation to minimise the chances of insurrection. What was perceived as the communist threat gave rise to the political corporatism of fascism in Europe. Like the communist economic system, fascism favoured greater isolation from international trade and a larger role for the state in economic life. The peripheral states of the Russian Empire – Poland, Lithuania, Latvia, Estonia and Finland – all gained independence with the breakdown of central Russian authority. (Poland was also helped by the dismemberment of Austria-Hungary and the German Empire).

Ironically, the nineteenth-century 'sick man of Europe' lasted one year longer than the Russian Empire. The Turkish Empire had been supported by western Europe as a bulwark against Russian expansion. Under the peace settlement, the former Turkish Empire in the Middle East was divided between Britain and France as League of Nation mandate territories (Hardach, 1977, p. 243; Aldcroft, 1977, pp. 26–30). The search for oil in these territories became yet another source of international friction.

The break-up of neither the Russian nor the Turkish Empire was of great importance to international economic relations, but the treatment of the German and Austrian Empires at the Versailles peace settlement certainly was. The victorious powers were determined not to repeat what they saw as the mistake of the Treaty of Vienna in 1815 at the end of the previous European war – the neglect of nationalist aspirations for independence (Carr, 1961).[4] In avoiding this mistake and dividing Europe into a large number of small nation-states, they made a different error: they neglected the economic unity of European regions, which they disrupted by creating the new states, hedged with barriers to trade and investment (see Figure 10.1). Out of the Austro-Hungarian Empire were created the states of Austria, Hungary and Czechoslovakia. Austria was in particular difficulties, its economy and railway network having developed as the centre of a great empire. Of its textile industry, the yarn-spinning and finishing mills remained, but the weaving mills were in Czechoslovakia, which protected its own infant spinning industry (Mitrany, 1936, pp. 172–3; Aldcroft, 1977, chs. 2 and 3). Although Austria's major coal deposits were given to Poland and Czechoslovakia, the Alpine ironworks that needed the coal were still in Austria. Similarly, Hungary's great flour mills lost their sources of supply and the market for their products.

Germany's fate was determined primarily by a desire for retribution. Alsace-Lorraine was returned to France. By redefining products, the boundary change increased French exports of iron ore and German imports. Under the terms of the peace treaty, Germany ceded 13 per cent of its territory, and with it 10 per cent of its population. It lost three-quarters of its potential output of iron ore, one-third of its potential output of coal, and 15 per cent of the area under cultivation. Because income per head was lower in the east, the loss of the poor agricultural territories there may have actually raised average German living standards, again purely by the national redefinition of boundary changes (Hardach, 1977, p. 244; Aldcroft, 1977, p. 23). Some of eastern Germany was transferred to the new Polish state. Poland was granted access to the Baltic at Danzig by a corridor which separated East Prussia from the rest of Germany. This was to provide a proximate cause of the next world war.

Had the opportunity been taken to create a sounder world economic order at the same time, the multiplication of national boundaries might not have had adverse repercussions. A common view blamed the accompanying insistence on unfettered national economic sovereignty, and belief in the recuperative power of private enterprise and automatic economic forces, for the failure of international economic recovery in the inter-war years (Arndt, 1972, ch. 9, pp. 296–7).[5] With the policy implications of state central planning and import controls, such views conferred on the inter-war economic debate a much wider relevance. The

International trade in the twilight of liberal capitalism

Figure 10.1 European frontiers after Versailles

apparently powerful growth of the centrally planned Russian economy, largely independent of economic relations with the market economies throughout the 1920s and the depression of the 1930s, added more grist to the anti-liberal capitalist mill. After 1945 the prestige so acquired for central planning and control of international trade was a major stimulus to the import-substituting industrialisation policies of the less developed countries (see Chapter 13).

The decline of international trade

Purely because of the multiplication of European national boundaries, the volume of international trade and investment should have increased, at least relative to domestic economic activity. In fact, the pursuit of nationalist commercial policies reduced the volume, both absolutely and relative to national income. The mean trade–income ratio for Great Britain, Italy, Sweden, Norway and Denmark fell from 43.7 per cent between 1905 and 1914, to 35.7 per cent between 1925 and 1934, and to 26.7 per cent between 1935 and 1944 (Grassman, 1980).[6] The decline in the American trade–income ratio began from a lower level, and the rapid phase occurred earlier. From 12.4 per cent between 1915 and 1924, the mean American trade–income ratio reached 7.7 per cent between 1925 and 1934, and 6.8 per cent between 1935 and 1944.

Not all of this fall by any means was caused by commercial policies. The rest of the world was becoming more independent of Europe as a source of supply of manufactures, and especially of some traditional types of export product such as cotton textiles (Svennilson, 1954, p. 22).

Cut off by submarine warfare and military demands on shipping space, overseas markets during the war were stimulated to start industries of their own. Industrial development of overseas competition, especially in the United States and Japan, was given a boost. Even so, from the 1870s until the end of the 1920s, industrialisation had increased such countries' ability to export and to buy more imports of manufactures (League of Nations, 1945).[7] The United States by now was anyway largely economically independent of the rest of the world, except in certain tropical products. Consequently, the stagnation of the open European economies while the United States grew in the 1920s inevitably lowered the world trade–income ratio.

Agricultural raw material trade rose little for two reasons. First, the materials were often substitutes and so competed against each other. Second, accidents of location of supply and demand in relation to national borders contracted international trade. Rubber displaced leather in some uses, and wood pulp in the form of rayon competed against and partially displaced cotton. Competition for minerals was beginning, with light metals replacing wood in many constructional jobs, petroleum-based compounds being made into nylon or synthetic rubber, and dyes and drugs being manufactured from synthesised chemicals, rather than from the juices of plants (Yates, 1959). Whether these technological changes expanded or contracted the demand for imports depended upon whether they brought into use materials plentifully available within the frontiers of the major manufacturing countries. The United States lacked natural rubber, but possessed raw materials for synthetic rubber, and wood. Britain depended almost entirely on imports for pulp and pulp wood as well as for rubber. Britain's propensity to import these items therefore exceeded that of the United States in the inter-war years. Similarly, the growth of the Indian cotton textile industry did not increase world raw cotton trade in parallel because raw cotton supplies were available within the country. In the impact upon international trade, however, Japanese cotton was very similar to the earlier European industry.

Most contemporaries blamed commercial policies for the decline of trade, the problems of the inter-war years and the causes of the 1919 Depression.[8] Others focused their attention on deficient demand. According to this view, tariffs and the decline of trade were inevitable concomitants of industrialisation. They need not have greatly harmed international economic relations under 'normal' conditions (Lewis, 1949, p. 164; Arndt, 1972, p. 271). The paralysing restrictions of the inter-war years were attempts to remedy instability and unemployment, which had their root cause in domestic economic mismanagement. Indeed, restrictive trade policies that attempted to restore domestic employment, such as Schacht's 'New Plan' in Germany of 1934, were amply justified in the absence of international reflation. The central problem of the international trade of western Europe was that trade in manufactures was low because the industrial countries were buying too little from the primary producers and paying too low a price. Expansionary fiscal and monetary policies in industrial Europe, by implication, would have boosted demand and prices even with an unchanged level of protection.

Although the United States tariff policy has often been blamed for exacerbating inter-war trade policies, the Fordney–McCumber duties of 1922 which raised tariff levels steeply did so on products which would not have been imported in large quantities anyway. On sewing machines a tariff of 15–30 per cent was levied, yet under free trade the previous year exports totalled $7.3 million, and imports a mere $0.4 million. The development of the vast resources of the United States meant that the economy had little need to import manufactures from Europe, or primary commodities from the temperate zone exporters.

Would a reduction in tariff protection or a revaluation have much reduced the American trade surplus with Europe? And was that objective desirable in any case? It was desirable for the stability of European balances of payments, for the United States was accumulating short-term claims on Europe which were likely to be withdrawn at any time when interest rates or currency risk altered (Falkus, 1971).[9] Quantitative controls on European imports would have solved that problem at the expense of further reducing international trade. Such a policy in the early and mid-1920s might have reduced economic dislocation in Europe at the end of the decade when American funds were withdrawn. Alternatively, more expansionary US domestic economic policies would have achieved the same ends at a higher level of world employment. They would have drawn in imports, especially luxuries, probably more than proportionately to the growth of income, and eased the chronic imbalance in the international accounts with Europe.

As with the Fordney–McCumber duties, so too the British lapse from free trade with the Safeguarding of Industry Act 1921, and the continuation of the McKenna duties, cannot be regarded as a major contributor to the contraction of trade. These duties covered only a small number of commodities, and the tariff rates were not prohibitively high. Dyestuff imports were prohibited for ten years except under licence because of their relationship with explosive technology, the dominance of Germany, and the weakness of British industry in the field. Duties imposed on German imports in part-payment of war reparations averaged 26 per cent, yet by 1930 only 17 per cent of imports by value were dutiable. Protective tariffs, in contrast to low duties designed merely to earn revenue, did not affect more than 2 to 3 per cent of imports (Pollard, 1962, p. 194). With low price elasticities for British trade, these tariffs would have had little effect.[10]

Contrary to the predictions of the simple factor price equalisation theorem, but consistent with the analysis of earlier chapters, declining trade was matched by diminished factor movements between the world wars (cf. Mundell, 1981). The great intercontinental migrations before 1914 did not diminish the need for trade, and the contraction of trade in the 1930s did not increase factor mobility. Intercontinental migration declined to around 6.7 million in the 1920s, little more than half the outflow of 1901–10 (11.6 million), and to 1.9 million in the 1930s (Woytinsky and Woytinsky, 1953, pp. 75, 82, 95–7). The United States recorded the greatest drop in immigration numbers in the 1920s, at least partly because of the quota restrictions imposed in 1921 and 1924. Thanks to very high American unemployment, during the 1930s migrant numbers did not reach the quota limits; there was actually net emigration from the United States between 1930 and 1934. Southern and eastern Europe, as before 1914, continued to provide more than half the immigrants, although the UK remained the largest single source. Within Europe about 7 million people were transferred or fled as a consequence of wars, persecutions and boundary changes. In may instances, it became difficult to distinguish flights of refugees from normal migration.

Quota limitation reflected the ending of opportunities for extensive expansion in the United States, the recognition of the effect of immigration on wages and the growing separation of US interests from Europe after the war years. If the quotas were an important cause of the reduction in migration, trade could be expected to increase, other things being equal, according to the factor price equalisation theorem.

Economic performance in the inter-war years

Despite the decline in international trade and factor mobility, the unemployment and the instability of the period, real incomes in most countries rose quite considerably, so that those in employment (and sometimes the unemployed) achieved higher living standards in 1938 than in 1913.

Greater prosperity was based upon a variety of technical developments which were so fundamental that they were comparable with the techniques of the first Industrial Revolution. The most radical developments were in new applications of electricity and the internal combustion engine, which transformed mechanical motive power sources for transport, agriculture and industry. The basic innovations were achieved several decades before the war. But continual progress after the war halved in a decade the coal required to generate a given output of electricity, and greatly widened the radius around primary energy sources within which electricity could be cheaply supplied, by the development of high-voltage transmission (Svennilson, 1954, pp. 20–1). More flexible transport of goods and people was made possible by the motor vehicle in the form of the lorry, the bus and the private car. Between 1919 and 1928 the number of motor vehicles, excluding lorries, registered in the United States tripled to over 21 million and the number registered in the United Kingdom rose fourfold to 1 million. In the three years before 1929, African motor car registration almost doubled to about 350,000 vehicles.

The consumption of other goods rose with the higher incomes created by the new technologies during the late 1920s. In 1927 there were 2.5 million wireless sets in Great Britain, 2 million in Germany and over a quarter of a million each in Japan and Australia. The diet of the West became lighter and more varied, including more fruit and dairy produce (Ohlin, 1931, p. 19). State education and insurance against unemployment, sickness and old age diverted a part of national income from investment to consumption. An ageing population, as population growth declined, increased the proportion of the population that earned a living in the market, as did the greater female labour force participation. Hence family income and consumption rose by more than wages.

Inter-war growth experience varied greatly between nations. Much depends on what periods or sub-periods are considered. There is a case for measuring 'inter-war' growth across the years 1913–50. That takes into account 'catching up' after war years as well as output losses during hostilities. A break point in 1929 distinguishes possible pre- and post-Great Depression phases. For the United States and the regions of recent European settlement, the 1920s were rather prosperous, and the 1930s a period of acute depression. Most of the former belligerent states of Europe took longer to recover from the war. Civil war with outside intervention continued in Russia until 1922, by which year Greece was again at war with Turkey. In Hungary and in Ireland too there was civil strife. France's rapid recovery from the war can be explained by an undervalued franc, an expansionary fiscal policy and the benefits of the Versailles Treaty. Table 10.1 shows the contrast in growth rate of France and the USA on the one hand, and Germany, Hungary, the United Kingdom and the USSR on the other, between 1913 and 1929.

Table 10.1 allows an assessment of the poor reputation of the inter-war period.

Table 10.1 Annual average growth rates of real gross national product per head, 1913–50 (%).[a]

	1913–29	1929–50	1913–50
France	1.9	0.5	0.9
Germany	0.8	0.7	0.7
Italy	1.0	0.5	0.7
UK	0.3	1.3	0.8
Russia/USSR	0.4	3.7	2.3
USA	1.6	1.5	1.5
Australia	−0.5	1.2	0.9
Canada	1.1	1.8	1.5
Argentina	0.9	0.6	0.7
Japan	2.4	−1.1	0.9
India	0.1	−0.5	−0.3
Mexico	0.1	1.6	0.9
Brazil	1.4	2.3	1.9

[a] Measured in international dollars at 1980 prices.
Source: Calculated from Maddison (1989, table 1.3).

Growth performance may be compared with some aggregates before the First World War. In the first thirteen years of the twentieth century, the group of industrialised economies that were later to form the OECD expanded on average at 1.5 per cent a year. The largest Latin American economies averaged 2 per cent and the Asian economies attained 0.8 per cent. In drawing conclusions from Table 10.1 it is important to be aware of the way in which the measurement of growth rates may be affected by the choice of initial or terminal years. If the terminal year for one country is a recession, whereas for the country with which it is being compared that terminal year is a boom, then even if the two economies were growing on trend at identical rates, the second country would seem to have grown faster than the first. Similar considerations apply to the comparison of different periods for the same country. Bearing in mind these caveats, France's economic growth was high in the 1920s at 1.9 per cent 1913–29, but fell badly in the 1930s, reducing the average for 1929–50 to 0.9 per cent. For all four European economies in the table, the growth over 1913–25 was considerably less than 1.5 per cent. Only Canada and the United States among OECD members matched the pre-war average. The United States achieved a similar growth rate to France's between 1913 and 1929, but unlike France, maintained that performance in the second period. Among the agricultural regions of recent European settlement, Canada grew quite rapidly, Argentinian growth resembled Germany's or Italy's, and the Australian economy suffered badly in the 1920s.

Outside Europe and the European offshoots, growth experience was mixed. The Indian figures make the much criticised nineteenth-century growth look

Table 10.2 Growth rates per annum in the dollar value of exports, 1913–37.

	Total	Per head		Total	Per head
Japan	5.2	3.85	USA	1.4	0.25
Malaysia	4.2	n.a.	World	1.4	
Canada	3.7	2.2	Belgium	0.7	0.35
Indonesia	3.0		Italy	0.5	−0.15
China (inc. Manchuria)	2.4		UK	0.1	−0.3
Thailand	2.4				
Australia	2.2	0.7	Germany	−0.1	−0.6
			India (inc Burma)	−0.1	
Netherlands	1.8	0.4	France	−1.3	−1.3
Argentina	1.6	−0.7	USSR	−12.4	−13.2

Source: Maddison (1969, table 10 and appendix C).

respectable in comparison, although the second figure is pulled down by the disruptions after independence in 1947. Other Asian economies, China and Thailand, like India seem to have earned no higher incomes in 1950 than in 1913. Brazilian economic growth exceeded Japan's over the whole period 1913–50 because Brazil avoided the devastation of war. Disastrous earnings of the major export crop, coffee (see below), proved much less harmful (Fishlow, 1980). The coffee sector accounted for little more than 10 per cent of national income, and the ratio of exports to GDP was only about 15 per cent. Growth in the Brazilian manufactures sector (and that of Indian manufactures) was based upon massive substitutions of domestic goods for imports.

Export performance (Table 10.2) varied between countries as much as income growth. Japan experienced the fastest increase in output before the Second World War of the economies listed, and also the greatest expansion rate of exports. Japan and probably Malaya were exceptional in the inter-war years in experiencing export-led growth; their economies became increasingly open and grew rapidly. Otherwise there was an inverse association between growth and export orientation.

Both Germany and the UK grew faster in the 1930s, when world trade and their participation in it was greatly reduced but a substantial component of this growth was increased capacity utilisation, a cyclical or policy-induced upswing. The USSR is the most extreme example of this phenomenon.[11] Most of Europe was less dependent on trade in 1938 than in 1913. Apart from Canada, the top six export growth economies earned low incomes, and initially were all mainly exporters of primary products. The large economies of industrial Europe which exported manufactures – the UK, France and Germany – by contrast all suffered declines in their dollar values of exports per head. A cursory examination of the evidence, then, suggested that the effect of trade in primary products in the inter-war years was rather different from that of trade in manufactures. This experience contributed much to the later suspicion of primary product trade as a means of economic development.

Table 10.3 The composition of world exports, 1913–37 (Index nos 1913=100).

Year		All manufactures	All primary products	Cereals	All food	Agricultural raw materials	Fuel	All minerals
1913		100($6,855m)	100($12,248m)	100($1,784m)	100($5,535m)	100($4,040m)	100($919m)	100($2,673m)
1929	(a)	129	138	107	136	128	173	155
	(b)	180	164	115	153	161	224	191
1937	(a)	107	129	98	133	111	171	151
	(b)	133	131	85	113	122	205	184

(a) = 1913 constant prices; (b) = actual values, $m
Source: Yates (1959).

Trade in primary products

The distrust was largely based upon misinterpretation. Although technical change did adversely affect some primary products, especially coal, the main problem was the low demand from the industrial countries because of their increasing agricultural protection and their high and persistent levels of unemployment. Introduction of agricultural price support policies by the primary product exporters exacerbated their problems.

Within the category of primary products there were substantial differences in growth rates that reflected the buoyancy of the demands of industrial economics. As Table 10.3 shows, the volume of total primary exports grew faster between 1913 and 1937 than did manufactures. Trade in cereals, however, actually declined, while the value of trade in fuels doubled. The greater expansion of primary product export volume compared with manufactures export volume, when they increased similarly in value, shows that the commodity terms of trade moved against primary products and in favour of manufactures between 1913 and 1937.

Open 'non-industrial' countries were dependent on the demand from industrial countries, to which 85 per cent of their exports were sent in 1913, but the industrial countries depended on other industrial countries for two-thirds of their export demand (Yates, 1959, pp. 56–7, 62).[12] A stagnation or decline in industrial countries therefore greatly hurt primary exporters. The industrial countries in the inter-war years wanted to prevent a further decline in their agricultural sectors and introduced measures which reduced their primary imports. Adopting agricultural protectionism was in part a response to the disruption of supplies caused by the war, which had shaken the assumptions on which the nineteenth-century pattern of food production was based. Britain began a state-aided sugar beet industry in 1925, and the same year Italy opened its battle for self-sufficiency in wheat.

Technical progress in temperate zone agriculture further worsened the terms of trade by increasing supply. The use of the tractor and the combine harvester expanded rapidly in the late 1920s. In 1928 over 140,000 tractors were sold in the United States or exported (Ohlin, 1931, pp. 91–4). Two-thirds of the 68,000 American combine harvesters in use in 1929 had been bought within a year. Argentinian annual purchases of these harvesters doubled to over 6,000 between 1928 and 1929. Over 20,000 tractors were sold in Canada in 1928. Nevertheless, cereal exports were lower in 1937 than in 1913, so a demand-side explanation for the decline in the terms of trade is more plausible than a supply-side explanation.

Colonial economic policy accentuated the increase in primary product export supply. Rapid expansion of cocoa, oil seeds and fats sales raised Africa's share of world exports. In the French African colonies, the use of administrative measures to enforce the planting of coffee, cocoa and cotton was common, but not very successful (Suret-Canale, 1971). Imperial tariff preference on exports to France offered another incentive to increase agricultural production for export, as did the African peasants' need to earn the cash to pay taxes. French colonial public expenditure was also directed towards agricultural exports. By 1942, 8,455 million francs had been allocated by the central government for the development of the colonies (£113 million at the 1935 exchange rate), of which 1,153 million francs represented unstarted projects (Coquery-Vidrovitch, 1981). Forty per cent went to French West Africa. Most projects were public works or agricultural processing, mainly peanut shelling.

Precisely what contribution the British African colonial governments made to export growth, other than through the maintenance of law and order, is still a matter of dispute. So is the desirability of that pattern of development (Ehrlich, 1973; Drummond, 1974, ch. 9; Abbot, 1973; Meredith, 1975). The construction of a railway in Kano, Nigeria, in 1911 allowed a groundnut boom which tripled the tonnage exported between 1916 and 1929. In 1920 the Empire Cotton Growing Corporation was established and cotton cultivation for export was encouraged in Uganda. In Kenya, Africans were denied access to land which would allow the profitable cultivation of coffee for export, on the grounds that the fertility of the soil had to be protected to prevent the emergence of another US 'dustbowl'. This restriction also kept down the wages of labour employed by the white farmers. The Colonial Development Act of 1929, permitting the disbursement of up to £1 million a year, provided assistance mainly to small-scale transport and public health schemes, 60 per cent in the form of grants.

If colonial policy prevented colonial industrialisation by requiring colonies to maintain an 'open-door' tariff policy, despite the UK's abandonment of free trade, then it also helped increase the supply of primary exports. This is not necessarily inconsistent with an optimum policy even so. Although often asserted, a colonial open-door policy has not been demonstrated harmful to colonial development. A decline in the commodity terms of trade because of increased supply in any case does not inevitably imply a loss, only a decrease in the share of the gains from trade. Objections to primary product exports, rehearsed in Chapter 6, are based upon other, usually erroneous, beliefs and often go hand in hand with a distrust of international economic relations altogether, linked to a fear of dependency. Birnberg and Resnick (1975) in this spirit formulated and estimated econometrically an ingenious model of colonial development based upon exports, to explain primary product exporters' experience during the inter-war years (and, in most cases, from the end of the nineteenth century). The colonies for which they estimated their model were Ceylon, India, Nigeria, the Philippines and Taiwan. They also included countries which were not formal colonies – Chile, Egypt, Thailand and Cuba – because they believed that foreign influence, although exercised more subtly through international trade and finance, subjected these countries to the same internal and external forces which determined their pattern of development. The argument can be reversed: formal political control did not involve distortion of the colonial economies, as the similar development of independent economies shows.

The structure of Birnberg and Resnick's model is given in Figure 10.2. The

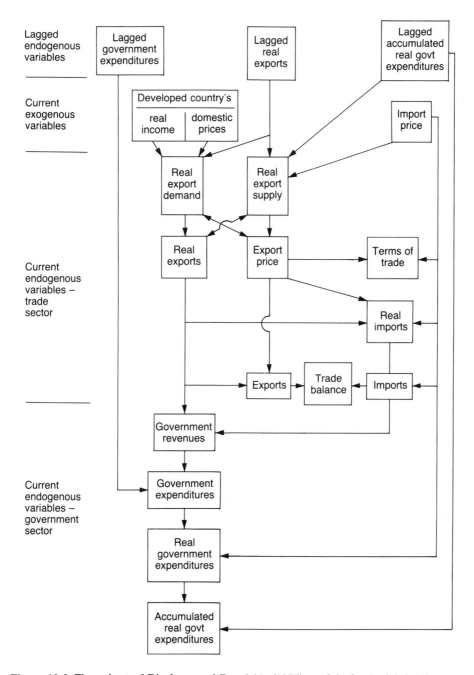

Figure 10.2 Flow chart of Birnberg and Resnick's (1975) model of colonial development

volume of colonial exports is determined by increases in accumulated real government expenditures oriented towards export growth. These expenditures include the provision of transport facilities, irrigation, education and public health. Colonial export volumes depend also upon import prices, reflecting the costs of export production (the prices of capital goods, transport equipment, and fertilisers). The demand for colonial real exports depends upon the real income, domestic prices and trade policies of the developed country to which the colony is 'tied'. Export price is determined by the balancing of colonial export demand and supply. The colonies did not balance their commodity trade, so colonial import demand is determined by exports, import prices and export prices. Nominal government revenues are generated from the direct and indirect taxation of real exports and nominal imports. Nominal government expenditures depend upon government current revenues and lagged expenditure.

Birnberg and Resnick (1975) obtained low supply and demand price elasticities for colonial export volumes. Thus changes in supply or demand conditions caused large changes in export prices. The average real income elasticity of demand for colonial exports was approximately equal to one in the long run, implying that the export growth prospects were reasonable. Birnberg and Resnick found a wide range of estimated coefficients of the response of exports to long-run accumulated government expenditure. They maintain that these values are as important for explaining the low export growth of India and Egypt as for explaining the high export growth of Ceylon and Philippines. The response coefficients, they believe, take their particular values because of the uses to which government expenditure was put. Almost one-half of the Indian budget was devoted to military expenditure, and Egypt's government expenditures were largely allocated to paying interest on previous international loans. Hence the low productivity of total government expenditures in boosting Indian and Egyptian exports. However, a measure of the openness of the economy to trade – for example, proxied by population size – is likely to alter this conclusion markedly. A very small proportion of Indian output was exported, so government expenditure mainly increased the ability to supply at home, not abroad.

Two colonial characteristics – the developed country to which a colony was tied, and the productivity of government expenditures directed towards development – are key explanations for the similarities and differences of the colonial development process, Birnberg and Resnick maintain. The economic losses from the First World War were greatest for the UK bloc. Their terms of trade deteriorated because of the post-war rise in import prices and the weakness of the British economy. Those linked to the United States, and Ceylon, suffered most from the Great Depression because of the decline in the demand for rubber. The inclusion of rubber-exporting Malaya in the sample would have shown, as Ceylon did show, that political links were not necessarily coterminous with economic links.

Probably the most adverse effects of primary product export-oriented development in the inter-war years were experienced not by any of Birnberg and

Resnick's sample, but by Japan. The collapse of Japanese silk exports and the political repercussions in the 1930s eventually resounded around the world (Allen, 1981, pp. 120–1). Raw silk was second only to rice among Japanese agricultural crops. Between 1914 and 1929 production tripled to supply the American demand for silk stockings. With the collapse of the United States market, the export price of silk fell by 1932 to almost one-quarter of the 1929 level. Farmers' cash incomes from this source were thereby greatly reduced just at the time that the price of rice (the other peasant income earner) had fallen. The peasants attributed responsibility to the politicians and to the zaibatsu (the financial/industrial conglomerates of the modern sector). The army which was largely recruited from the agricultural sector, expressed their discontent. Silk prices thereby contributed to the overthrow of the liberal government and the transfer of power to those who favoured military aggression. In 1931 Japan invaded Manchuria, and in 1937 the Sino–Japanese War broke out. The Japanese economy was traumatised because of export concentration rather than because of the nature of the product itself. But like the late nineteenth-century West Indian experience with sugar, Japanese specialisation served to underline the element of uncertainty and instability in international trade which the classical analysis of the gains from trade had neglected.

The fuel trade

The secular impact of a weak demand for a primary product was felt strongly in the inter-war years by one of the foremost industrial powers, the UK. In contrast to the pre-war increase of 4 per cent per annum, the demand for coal remained stable in the 1920s. Oil and hydroelectricity reduced the demand for coal and prevented rising industrial production boosting coal sales (Ohlin, 1931, p. 60). In 1913 international trade in coal and coke ranked third behind cotton and wheat. Even at the lower 1938 level, the coal trade was surpassed only by cotton. The UK, Germany and Belgium provided the import needs of the Netherlands, France, Italy, Austria and others. The United States exported a little to Canada, and the European surplus, together with small amounts from Japan and Australia, supplied the needs of other continents (Yates, 1959, p. 150). These were, however, relatively small because coal was a motive force for industry, steam-ships and railways; the industrial countries therefore consumed 93 per cent of the world's coal output. Coal prices, especially British coal prices, rose relative to commodity prices after 1928, reinforcing the switch to oil for power and heat that was already occurring. This was one of the reasons for the concentration of long-term unemployment in the British coal industry.

Despite the decline of coal, Table 10.3 shows a substantial increase in the fuel and mineral trade between 1913 and 1937, mainly because of oil. The politics of the oil trade in these years in some respects was a dummy-run for the era after the Second World War, and in other respects laid the ground for the post-war trade.

Although small in 1919, the trade was regarded as a matter of extreme national importance. Oil played a vital role in the war, and the British recognised that the nineteenth-century technology which had made its indigenous deposits of coal so valuable was nearly obsolete (Tugendhat and Hamilton, 1975; Longrigg, 1961; Penrose, 1968).

At the San Remo Conference in April 1920, Britain and France divided the Arab territories of the Turkish Empire between them as League of Nations mandates, a division which was to give them control over much of the Middle Eastern oil. Throughout 1919 British officials refused to allow Standard of New York to resume its pre-war explorations in Palestine while (Anglo-Dutch) Shell geologists were at work in Mesopotamia. The US State Department believed that the British were planning to use their military presence in the Middle East to exclude US interests from promising oil areas. Between 1914 and 1920 US domestic demand more than doubled, and the tripling of crude oil prices since 1913 convinced many Americans that their domestic fields would soon be exhausted. The Americans were therefore extremely concerned to find new sources of supply. Eventually an agreement was reached in 1922 which allowed the American firms to participate in the development of the Mesopotamian oil fields.

By then the oil crisis was almost over. In December 1922 a massive oil discovery was made in Venezuela near the shores of Lake Maracaibo, which enabled Venezuela to replace Russia as the world's second largest producer in 1928. New reserves were discovered in the United States where, from 1922, more oil was produced than the refineries could handle.

Commodity control schemes

The small number of companies engaged in the oil trade made the creation of a cartel or price agreements relatively easy. To end an international price war, in 1928 the companies signed the Achnacarry agreement. The weakening of primary product prices and the larger number of producers in other sectors increased the electoral advantage from governments' introducing price support or output control schemes during the 1920s. Their efforts ensured that, for many of the inter-war years, market prices were not necessarily market-clearing prices. Stocks were accumulating during the late 1920s and the impossibility of financing them any longer contributed to the onset and severity of the Depression after 1929. Stocks were only a problem for commodities such as sugar, wheat and coffee, for which there was a strong increase in supply relative to demand, because of technical progress or, in the case of coffee, because of a price support scheme (Rowe, 1965, ch. 7). The area planted with coffee in Colombia increased by one-third between 1926 and 1928 because of the Brazilian scheme (Ohlin, 1931, pp. 51–4). The wheat fields of the United States and Canada especially were being mechanised, while sugar costs in Cuba were reduced by the introduction of large-scale mills, and in Java by improved varieties of sugar cane.

For many products – for example, copper, tin, petrol, meat, dairy products and wool – the evidence of imbalance pointed the other way.

One of the more successful attempts to raise primary product prices was the Stevenson rubber output restriction scheme of 1922. When the objective of increasing the dollar earnings of the British Empire was achieved, the United States was upset. Its motor industry was the major user of rubber, and by 1919 it was extremely anxious about the supply of raw materials. During and immediately after the war, the allied governments operated a number of control schemes for the benefit of consumers. They were the forerunners of post-war cartel agreements in copper, tin and rubber (Rowe, 1965, ch. 10). A rubber tree takes about seven years to reach a tappable size. Those trees planted in the boom of 1909–10 were coming into production in 1917, a time when the war was causing a shortage of shipping space. A voluntary scheme to restrict output to 75 per cent of plantation company capacity was introduced for twelve months. With the ending of the post-war boom of 1920, the price of rubber more than halved in four months, and voluntary restriction was again introduced. After a year the lack of support for the scheme's continuance led to a compulsory restriction scheme for both Malaya and Ceylon in 1922. By November 1925 the restriction scheme had doubled the 1920 peak price.

A Brazilian scheme in 1923 regulated the flow of coffee on to the market at a rate estimated to stabilise the price independently of good and bad crops. By 1925 the price had risen substantially, mainly because of market conditions rather than the valorisation scheme. Bumper crops occurred in 1927, 1928 and 1929. In that last year the scheme for stockholding ran out of money and prices collapsed. After an abundant harvest in 1926, the Cuban government restricted sugar output and so held up the price of sugar until 1929.

In contrast to rubber, coffee and sugar, governments played no part in the establishment of the international copper cartel in 1926. This was an attempt to gain monopoly profits for the American capital controlling about 75 per cent of world mine production and for the Union Minière du Haut-Katanga in the Belgian Congo, then supplying about 15 per cent of world exports. Within three years, by rationing the European market and restricting output, the cartel doubled copper prices. Buyers reacted with a strike in March 1929. The cartel's price policy undoubtedly encouraged the development of the new copper fields in northern Rhodesia. There had been a similar response to the British rubber control. Production in the Netherlands East Indies expanded to take advantage of the high prices. The formation of a centralised buying pool by the large American consumers in 1926 forced down prices after that date, until the abandonment of the scheme in 1928.

Nevertheless control schemes continued to proliferate. The Canadian wheat pool in 1928, faced with a large crop, decided to hold back part of it to prevent prices falling unduly low, just as the Brazilians had done. These stocks were successfully financed, but although the 1929 crop was smaller, they were not reduced in August 1930. The collapse of wheat prices made further private finance impossible, and the Canadian federal government took over the selling side of the pool's business. In the United States, the fall in wheat prices in 1930 led to heavy government purchases, which kept Chicago prices above the world level, and eliminated exports. Direct acreage control was introduced by the Roosevelt administration in 1933 to maintain prices by restricting output. So effective was the scheme that the drought in 1934 turned the USA into a net importer of wheat for the first time.

Perhaps the most interesting commodity control scheme of the 1930s was for tin. Certainly it had more claimed for it than most such schemes. 'If such machinery had existed in 1929 there would have been no major slump, much higher living standards all round, and no Second World War' (Lewis, 1949, p. 174). The machinery in question was the tin buffer stock of 1934, financed by the producer governments. Tin was to be released on to the market when demand was high and accumulated when it was low, holding the price on its long-term trend or level. If the price trend had been judged correctly, most buying and storage costs could have been covered by the earnings from selling at an earlier or later period. Buyers would have been better off through the reduction in price fluctuations. Unfortunately, the buffer stock was too small (Rowe, 1965, p. 143). When prices rose in 1935, the stock was released and the scheme ended. In mid-1938, with falling prices, a stock was again accumulated and held until the war scramble for raw materials the following year. The sale of all the stock was then quite insufficient to have any effect on prices.

The commodity control schemes of the 1920s generally suffered from an inability to control total output. They therefore tended to raise world output by temporarily raising particular commodity prices, and to introduce more volatility into prices when the schemes collapsed. This often made the forming of correct expectations about future prices very difficult, and exacerbated speculation.[13]

Trade in manufactures

As with primary products, some branches of manufactures trade grew rapidly while others declined. Those countries with a comparative advantage in the dynamic sectors gained accordingly. The United States increased competitiveness the most between 1913 and 1929, in the growth sectors of machinery, transport equipment and chemicals (Maizels, 1963, ch. 8) (see Table 10.4). In 1913 a far higher proportion of manufactured exports of the United States (over 40 per cent) and of Germany (about 35 per cent) than of other industrial countries consisted of the 'expanding' commodity groups. This reflected to a large extent the more dynamic character of the economies of these two countries, and their higher

Table 10.4 Changes in the volume of exports of manufactures, 1913–37 ($bn, constant prices)

	UK	France	Germany	Other western Europe	Canada	USA	India	Japan	Total
1913 exports	1.96	0.79	1.73	0.83	0.04	0.85	0.15	0.15	6.50
Change to 1929 attributable to:									
Market growth	+0.74	+0.30	+0.65	+0.31	+0.02	+0.32	0.06	+0.06	+2.44
diversification	−0.01	−0.03	−0.09	−0.08	+0.02	+0.26	−0.04	−0.02	−
competitiveness	−0.86	−0.01	−0.43	+0.26	+0.24	+0.65	+0.03	+0.14	−
Total	−0.14	+0.25	+0.12	+0.49	+0.27	+1.22	+0.05	+0.18	+2.44
1929 exports (values at 1913 prices)	1.82	1.04	1.85	1.32	0.31	2.07	0.20	0.33	8.94
1929 exports (values at 1955 prices)	5.18	2.76	4.76	3.95	0.93	5.09	0.42	0.76	23.84
market growth	−0.86	−0.46	−0.79	−0.65	−0.15	−0.84	−0.07	−0.13	−3.94
diversification	+0.21	−0.08	−0.10	−0.05	0	−0.04	+0.02	+0.06	−
competitiveness	−0.50	−1.07	−0.63	+0.62	+0.38	−0.13	+0.10	+1.20	−
Total	−1.14	−1.61	−1.52	−0.08	+0.23	−1.02	+0.05	+1.13	−3.94
1937 exports	4.03	1.15	3.24	3.87	1.16	4.08	0.48	1.89	19.90

Source: Maizels (1963).

technical level. At the other extreme, only 5 per cent of Japanese manufactured exports were in the 'expanding' groups in 1913.

Turning to the main markets for imported manufactures, in Canada the rise of the United States' market share was accounted for by America's predominant position as a supplier of the three growth sectors. The rise in Canada's share of the American import market was almost wholly in newsprint and non-ferrous metals. The fall in the UK share of that market was mainly in textiles, but the decline in the German share was spread across all groups, reflecting US import substitution. In the British import market, the rise of the Canadian share was, as in the United States, due to the greater sales of resource-based manufactures, and the sharp fall in the French share was mainly in textiles, this last being associated with the devaluation of sterling in 1931 and the Import Duties Act of 1932. France's traditional exports of silk fabrics were facing a secular decline because of competition from the textiles, especially cotton and rayon (developed in the 1930s particularly by Germany, Italy and Japan to lessen dependence on imported raw cotton). A second reason for the French decline was the overvaluation of the franc during the 1930s.

Before the First World War, Britain was the major supplier of manufactured goods to Japan. Many of the Japanese warships that destroyed the Russian fleet at Port Arthur in 1905 were built on the Tyne, and over half of the Japanese imports of manufactures in 1913 originated in Britain. By 1929 this component had fallen to one-quarter, the decline being taken up by American exports. A similar collapse occurred in the British share of manufactures exported to the southern dominions of Australia, New Zealand and South Africa. The share dropped from three-quarters in 1913 to three-fifths in 1929, though imperial preferences may have reduced the decline during the 1930s (see below).

A similar pattern was traced out by the British share in the Indian market for imported manufactures during the inter-war period. This trend is evidence that India was not a captive market for British goods. The UK did not use its influence over the Indian Tariff Board, which was implementing a policy of selective protection, granting no preferences for British goods during the 1920s and very few during the 1930s (Drummond, 1974, p. 428). Britain supplied 80 per cent of India's imports of manufactures in 1913, a similar proportion to that in the politically autonomous southern dominions. The composition of these exports was different, however, the greater part supplied to India being textiles. The growth of factory production of cotton textiles in India in the inter-war years reduced the level of imports, mainly to Britain's loss. By 1937 British textile exports to India were only one-seventh of their 1913 volume.

Changes in competitiveness

British export decline between 1913 and 1929 was due almost entirely to the loss of market share shown in Table 10.4, conventionally attributed mainly to the overvaluation of sterling. Canada and the smaller industrial countries of western Europe made significant gains in the share of individual markets between 1913 and 1929, and this trend continued into the 1930s. By 1934 the yen had been devalued to 36 per cent of its 1929 gold parity compared with a devaluation to 60–2 per cent of parity for the UK and the USA. Hence Japan achieved a major increase in its market share by 1937 (Maizels, 1963, p. 207). The main losers in 'competitiveness' were France (the franc was overvalued), the UK and Germany, which suffered from the UK's tariff as well as from overvaluation during the 1930s.

Japanese international competitiveness especially in textiles was attenuated during the 1920s by the Japanese ability to finance raw material imports largely from the proceeds of growing sales of raw silk to the United States, and by drawing upon foreign currency holdings which had accumulated during the First World War. Late in the decade, when the price of raw silk collapsed in the USA and Japanese currency holdings were exhausted, Japan became increasingly dependent upon the export of manufactured goods. They would sell in sufficient volume only when the yen depreciated. This competition was erratic and could not be offset by price adjustment in the short run. Even so, there may have been beneficial secondary effects. Between 1929 and 1935, the Dominican Republic, Haiti and Honduras switched from the United States or the UK to Japan as the principal supplier of cotton piece goods, the Japanese share of imports in Haiti rising from 0.1 to 17.7 per cent (League of Nations, 1945, p. 111). Although trade in general was declining and the quantity of most manufactured goods imported into these countries also declined, the low Japanese prices stimulated a rise in cotton piece good imports. The higher demand for raw cotton meant that the pressure on raw material prices was reduced, and the income of primary producers, and their demand for manufactured imports, was enhanced.

By and large, however, it was more difficult during the 1930s for older industrial countries to profit from their sales to primary producing regions, and previous gains were often wiped out. The United Kingdom was particularly dependent on selling manufactures to countries which depended on exporting primary products to all parts of the world. The collapse of international lending and multilateral trade greatly reduced this option.

Multinational companies and trade in manufactures

Large companies which formerly exported to foreign markets were often encouraged by tariff policies such as India's to invest in them during the inter-war years. One of the biggest of these companies was Unilever, a manufacturer of detergents and edible oils. The deterrent to direct investment in most colonies was the limited size of the market. In India the lack of local competition with imports left Unilever reluctant to incur the expense of investment until the 1920s, and then it was a misjudgement. The post-war depression pulled down sales, and profits slumped between 1921 and 1922. William Lever wrongly attributed his firm's misfortunes to Gandhi's call for a boycott of imports from the UK (Fieldhouse, 1978, ch. 4). So in 1922 Lever acquired a controlling interest in two local companies. The boycott ceased and, because soap was not tariff protected, soap imports rose to an all-time peak in 1928. Production in Unilever's Calcutta factory remained small, limited to local not Unilever brands. The boycott revived during the slump, soap received a 25 per cent tariff in 1931, and sales of imported soap almost halved between 1929 and 1932. Early in 1933, these events persuaded Unilever to build a substantial factory in Bombay, alongside another factory making ghee substitute, which had been begun the year before on the same grounds – that the established market would otherwise be lost.

American exports in the dynamic sector of transport equipment were particularly prone to be replaced by multinational production abroad. Two of the largest US multinationals, General Motors and Ford, originated in the motor industry. In the first quarter of the twentieth century, the US motor industry built up a massive technological lead over the European industries, at least as judged by output per man, despite the invention of the petrol-driven motor vehicle in Europe in the 1860s and 1870s. General Motors and Ford, therefore, possessed an established advantage over their European rivals when they began production overseas. As with Unilever's decision to produce in the Indian market, it was tariff barriers that motivated setting up production plants rather than exporting, although Ford assembled in the UK before tariff protection (Foreman-Peck, 1982).

Almost from the beginning of the industry, the bulkiness of motor vehicles justified assembly plants overseas when the market was large enough. Shipping vehicles 'completely knocked down' was so much cheaper. By the end of 1926 General Motors (GM) had invested $30 million in plant, equipment inventories and working capital (overseas sales were $100 million). The following year GM increased this sum with the acquisition of Opel, which already held 40 per cent of the German market. GM was more sensitive than Ford to the trend of economic nationalism, attempting to acquire established national names as façades for their products. This GM had already achieved in the UK with the purchase of Vauxhall and Bedford. In Australia it formed an alliance with the indigenous body-building firm Holden. Even in India, General Motors (and Ford) established subsidiaries to get round the tariff barrier and to receive advantages in Indian government orders. To the chagrin of the British Board of Trade, Morris, the largest British motor manufacturer, declined to do the same (Foreman-Peck, 1986). Ford was initially a larger investor overseas than General Motors. By 1923 Ford's assets in foreign plants totalled almost $50 million, but these assets were almost entirely paid for from overseas profits. American multinational motor vehicle production abroad in 1928 at 309,000 in total was greater than the output of any single national motor industry outside the United States.

Not all states that wanted to use their technology for import substitution were favourably inclined to the permanent presence of American firms. Ford's exports to the USSR were sufficient for the company to boast in 1927 that 85 per cent of the trucks (lorries) and tractors in the country were Fords (Nevins and Hill, 1957, appendix 1). Two years later Ford contracted to supply the technology for a factory at Nizhni Novgorod (Gorki) to produce 100,000 units a year. In return the Russian government agreed to buy a total of 72,000 Ford units on a four-year schedule. What happened to this contract provides an insight into why multinationals have been unwilling to take out licensing agreements in the face of economic nationalism. Ford lost over $21 million on the $18 million sales, and Ford officers thought that the products were below standard. Sub-standard Ford products could have adversely affected sales elsewhere had the Soviet economy been less closed to international relations.

Commercial policy and the Depression of 1929

Ultimately far more important than multinational production for the trade in manufactures were the 'beggar-my-neighbour' commercial policies of the 1930s that contracted income and employment. The decline in agricultural prices during the 1920s in the United States tempted Hoover, during his presidential campaign of 1928, to help farmers by proposing to raise tariffs. High tariffs were anyway a bargaining advantage in a most favoured nation (MFN) system without discrimination between countries according to whether or not they offered reciprocal concessions. There was more to offer to give up. Political lobbying extended the promised Smoot–Hawley Tariff Act of 1930 to non-agricultural products. Although some countries had already raised tariffs, the passage of the Act caused a wave of foreign protest and reprisals by sixty countries. France retaliated with quotas, Spain withdrew MFN status from the USA and imposed tariffs of up to 700 per cent on US goods, while Switzerland encouraged informal boycotts. The timing of the Act was most unfortunate. In the middle of a world depression it exported unemployment, but the direct impact was small: for the United States the tariff and retaliation cost only 0.44 per cent of national income (Brunner, 1985; Kindleberger, 1973, pp. 133–4). What mattered more was the indirect effect, signalling that the United States was not taking on the leadership of the world economy that its economic importance required.

Between 1929 and 1931 agricultural prices fell by about one-half. Wheat export prices alone dropped by more in the twelve months after December 1929 (Rowe, 1965, p. 85). With specific duties in force – that is, duties fixed in money terms – the actual tariff percentage on agricultural products rose as prices declined. Nominal tariff levels and trade restrictions increased markedly, raising trade barriers even higher. Just as the protectionist Meline Tariff of 1892 reversed the increase in agricultural imports of the 1870s and 1880s, so the French government again took drastic action in 1931 when the proportion of agricultural imports rose above 25–30 per cent of the total, threatening agricultural interests. From 22.8 per cent in 1928–30, agricultural imports as a proportion of the total rose to 34.5 per cent in 1930–1 and, after the introduction of quotas, declined to 25.3 per cent in 1937 (Weiller, 1971).

Quotas were preferred to tariffs because they gave better control of imports in the face of the massive price declines. Although they better protected internal economic activity, the main reason for the introduction of quotas (their use as an instrument for bargaining with other states) was less satisfactory (Heusser, 1939). Whereas in a tariff war between two countries the welfare of the country initiating the war may be improved in the end, this does not happen in a quota war (Rodrigues, 1981). Successive retaliations to the other country's quota limitations lead asymptotically to the complete elimination of all trade between the two states. Italy was the first important country to use quotas as reprisals, on French exports of wines, liquors, perfumes and soaps (Heusser, 1939, p. 43). The French retaliated in July 1932 with limitations on Italian exports of fruit and vegetables. The Italians replied with restrictions on their imports of French cars and clothing. In June 1933 the French reduced the Italian meat quotas. A week later French cotton, yarn, lace, tools, machinery and hides exports to Italy were further cut.

The vast amount of administration involved in the allocation of quotas at first required the delegation of powers to distribute import licences. In Greece the Chambers of Commerce distributed them, while in France 'Inter-Professional Committees' of traders and producers were consulted about the distribution.

Some co-operation facilitated the formation of import monopolies, which in order to maximise profits could sometimes find it advantageous to import less than the full quota, limiting trade even further (Heusser, 1939, pp. 52, 128). In Germany imports were restricted by tariff increases, quotas and exchange control, introduced originally to prevent capital flight. The foreign exchange which importers could buy was reduced by an amount deemed to correspond to the decrease in foreign prices and domestic incomes since 1931. Regional Chambers of Commerce were consulted about currency quotas by the Ministry of Finance acting for the Reichsbank.

Before the Great Depression, Europe was divided into two main trading blocs: the industrial countries of Europe trading with each other, and industrial Europe trading with agrarian 'border Europe' (Liepmann, 1938, pp. 343–4). The UK's departure from the gold standard (while France, Switzerland, the Netherlands, Italy and Belgium did not devalue) and its new tariff barrier of 1931/2 reduced its usefulness as a market in the first bloc. The British emergency Tariff Act 1931 allowed *ad valorem* duties of up to 100 per cent. The Import Duties Act 1932 did not apply to empire goods at that time, or to important industrial raw materials or food. Otherwise a general 10 per cent tariff was imposed and an Import Duties Advisory Committee was appointed to suggest alterations (Richardson, 1938, p. 129). The recommendations of the Committee, in contrast to US tariff-making bodies, do not seem to have been greatly influenced by the lobbying of interested industrial groups. (Capie, 1981). After alterations agreed at the Empire Conference at Ottawa, about one-quarter of imports remained duty free (though many were restricted by other methods), on one-half duties of 10–20 per cent were levied, 8 per cent paid over 20 per cent, and the remainder were subject to the old duties of the 1920s.

Similarly drastic Italian and Czech protection also reduced the viability of these countries as markets in the first bloc. Among the remaining states, industrial exports, which were still liable to moderate duties, continued to sell fairly well. Any substantial agricultural exports of these countries had already been largely eliminated by 1931. The connections between central Europe (Germany, Austria and Czechoslovakia) and 'border Europe' were threatened by the central agricultural protectionism and by the industrial protectionism of eastern and south-eastern Europe and Spain. The trade of northern and north-eastern countries and Holland with Germany and the UK, based mainly upon the exchange of timber and timber products, dairy produce and meat for industrial products, was well maintained up to 1931, duties on these articles remaining moderate. Trade between the Mediterranean border states and industrial Europe (England, Germany and France) was also affected little because the major export, fruit, was less badly hit by duties.

The worst happened between 1932 and 1935 when quota and tariff barriers escalated. By 1935 world trade had fallen to one-third of the 1929 level in gold dollars, and even measured in paper pounds trade had lost 45 per cent of the 1929 value (Hardach, 1977, chs. 2–6). Proposals for tariff reductions at the Economic

Conference at Geneva in 1930 failed because the bilateral negotiations violated the 'most favoured nation clause' of the principal commercial treaties.[14] A proposed Austro-German customs union failed because of French and Italian objections, based nominally on this clause, but really motivated by political considerations (as was the proposal). Earlier commercial treaties were renounced, and exchange control and bilateral agreements replaced free-market trade. Barter agreements became a major means of conducting trade: in December 1932, for example, Hungary agreed to exchange eggs and pigs for Czech coal. The US Reciprocal Trade Agreements Amendment 1934 was merely a tactic to reduce foreign retaliation while maintaining high tariffs (Conybeare, 1987, pp. 250–7).

The general increase of trade restrictions after 1929 could have been beneficial: by cutting imports they might have arrested the secondary decline in domestic income and employment, and so prevented production falling in the same proportion as exports (Lewis, 1949, p. 10). But a test of this hypothesis, by modelling the interaction of foreign trade and national income among the European economies, refuted it (Friedman, 1978). For the UK, tariff protection was rather helpful, raising 1938 national income by perhaps 2.3 per cent (Capie, 1983, p. 107; Kitson and Solomou, 1990; Foreman-Peck et al., 1992). Generally, however, the effects of lower levels of trade on income swamped the tendency of national policies to stimulate domestic economic activity. This conclusion is strengthened when industrial and agricultural European economies are compared, and when the effects of alternative policy measures are considered.

After the deluge

Those European nations with empires – the UK, France and Holland – could respond to these policies of trade destruction by tightening their commercial links with their overseas territories, as they had already been doing. France's exports to its empire rose from 12.4 per cent of the total in 1913 to 15.2 per cent in 1928–30. The British political move towards closer economic ties with the empire can be dated to 1917 when Britain committed itself to encourage empire settlement and to introduce preferential tariffs. Significantly, the initiative for the policy came from the Prime Minister of New Zealand, and subsequent moves towards closer union also originated from the dominions. A Canadian lead of 1929 gave rise to the Imperial Economic Conference the following year, which in turn led to the Ottawa Conference of 1932 (Drummond, 1974, pp. 25–6, 31). The conferees extended to each other reciprocal tariff preferences intended to increase trade between empire countries, and proclaimed that the consequent increase in empire purchasing power would also expand the trade of the rest of the world.

Throughout the inter-war period, the empire drew a decreasing proportion of its imports from the UK. However, in 1932–5 the proportion was slightly higher than might be expected from the trend. If this was the result of Ottawa then the agreements raised British exports by 3.5 per cent in 1933 and by 5.4 per cent in

1937. With a multiplier of 2, output and employment might have been 0.5 and 1 per cent higher in the two years. British imports by a similar method were 7.2 and 10.3 per cent higher in 1933 and 1937 as a consequence (Drummond, 1974, pp. 286–7). Much of the trade diversion did not involve a loss of income or employment to foreign countries because, when empire countries sold more primary products to the UK, they sold fewer in foreign markets, allowing room for important substitution to raise foreign income and employment. In any case, empire imports of foreign goods rose by three-quarters in value terms between 1933 and 1937, and British foreign imports increased by half.

In view of the measures taken to reinforce empire economic ties in the face of increasing economic nationalism in the 1930s, it is not surprising to find that the share of trade and investment of imperial countries in their colonies was higher than the share of the countries in all LDC trade and investment by 1938. The foreign direct investment share was considerably higher than the trade ratios. In Madagascar, Indo-China and British Crown colonies, foreign investment in extractive industries was banned. The investor countries, the UK and France, also imposed restrictions on direct investment in non-associated countries. In British, American, Belgian and Dutch colonial territories, few outright limitations were placed upon investment from third countries in non-extractive sectors (Svedberg, 1981). After decolonisation in 1967, compared with 1938, the British share in foreign direct investment in ex-dependencies actually increased, while in French former colonies, the French share fell slightly. One inference that might be drawn is that there was no imperial distortion of investment in British colonies during the 1930s. Two other possibilities remain however: influence was later retained through 'neo-colonialism', (or inertia), or exchange controls in former colonies after 1967 precluded a search for new opportunities by the investors of the former imperial power, who instead ploughed back their retained profits.

Those countries, such as Argentina and Denmark, outside the British Empire but for which the UK was an important market, negotiated the best trade agreements they could without the leverage of Commonwealth membership.[15] Agricultural 'border Europe' was drawn further into the German sphere of influence to form another quasi-imperial grouping. Under Schacht's 'New Plan' of 1934, twenty-five supervisory centres allocated foreign exchange for important transactions approved by the state. Bilateral clearing agreements were concluded especially with central Europe and South America, in which German purchases were credited against offset purchases by foreigners in German markets. By the spring of 1938 more than half of Germany's foreign trade was conducted through these agreements.

If Germany used its market power to exploit agrarian border Europe through these arrangements, its terms of trade should have risen; instead they fell. A monopoly/monopsony framework is not appropriate in the conditions of widespread unemployment of Germany and south-eastern Europe (Neal, 1979).[16] The foreign goods that Germany bought from Europe were 20–40 per cent above the world market prices, and on average Germany paid more for the same commodity from a clearing agreement country. Neither did Germany secure higher prices for its exports; even when it reached full employment, German purchasing agencies offset the effects of higher prices by offering Yugoslavia very generous deferred purchase terms with almost negligible down-payments. It is possible that future economic or political gains were being purchased at the expense of present gains, but then this investment was undertaken for at least five years and the monopoly position was never exploited.

Most probably it was Germany's desire to obtain foreign exchange by any means possible that was responsible for these generous arrangements. Effectively, a two-tier exchange rate was introduced, with German ASKI (*Auslander-SonderKonten für Inlandszuklungen*) marks received by an exporter to Germany being sold to an importer from Germany at a discount, so that, although the overvalued gold mark was still used to make payment on foreign debt, the ASKI mark floated in these 'private organisation agreements'.

More important for the states of south-eastern Europe were the clearing agreements. Germans exported little but imported much, so that Germany accumulated large debts in blocked marks in these economies, which exporters tried to sell in their own countries. If the central bank supported the exchange rate agreed upon in the bilateral agreement, it paid out domestic currency at that rate for claims on blocked marks. In one sense this was an extension of credit to Germany and raised the domestic price level of employment (the Hungarian, Greek and Bulgarian cases). Alternatively, the central bank could refuse to buy blocked marks from exporters until they received a request for the marks from domestic importers of German goods (as in Romania and Yugoslavia). This amounted to a devaluation of the mark, or a revaluation of the domestic currency, which had no expansionary effects.

The increasing politicisation of international trade associated with these bilateral agreements was matched by the first important collective use of trade sanctions as a political weapon during the 1930s. On 11 October 1935 fifty of the fifty-four League of Nations members agreed to apply sanctions against Italy for its invasion of Ethiopia (Le Roy Bennett, 1980, ch. 2). The members stopped arms sales to Italy but not to Ethiopia, cut off credit to the Italian governments and to Italian businesses, prohibited imports from Italy and placed an embargo on the export to Italy of strategic raw materials (including rubber, tin, aluminium and manganese), which were substantially controlled by the boycotting states. Within a few months Italian gold reserves were dangerously low and the lira was devalued. Italian foreign trade was greatly reduced. The measures were insufficient to prevent the conquest of Ethiopia, however, and more effective sanctions, such as embargoes on food, coal, steel and oil, and refusal of Italian access to the Suez Canal, were rejected. Two months after Mussolini's announcement of victory, sanctions were abandoned in July 1936 at the initiative of the British. The British wished to secure Italian support in resisting Germany in Europe, following Hitler's remilitarisation of the Rhineland in 1936.

Sanctions were prevented from achieving the success of the commercially motivated trade quotas because political stability in Europe was judged of overriding importance. The history of Europe in the inter-war period suggests that the commercial policies pursued were equally unconducive to political harmony, encouraging the building of 'siege economies', aggressive foreign policies and restrictions on personal freedom.

Summary and conclusion

The First World War coupled with the Versailles Treaty proliferated the traps into which economic relations could fall. The multiplication of independent policy-makers with divergent interests in Europe and political instability held back recovery for Europe as a whole. Meanwhile the United States developed rapidly, becoming increasingly self-sufficient in any products that Europe could supply. This required an adjustment in European trading relations, and in those of the temperate zone primary commodity exporters outside Europe and the United States. Such an adjustment, however, was not made. Had America pursued more expansionary monetary or fiscal policies, or revalued or reduced tariffs, the necessity for adjustment would have been greatly reduced. But equally, had Europe avoided war the imbalance between the two sub-continents would have been much smaller.

Economic instability, unemployment and the decline of trade conferred a poor reputation on the inter-war years. Yet many countries substantially increased real incomes per head over pre-war levels. They did so despite tight restrictions on trade and factor flows by finding substitute channels of economic activity. Incomes in most cases would have been higher in an unchanging economic environment if trade had been less constrained. But in the world economy between the two world wars, subject to violent shocks, adjustment to which inevitably took time, transitional arrangements mattered more for well-being than those policies that maximised welfare in the long run. By suddenly cutting off foreign markets, the trade collapse from 1929 caused great hardship and an increase in political extremism.

Japan suffered as much as most countries from the trade effects of the Depression, yet Japanese exports grew faster than national income, which in turn expanded more rapidly than those of other economies. During the 1930s this export growth was stimulated by the pronounced depreciation of the yen, all the greater because of Japanese dependence on imported raw materials. Japanese growth was to some extent at the expense of employment in other economies, whose policies and economies were less quick to change.

The now massive US economy was self-sufficient in all but a few raw materials. About the time of the war, the USA ceased to be a net exporter of natural resources as measured by the resource content of its trade, and became a net importer. Rapid 'motorisation' of the USA demanded huge quantities of oil and rubber, precipitating the oil crisis of the early 1920s, and creating the opportunity for the rubber output restriction scheme to push up rubber prices markedly. Other primary commodities, especially cereals, were not so favoured, and wheat-exporting countries, such as Canada, accordingly stagnated. Increased agricultural protection by the industrial countries, unwilling to accept the structural readjustments required by an expanding international division of labour, further constrained international trade in food. Protection became very heavy from 1930 and added to the severity of the Depression. Quotas, rather than tariffs, were especially destructive of trade once quotas became instruments of economic warfare and bargaining.

Price control schemes to maintain incomes of producers of some primary commodities countered price weakness. Although successful for a few years, these schemes usually stimulated the expansion of supply elsewhere and the accumulation of stocks during the 1920s. Consequently, when the stocks could no longer be financed during the Depression, the fall in prices and incomes of the primary commodity exporters was the more acute. During the 1930s, control schemes were reintroduced and extended, generally with government support, but only the prices of tin, tea and rubber were decisively influenced.

The volume of trade in manufactured goods declined more than the volume of primary products. Production of manufactures was more 'footloose' in response to barriers to trade. Companies which formerly exported to a large market often preferred to manufacture there rather than be excluded by tariffs and exchange controls. Also the industrialisation of lower-income countries such as India and Brazil led to import substitution, which was not offset in the 1930s by the increased imports generated by higher incomes. The greatest transformation of manufactures trade since the pre-war era was the removal of the German challenge to British supremacy. But that was overshadowed by the unprecedented achievements of United States manufacturing. By the end of the 1920s, the United States was exporting more manufactures than the UK, largely by virtue of its comparative advantage in the rapidly growing sectors of chemicals, machinery and transport equipment. British industrial structure, heavily weighted towards textiles, suffered from the import substitution and international competition of the newly industrialising countries.

The rise in trade barriers during the Depression after 1929 prompted a tightening of economic ties within empires. The 1932 Ottawa Agreement probably halted the decline in Britain's share of the dominion markets temporarily, and raised UK imports from the empire possibly by as much as 10 per cent of total imports in 1937. Although creating difficulties with outsiders, especially the Americans, the net result of the agreement was probably expansionary for the world as a whole, because the higher British import of primary products meant that the empire countries exported less to foreign markets than they would otherwise have done. This created room for import substitution to raise foreign incomes and employment. Between countries outside empires, bilateral agreements were often the alternative. Germany established bilateral trading agreements with central Europe and South America which mattered a great deal to those economies. The effects of the agreements depended on the central bank policy of the dependent states. Some countries, notably Hungary, Greece and Bulgaria, benefited substantially from the boost to aggregate demand arising from their exports. Other economies, Romania and Yugoslavia, did not gain because their banks failed to expand the money supply.

International economic policy bolstered the system of alliances in Europe before 1914, and thereby contributed to the increase in tensions that eventually resulted in war. Similarly, the abandonment of the market for government-controlled international economic relations in the inter-war years also became a source of friction.

Notes

1. The report of the British Cunliffe Committee of 1919 exemplifies this tendency, with its advocacy of a return to the pre-war gold standard at the earliest possible opportunity. The majority report of the Gold Delegation of the Financial Committee of the League of Nations in 1932 emphasised the changes in economic structure caused by the war. See also Keynes (1920, p. 10) for nostalgia.
2. It would be out of place to discuss in detail what were the proximate causes of the war, but Taylor believes the most responsible person was Schlieffen (already dead by 1914), who had prepared Germany's war contingency plans, which made certain that once Germany began to mobilise, war was inevitable, and that any war in Europe would be a general war (Taylor, 1967).
3. Mendershausen (1941, p. 260) cites a world total mortality from influenza of 15 million. Adding this to civilian deaths, he makes a civilian total of 28 million, and a civilian and military total of 41 million.
4. Carr was present at the Versailles negotiations in a diplomatic capacity.
5. The Royal Institute of International Affairs recognised the political importance of Arndt's interpretation and added an appendix to his book maintaining that it was the strangling of market forces by the post-war boundaries settlement, as described above, and war debts and reparations (see Chapter 11) that were the root cause of inter-war economic difficulties.
6. Trade, in these ratios, is the sum of exports and imports.
7. Subsequent, more detailed analysis by Maizels (1963) confirmed these conclusions.
8. The World Economic Conference of 1927 was unanimous in condemning tariffs.
9. The relationship between the relevant elasticities which must be fulfilled if a revaluation of the currency is to diminish the current account surplus is the Marshall–Lerner condition. If export and import supply are perfectly elastic then the condition is that the sum of the absolute value of the price elasticities of demand for imports and for exports should exceed one, a condition satisfied according to Falkus (1971) by the US and UK elasticities. Because it would have only increased American imports, not contracted American exports, a 10 per cent reduction in the United States tariff would have had a smaller effect than a 10 per cent revaluation.
10. Elasticity pessimism – a belief in the low responsiveness of international economic relations to relative price changes – was widespread between the world wars (see, for example, Eichengreen, 1981, p. 25). That such pessimism has not been confirmed in more recent years suggests that the estimates of low price elasticities were due to faulty techniques.

11. It should be noted that the valuation of output of a centrally planned economy such as the inter-war USSR creates special difficulties for the comparison.
12. The industrial countries of 1913 were the USA, Britain, Belgium, France, Germany, Austria, Hungary, the Netherlands, Italy, Sweden, Switzerland and Japan, in this calculation.
13. Davis (1975) believes that the irrationality of expectations, in their optimism to 1929, and their pessimism in the 1930s, have been underestimated in explanations of the inter-war economy.
14. The MFN clause stated that any tariff reduction granted to one state should equally be granted to all others.
15. The Runciman–Roca Agreement between the UK and Argentina in 1933 is discussed by Gravil and Rooth (1978).
16. For an econometric model of the monopoly/monopsony approach, see Friedman (1976).

11 The disintegration of the gold standard

International monetary relations between 1919 and 1939 were profoundly influenced by economic policies adopted during the First World War. The two keys were budgetary strategies and intensity of national mobilisations. These rearranged international costs and prices, and redistributed debts and assets between nations. Then with peace came social upheavals that put unprecedented additional pressures on fiscal and monetary policies. Despite a widespread desire to return to the 'normality' of the fixed exchange rate gold standard immediately after the war, most former belligerents were unable to do so and instead floated their exchange rates. The political problems, budget deficits and inflation that necessitated floating rates form the subject matter of the first section of the chapter.

In the UK, the keystone of the pre-1914 system, the objective was not only to return to the gold standard, but also to resume the pre-war parity. The political and economic reasons for the decision to revert to a fixed rate, and the success of the floating regimes in facilitating the return to gold, are discussed in the following section. Towards the end of the 1920s most of the world had adopted a form of gold standard, but one in which, it has frequently been asserted, exchange rates were inappropriate to prevailing costs and prices. Moreover, central banks supposedly pursued policies which prevented the classic gold standard adjustment. The UK's departure from the gold standard in 1931 was the culmination of their policies superimposed on structural distortions. These allegations form the third topic of the chapter.

The depression that began in 1929 was the severest that the world has ever seen. Although the primary product exporters were clearly heading for recession, the severity of the downturn depended on the US economic collapse. The role of the gold standard, and the world economic environment of the 1920s in transmitting this depression and its reverberations, are analysed in the fourth section.

The world economy which emerged in 1933 was divided into three monetary blocs: (a) a sterling area centred on the United Kingdom and (b) a dollar area led by the United States, both floating against the rest of the world, and (c) the countries that remained on the gold standard, to a greater or lesser extent with the help of exchange controls and import quotas. These were the years of competitive depreciation and international monetary chaos, according to the later architects of the Bretton Woods system. The final section assesses the truth of that judgement.

Floating exchange rates and post-war inflation

During the war, the belligerent states maintained the gold standard's façade, but the monetary edifice was gutted. Allied exchange rates were pegged at their pre-war gold standard parities by official purchases and sales of the currencies. The Russian economy was least able to withstand the strains of war. Thus the Russian budgetary explosion and exchange rate collapse were among the earlier warnings to contemporaries of the dire consequences of the departure from fiscal rectitude (policy-makers' perceptions of the post-war inflations played a key role in the events of the later 1920s). As a proportion of total government expenditure, the Russian budget deficit rose rapidly during the war. In 1915, 31 per cent, and in 1916 one-quarter, of the deficit was financed by an increase in the money supply (Katzenellenbaum, 1925). This rate of monetary expansion was inconsistent with stable prices. Prices rose by 20–30 per cent per annum in 1914 and 1915, while the annual monetary expansion was around 80–90 per cent. From 1916 prices began to soar faster that the money supply for two reasons. First, and primarily, the Russian public was increasingly unwilling to hold money which was declining in value so rapidly. Second, the volume of goods being placed on the market was falling.

With a greater average income per head and more developed capital markets, the United Kingdom did not have the same budgetary problems as Russia. None the less, the British government was so reluctant to impose additional taxes to pay for the war that revenue covered only 27 per cent of expenditure in the financial year 1915/16 (Morgan, 1952). Prices therefore climbed, retail prices perhaps doubling by 1918. During the immediate post-war monetary expansion until 1920, wholesale prices increased by about a third, and retail prices rose by approximately one-quarter.

Japan experienced inflation transmitted from the industrial economies. Japanese foreign exchange balances, virtually all held as sterling in London, counted as part of the specie reserve against the note issue. During the war, as claims on foreign exchange accumulated from exports, the Bank of Japan increased the note issue commensurately. Wholesale prices more than tripled by 1920 (Inouye, 1931).

Exchange rates could remain unchanged with countries such as the United States, which experienced a lower rate of inflation, only if they were supported by governments, or if there were restrictions on the international movement of goods and gold. The British government spent over 2 billion dollars supporting the dollar–sterling exchange rate after 1915 (Moggridge, 1972, p. 17). Once peace was restored, the even greater level of official intervention required to maintain 1914 parities was clearly impossible, and the pound was allowed to find its own level on the foreign exchanges. It fell against the dollar, reaching a low of $3.378 in 1920.

The collapse of the German exchange rate as inflation soared was the most dramatic and influential immediate post-war-monetary event. In contrast to Britain, there was no attempt to reduce the budget deficit after the war. A coalition between industrialists and workers maintained government expenditure in 1919 (Balderston, 1981). Transfer payments to the wounded, to dependants of the dead and to the unemployed, and payments from the Bismarckian social insurance institutions were all raised. Armaments orders were prolonged, and compensation for assets lost through the treaty consumed large sums.

An inefficient tax system, public enterprise underpricing and official wishful thinking in the face of reparations payments dragged the economy into hyperinflation. Tax revenues remained low because the states, rather than the central government (which faced the greatest demands on its expenditure), controlled direct taxation. Political disturbances in any case hindered the efficient collection of taxes. Revenue requirements were inflated by the massive subsidies that public enterprises needed. The mark exchange rate depreciated, internal prices rose, the government increased the note issue and the exchange rate fell further.

The official German position was that the unfavourable balance of payments was the cause of the inflation that accelerated during the early 1920s; the government budget deficit responded passively. Certainly a deterioration of the balance of payments was on occasion associated with the payment of war reparations, hence the attraction of the official view of inflation. However, inflationary government financial policies always preceded depreciation. Towards the end of October 1923, the special paper used for notes occupied thirty paper mills. On 25 October, notes to the total of 120,000,000 billion (12×10^{16}) paper marks had been stamped, but demand during the day reached 10^{18} marks (Bresciani-Turoni, 1937; see also Graham, 1930).[1]

Speculation probably did not destabilise the mark because of the dispersion of views about future changes in its value. In Germany, as in Russia earlier and France later, the normal demand for foreign exchange to make payments was supplemented by a demand for foreign exchange as a stable store of value for savings. Many foreigners, unable to believe that the mark could permanently depreciate so much, bought the currency expecting it to rise in value. In the middle of 1922 foreigners held about 60,000 billion paper marks, on which they must have lost approximately 5,000 billion gold marks as the exchange rate fell, a sum triple that paid by Germany in foreign exchange for reparations (Bresciani-Turoni, 1937, p. 252). After stabilisation in 1924, the German profit from the sale of marks between 1918 and 1924 was estimated at between 7.6 and 8.7 thousand billion gold marks. Against this profit must be offset an unknown sum for the purchase by foreigners of German assets such as houses, under the stimulus of the low mark.

French budgetary policy and the course of the franc exchange rate can be explained by the French maxim that the Germans would pay. Conveniently running a massive budget deficit, supposedly to be financed by German reparations, could at first disguise French domestic political conflict (Moggridge, 1989; Aliber, 1962; Wolfe, 1951; Eichengreen, 1982). Only when these expectations became obviously unrealistic in the second half of 1922 was there unusual volatility of the franc exchange rate, apparently unjustified by considerations of relative purchasing power. It was especially the experience of the floating franc that bolstered the view that speculation was destabilising and that fixed exchange rates would have been more satisfactory. During the winter of 1923–4, the franc lost nearly half its dollar value in less than twenty weeks. In a second speculative attack in the summer of 1926, the franc fell another 20 per cent.

These movements in fact may easily be interpreted as reactions by speculators to new information about future equilibrium exchange rates, and an anticipation of the franc's inevitable move towards that new equilibrium. From June 1922 France broke with the UK and its other former allies over the flexibility of the schedule for German reparations laid down at the London Conference of May 1921. As the realisation dawned that reparations on the scale previously expected were no longer feasible, the prospect of sound French public finance receded.

In 1923 the French occupied the Ruhr to enforce payment of reparations. The Germans responded with passive resistance financed by their government. The German hyperinflation that followed contributed to the depreciation of the franc by decreasing the likelihood that reparations payments could be successfully extracted from the Germans. The imposition of new taxes in March 1924 and a pledge on the part of the Bank of France's gold reserves allowed a foreign loan to be raised for supporting the franc. Facing difficulties raising taxes to cover the budget deficit, the left-wing coalitions that governed from May 1924 to July 1926 proposed a capital levy. Together with the disclosure of subterfuge in the Bank of France's 1924 balance sheet, this threat to capital frightened French bondholders into selling for foreign currency assets. The sales drove the franc exchange rate further down (Schmid, 1974).

During 1925 the easier course was followed of four times raising the limit of the Bank of France's advances to the government, and three times increasing the note issue. The franc continued to fall. Eventually the moderate left-wing deputies concluded that the government-induced inflation was becoming harmful to their supporters, who could no longer be protected by fiscal benefits. Only then were they prepared to cede power to a conservative-led coalition which required sacrifices in order to eliminate inflation and stabilize the exchange rate. In July 1926 Poincaré was returned and announced a sound fiscal policy. The 32 per cent depreciation between May and July cannot be explained by fiscal and monetary events, and must therefore be attributed to destabilising speculation (Sicsic, 1992). However, the moderate left were able to use their remaining political influence to prevent any substantial revaluation of the franc. The relatively low value of the franc (about 122 to the pound) established at the end of 1926 was a contributory factor to the international monetary crisis of the early 1930s, yet international considerations played no part in the choice of exchange rate. Contemporaries did not agree that the franc was overvalued, bearing in mind the possibility of speculative capital flows. In view of the difficulty of achieving agreement on domestic issues within the French political system, such an outcome is perhaps not unexpected.

Floating exchange rates and sterling's return to gold

Amid this monetary disorder, policy-makers at international conferences in Brussels (1920) and Genoa (1922) were determined to anchor their economies to the gold standard. In effect that also meant returning at the pre-war parity, as the Swedes showed in April 1924, and the British in the following year. In so doing they hoped they would restore pre-war prosperity. The post-war inflations discredited the floating exchange rates with which they were associated, and so ruled out continued adherence to a 'managed' currency (Moggridge, 1972, pp. 85–6). In any case, for the UK there were strong political pressures from the dominions. At the Imperial Economic Conference of 1923 they recommended the early return of the UK and the sterling area to an effective gold standard (Pressnell, 1978). The southern dominions objected to holding sterling for fear of depreciation, and were impatient to stabilise on gold. South Africa threatened to return independently.

In deciding when to return, as well as the appropriateness of the pre-war rate, purchasing power parity calculations then and subsequently played an important part. These calculations, and the role of speculation, have formed much of the basis of the condemnation of the floating rates of the 1920s. Unpegged exchange rates allegedly failed to reduce the international disparities in prices and costs that had opened up during the war. Supposedly they also encouraged speculative capital flows, which were responsible both for eventual stabilisation at rates that overvalued the pound and undervalued the franc, and for the collapse of the mark (Triffin, 1968, ch. 2).[2]

Relative prices were certainly not an infallible guide to exchange rates in the post-war world, although they offered a long-term indication. During the massive Russian inflation from 1919, the exchange rate did not match the fall in internal purchasing power (Katzenellenbaum, 1925). By mid-1920 in Moscow, the rouble–sterling rate was about twice as high as warranted by the relative purchasing powers of the two currencies. From the end of 1920, the fall in the exchange rate exceeded the decline in the internal value of money. By 1 January 1922 the purchasing power parity of the pre-war had almost been achieved at £1 = 1.25 million roubles. This alignment of the exchange rate with domestic and foreign prices was brought about by more financial arbitrage as fighting diminished, by the relaxation of exchange controls, and by falls in price levels in other countries.

Similarly, the British-Japanese wholesale price indices of purchasing power parity indicated that a fall in the sterling–yen exchange rate should have occurred between 1920 and 1922, and thereafter should have been stable. The fall in 1925 allowed the exchange rate to reach the level indicated by relative prices that would have been appropriate in 1922 (Inouye, 1931). Inouye, a former Governor of the Bank of Japan and Minister of Finance, explained that the yen was overvalued in the early 1920s because of the government's ability to spend its sterling balances accumulated during the war. Once these balances were exhausted, as they were by the costs of reconstruction after the 1923 Tokyo earthquake, the yen was bound to fall, unless Japanese prices were reduced.

While war and the spending of sterling balances caused temporary divergences of the Russian and Japanese floating exchange rates from purchasing power parity, national budgetary policies, as already shown, mainly explained the deviation of the French franc and the German mark from levels implied by relative prices. When responsible domestic policies were pursued, floating rates generally followed movements in relative purchasing powers and did not fluctuate widely. Floating rates also permitted a degree of freedom for monetary policy. Finland maintained a fairly stable internal purchasing power, while abandoning any attempt to stabilise the exchange rate (Ohlin, 1931, p. 25). The wholesale prices index of Finland, unlike that of almost any other nation, remained unchanged throughout the 1920s.

The wholesale price indices used in purchasing power parity calculations are often said to be suspect because they measure the prices of internationally traded goods, whose prices in domestic currency will *reflect* changes in foreign exchange rates reasonably accurately (see, for example, Moggridge, 1972, p. 101). The prices of non-traded domestic goods are greatly understated in the wholesale price indices, yet they are also important for *determining* the equilibrium exchange rate. However, including real wages as a measure of non-trade prices reinforces conclusions obtained merely with wholesale indices. Ninety-six per cent of the variation of the dollar–sterling exchange rate between February 1921 and May 1925 could be explained by relative prices and wages, relative money stocks, incomes and interest rates (Clements and Frenkel, 1980). This evidence is consistent with relative prices determining the exchange rate.

Only between October 1923 and July 1924 did the sterling exchange rate depart from the level expected according to fundamental current economic factors (Hodgson, 1972). During these months fears became widespread, with some justification, that the British government (the first Labour government) would abandon the deflationary policy previously pursued, and that the future exchange rate would fall. More pertinent to the usefulness of floating rates for British purposes in the 1920s is evidence from the forward sterling rate that speculation that the British authorities were going to return to gold at the pre-war rate from autumn 1924 did cause floating sterling to become overvalued (Aliber, 1962; L. B. Thomas, 1973).[3]

Did speculative capital movements allow the sterling–dollar rate in 1925 to be fixed at an inappropriate rate? Overvaluation offers a convenient explanation for why British unemployment (and that of other countries in north-western Europe) was so unusually high during the 1920s. The pre-1914 rate of $4.86, Keynes argued at the time, supported by others later, overvalued sterling by at least 10 per cent (Keynes, 1925; Moggridge, 1972, esp. appendix 1). The extent to which non-traded goods dominate different price indices employment in the calculation determines the extent of overvaluation (Redmond, 1984). More important is the necessity to take into account exchange rates and prices in all trading partners in the judgement about the return to gold. This multilateral test indicates that the pound was overvalued by between 5 and 25 per cent.

The overvaluation thesis adequately explains the heavy British unemployment centred in the export industries, but it requires further support to explain why the effects of the overvaluation were not offset by a monetary contraction. A subtle supply-side argument suggests that a lower exchange rate would have reduced unemployment by raising prices. Higher prices would have cut the real value of unemployment benefits with fixed money values. Lower real unemployment benefits would then have reduced unemployment, if the responsiveness of the unemployed was high enough to matter much (Matthews, 1986a, 1986b).

A monetary contraction necessary to support the exchange rate should also have reduced the domestic price level, restoring competitiveness to export and to import-competing industries. British prices did in fact decline between 1925 and 1929: the consumers' expenditure average value index fell by almost 5 per cent. But the decision to return at $4.86 was also based on the assumption that prices in the rest of the world, in particular the United States, would rise substantially, which they did not. Between 1925 and 1929 the US consumers' expenditure average value index fell very slightly, ensuring that, if the sterling–dollar rate was not to be too high, British prices would have to fall sharply. Such a decline could in fact be accomplished only slowly.

As much an error as the choice of the sterling exchange rate was the resumption of sterling area arrangements against gold in 1925 at the pre-war rate (Pressnell, 1978). A lower parity for the whole sterling area in 1925 would almost certainly have increased dollar earnings of the system as a whole, despite possible counter-depreciations elsewhere, because of the indirect nature of UK earnings from the outer sterling area. Sterling area raw material exporters often sold for dollars and spent in sterling. The southern dominions' currencies were supported by borrowing and were probably overvalued as well against sterling. Without a lower parity, by 1928 the trade and income of primary producers were deteriorating badly. The Australian exchange rate at the end of 1929 was as low as was possible while adhering to the gold standard and the next month Australia in effect abandoned convertibility (Eichengreen, 1992, pp. 232–6). Sterling countries as a whole contributed to the strain on the pound that preceded its collapse in 1931.

The reconstructed gold standard

The reconstructed gold standard lasted only a short while and ushered in the world's greatest depression, which it exacerbated. Although the late 1920s saw no repetition of the post-war inflations, in a number of countries heavy unemployment persisted; the expected pre-1914 prosperity did not return with the pre-1914 monetary system. In this section we discuss whether the difference was due to central banks no longer following the gold standard 'rules of the game', to greater rigidities in the system that made it less responsive, or to the changed conditions brought about by the war, to which the restored monetary regime was inappropriate.

The impact of the war on the monetary system

The war transformed the United States from a major debtor into a major creditor, and the sale of overseas assets greatly reduced the United Kingdom's international investments (Jones, 1933; James, 1992). London's pre-eminence as a financial centre was no longer so marked relative to New York and Paris. Neither of these two centres held the same stake as London in the health of international commerce, or possessed the same experience in international financial management. The uncertainty generated by the disturbances of the early 1920s, the restrictions on overseas investment from London (not necessarily ultimately British investment) and the lesser experience of New York in long-term foreign investment left short-term securities as the main instrument of saving, in contrast to the position before the war. Perhaps £2 billion of these potentially volatile funds were free to cross national borders at short notice, greatly increasing national liquidity requirements. The exchange rates at which so many nations had hurriedly returned to gold were not necessarily appropriate, requiring rapid adjustment in a number of economies. But one apparent divergence between the pre-war and the restored gold standards was not a difference at all. The use of foreign exchange reserves instead of gold reserves, the so-called gold exchange standard, was recommended at the Genoa Conference in 1922 as a means of economising on gold. In fact, the ratio of foreign exchange reserves to gold reserves shows that the gold exchange standard was as widespread in 1910 as it was in 1924 and 1925 (Lindert, 1969, p. 76). The real difference lay in the reduced liquidity of sterling as a reserve currency. Sterling was no longer quite as good as gold.

International debt

The redistribution of international debt during the war has frequently been identified as a constraint on the restored gold standard. The First World War left the allies with debts to each other totalling about $26.5 billion, mostly owed to the

United States and Britain, with France as the main debtor. In addition, the Reparations Commission in 1921 required Germany to pay $33 billion, mainly to France and Britain (Aldcroft, 1977, p. 74). The attempted international transfer of these sums of money was the largest 'transfer problem' in economic history until then, and the reparations payments destabilised Germany in various ways.

The basic elements of the transfer problem are that, for a real resource transfer to take place, the donor nation must be able, voluntarily or under compulsion, to give resources up, and the recipient nation must be willing to accept them. Germany was unwilling to make the transfer, hence the inflation of 1923–4 (Keynes, 1929; Johnson, 1975; Keynes, 1922; Mantoux, 1946; McIntosh, 1977; Balogh and Graham, 1979). Keynes calculated that the Germans would have to pay 43 per cent of their national income in taxes, a proportion that could not be extracted by 'the whips and scorpions of any government' (though that level has been exceeded in many countries since then). Even with perfectly flexible prices and wages, and full employment, an apparently manageable transfer might prove impossible. If the size of the transfer is fixed on the basis of pre-transfer prices, the increased supply of exports from the donor country may force their price down so far that raising the requisite volume of foreign exchange is impossible. The recipient in this case is 'unwilling to accept'.

With imperfectly flexible prices and wages, the classical assumption that full employment would be maintained is incorrect. In the donor country, real resources can be released for foreigners only by cutting domestic demand, but there is no reason to suppose there would be a corresponding increase in the foreign demand for these goods; imports are deflationary. If the recipient does not accept the imports, the fall in aggregate demand in the donor country will react back, depressing demand further in the recipient country. In the international economy as a whole, planned savings increase and planned investment falls. As it turned out, the transfer was not made and so these economic problems did not arise. Before 1924 we have seen that Germany received a transfer from foreign speculators in marks which offset the reparations paid. After 1924, American capital flowed into Germany, balancing German transfers abroad. For Germany the main problem of the transfer was the constraint that it imposed on policy, especially in 1931 (see below).

Similarly, the economic impact of the transfers associated with inter-allied war debts was less significant than the political repercussions (Haberler, 1936, ch. 8). The British suggested cancelling all war debts. When the United States turned this proposal down, Balfour stated in 1922 that the UK had no choice but to collect the debts owed to it, but would do so only up to the limits of the British debt to the United States (Kindleberger, 1973, p. 41). Consequently, the war debts and reparation issue caused political instability and government deflation throughout the 1920s, culminating in the German deflationary policies of the early 1930s that helped Hitler to power, and which perhaps contributed to the crucial failure in 1931 of the Austrian Credit-Anstalt bank.

In 1931 US net receipts from war debts were about 10 per cent of the value of its exports and 68 per cent of the positive balance of trade. For the UK, the net liabilities were 2 per cent of import value and 4 per cent of the negative trade balance. For France, the proportions were 7 per cent and 25 per cent respectively. Between 1918 and 1931, the United States received $2.6 billion from the allies, less than half the amount specified in the revised settlements of the 1920s (Aldcroft, 1977, p. 95). France received in reparations 3.5 times what it paid in war debts to the USA and the UK. The latter ended up with a negative overall balance.

Private international transfers – lending, borrowing and debt service – in the 1920s also played a part in the difficulties of the gold standard from 1929. After 1929 defaults on sovereign debt far exceeded the earlier otherwise comparable phase of the 1870s. The new lenders did not learn from the mistakes of the old. Foreign lending by the United States, at over $6 billion from 1924 to 1929, was about double that of the United Kingdom, but when the capital flows are considered net of war debt and reparation payments, the lending of the two countries was roughly equal (Aldcroft, 1977, p. 66; Ohlin; 1931). Germany ceased to be a creditor in 1924, transformed into the most heavily indebted country in the world after Canada by 1929. Other countries – Poland, Romania, Yugoslavia and Bulgaria – even after 1924 maintained such poor international credit ratings that their international borrowings were extremely limited. Australia was most successful at borrowing funds apart from Germany, followed as some distance by Argentina. Each of these countries began to experience balance of payments problems by 1929 when export prices sagged, American foreign investment dried up, and the servicing of their foreign debt became impossible.

Price levels and price rigidities

Rearrangements of relative prices during the war had still not been entirely reversed by 1928. Given the exchange rates adopted, there were inevitable difficulties for the restored gold standard.

Table 11.1 gives an indication of the extent to which national prices relative to gold had diverged since 1914.[4] Some of the differences might be explained by measurement errors, taxes and tariffs, international investment, or changes in exchange rates, but not all. The UK increased its prices relative to gold between 1914 and 1928 by more than France (despite France's more rapid inflation, because the French lowered the gold value of the franc by 80 per cent), but the franc–sterling exchange rate was very much lower than in 1914. Australia and India maintained their pre-war parities with sterling, but their prices measured in gold increased by more than those of the UK. Similarly, Norway and Denmark held their pre-war exchange rates, but increased their prices by more than the United States.

Belief in the flexibility of prices and wages prevented much consideration of relative prices in the choice of exchange rate, despite the emergence of apparently

Table 11.1 Wholesale price levels in 1928 (gold indices 1913 or 1914 = 100)

Australia	165	New Zealand	147	Bulgaria	133
British India	163	Finland	145	Austria	130
Norway	161	Switzerland	145	Latvia	129
Japan	159	Spain	144	France	126
Peru	157	Czechoslovakia	143	Belgium	122
China	155	UK	140	Estonia	121
Denmark	153	Germany	140	Egypt	120
Canada	151	USA	140	S. Africa	120
Dutch E. Indies	149	Hungary	135	Poland	120
Netherlands	149	Italy	134	Chile	119
Sweden	148				

Source: Ohlin (1931).

unprecedented unemployment in north-western Europe after the post-war boom of 1919–20 (Svennilson, 1954, table 3). A new rigidity in prices and wages might explain the greater persistence of the effects of deflationary shocks (Temin, 1989, esp. appendix A). The failure of the inter-war gold standard to secure the necessary adjustment in domestic economies, and the severity of the Great Depression, may then also be attributable to the economies' structure rather than to the shocks. Collapses in primary goods prices and bank failures were certainly not novel. During the 1930s unemployment rates in northern Europe were similar to those in the 1920s, whereas in the rest of the world rates were much higher: for France they were higher by a factor of fifteen, for example. The Scandinavian countries and the UK returned to their pre-war exchange rates, whereas most other European nations adopted a lower rate when they eventually stabilised. It is therefore tempting to attribute the unemployment of northern Europe to overvalued exchange rates and slow adjustment.

A direct comparison of unemployment before and after 1914 is problematical because the coverage of the statistics was greatly increased after 1913 with the extension of unemployment benefits to more of the labour force. To these benefits has been attributed the high inter-war British unemployment rates. Supposedly they reduced the willingness to search for jobs and tilted the choice between work and 'leisure' on the dole (Benjamin and Kochin, 1979). The payment of unemployment benefits may have reduced wage flexibility: eligibility for benefits did not require applicants to accept lower wages or worse conditions than in their former job (Wright, 1981). In the UK money wages declined very markedly from 1920 to 1923, while unemployment benefits rose in both amount and scope. Furthermore, the evidence for structural changes in the labour markets between the pre-war and inter-war periods is mixed. Collective bargaining was on the increase in Europe before 1914 and covered a large proportion of the workforce after 1945 (Thomas, 1992). Inter-war industrial economies may have suffered an unfortunate conjuncture of severe shocks in newly rigid economies not stabilised as much by government expenditure as were their successors after the Second World War.

On the side of prices, increased market power may have given firms a greater opportunity for rigidity in pricing despite supply or demand shifts. A major merger boom during the 1920s in the UK raised industrial concentration, and in the United States the rise of the huge multidivisional corporation gave credence to those who argued for inappropriate pricing policies as the root cause of higher unemployment (Hannah and Kay, 1977; Bell, 1940).

The major structural change in the industrial economies, if it happened at all, came during, not before, the 1920s, as shown by a comparison of the slump of 1920–2 and that after 1929. In the first slump all prices seem to have been very flexible, thanks to the use of paper currencies; slump-induced price declines were frequently reversed quickly, money wages fell by 30–40 per cent in two years, and the large drop in bond yields stimulated investment (Ohlin, 1931, pp. 165–7, 289–90).

By 1929 the price system was much more rigid. Because of the pervasiveness of tariffs and cartels protecting the domestic markets, prices in international trade declined very seriously, while domestic prices were fairly inflexible. The prices of about half the industrial raw materials and semi-manufactured goods, as measured by the employment in supplying them, were controlled by cartels or producers' associations in Germany. Prices of German cartelised products fell from an index number of 104 to 103 between December 1928 and July 1930, while the non-cartelised prices dropped from 103 to 79. Belgian prices showed a similar pattern, with 'stabilised' prices actually rising from an index number of 811 to 871 over the same period, but other prices falling from 898 to 680. The primary product exporting economies, as has been seen in the previous chapter, increasingly during the 1920s attempted to 'stabilise' prices: that is, maintain them in the face of increased supply or falling demand. The consequences for the economy were that downturns in economic activity were more severe, and adjustments to overvalued exchange rates took longer and were accompanied by severe unemployment.

Monetary policy

A third explanation, in an influential League of Nations report, for the failure of the restored gold standard is that central banks no longer followed the 'rules' of the gold standard 'game' (Nurkse, 1944, pp. 212–13). Divergences of national price levels were possible because central banks did not ensure that changes in gold reserves were accompanied by parallel or multiple changes in domestic credit in the same direction. Between the two world wars, gold and other international assets came to act as a buffer or cushion for disturbances originating abroad. The effects of gold on the credit base were increasingly sterilised, so the maintenance of fixed rates became difficult. But monetary policy alone, the report maintains, was insufficient to offset the multiplier effects of a large fall in foreign demand as in 1929–32. Sixty per cent of the annual observations on twenty-six countries between 1932 and 1938 were inconsistent with the rule that a loss of foreign exchange reserves and gold should be met by a policy of contracting domestic credit. The report goes on to argue that the greater mobility of short-term funds tended to produce inverse movements of a central bank's international and domestic assets, if the bank followed the 'rules'. A loss of reserves countered by a higher discount rate attracted foreign assets. At the same time, domestic assets declined in response to open market purchases of securities intended to make the rate effective.

As shown in Chapter 9, this central bank behaviour was not new, and therefore cannot account for the failure of the gold standard between the world wars. Before 1914 central banks were also concerned about domestic employment and income, as well as their gold reserves. The banks' behaviour depended on balancing the two objectives and on the relative phases of domestic and foreign economic activity. The examination below of American, French and British central bank policies shows little deviation from the pre-1913 pattern except that then there was no American 'central bank' (the Federal Reserve System was set up in 1913). American monetary policy between the wars was singularly inept, and the mistakes of American management were rapidly transmitted to the rest of the world. The slump of 1920/1 probably originated in the United States. Unprecedentedly high American discount rates attracted large gold exports from other countries, which were similarly forced to raise interest rates to staunch the outflow (Friedman and Schwartz, 1963, p. 369 *et seq.*). The 1929 stock market boom, which withdrew US capital from abroad just when many primary producers were beginning to experience a downturn and especially needed this capital for their balance of payments, was also the result of a policy error. In 1924 and 1927 domestic and international requirements coincided: lower interest rates were needed in the United States to prevent a recession, while they also encouraged short-term capital to move into sterling, reducing the United Kingdom's foreign exchange problems. In 1929 domestic and international requirements diverged, and domestic interests predominated.

Gold flowed into the United States in every year during the 1920s except 1927 and 1928. These flows were prevented from expanding American credit and raising prices and employment for two reasons: the gold imports were thought to be temporary, so sterilisation was a stabilisation policy; and a fear of inflation persuaded member banks of the Federal Reserve to use their newly acquired gold to repay Federal Reserve credit, which therefore declined by the volume of gold imports (Nurkse, 1944, p. 74).

France also absorbed substantial volumes of gold, but, as noted in Chapter 9, the extensive French use of gold was not a phenomenon that began in the 1920s. Rather the undervalued exchange rate was the more fundamental cause. The British Treasury economist, Hawtrey, likened the Bank of France to a boa constrictor swallowing a goat: once it started taking in gold, it could not stop whether it wanted to or not (Royal Institute of International Affairs, 1931, p. 208). France, with one-ninth the national income of the United States, needed one-half the gold because of legislation governing the central bank, together with the desire of the French public to hold their money in the form of legal tender paper, rather than bank deposits. A much larger proportion of the purchasing power of the public in France therefore needed a backing narrowly defined by the central bank statutes. The obligations of the Bank of France amounted to more than 100 billion francs; most of this sum required covering by gold or by foreign exchange. There was an upper limit to the amount of foreign exchange which the Bank of France felt could safely be held; once this was reached, an increased demand for money by the French public needed to be covered by equal gold imports.

Like the Bank of France, the Bank of England was obliged to accumulate gold, but did not deliberately neutralise all reserve movements much more than before 1914. Despite shortages of the appropriate assets for open market operations, the Bank usually reinforced the effects of reserve losses. In view of the general weakness of sterling, however, the Bank did tend to offset reserve gains (Moggridge, 1972; Nurkse, 1944, p. 76). Possibly the restrictive policy was insufficiently tight; the growth of the money stock slightly exceeded the growth of domestic output, but then there were the effects on income and employment to consider. The Bank was so concerned to protect the money market and credit for industry and trade that in 1929 it was even prepared to seek an increase in the fiduciary issue to allow a gold loss without deflation (Sayers, 1976, pp. 211, 223, 312).

Central bank co-operation in principle could enhance the efficiency of monetary policy. Because each economy's policy spilled over on to its trading partners, co-operation that took into account these repercussions could allow the attainment of collective policy objectives at lower cost. Montague Norman for the Bank of England, Benjamin Strong for the Federal Reserve Bank of New York, Hjalmar Schacht for the Reichsbank and Emile Moreau for the Bank of France did try to agree co-operative policies in 1928, to alleviate the deflationary impact of the scramble for gold reserves (Eichengreen, 1984). That they were not very successful in supplying co-operative policies underlines the value of the gold standard as a device for reducing the demand for co-operative policies. The gold standard shrank the area in which agreement was necessary. National policy-makers were concerned to avoid the inflation of the early 1920s. The gold standard was a means of doing so, an institutional device for achieving a form of co-operative equilibrium. This is why such expensive attempts were made to defend it. But the regime was not an objective in its own right, despite political rhetoric.

Another attempt at increasing the level of co-operation was the new Bank for International Settlements. Established in 1929 to depoliticise and to manage the payment of German reparations, the Bank set out to become an instrument of international monetary co-operation, by improving the collective management of the gold standard. The date of foundation was not auspicious for achieving this end, nor was the imbalance in the supplies of world gold reserves.

Liquidity and the onset of international depression

Central bank policies and constitutions ensured that by 1928 the United States held 37.8 per cent of official monetary gold reserves, while the gold bloc, centred on France, hoarded another 20.8 per cent (League of Nations, 1935, p. LXXXI). Concentration of gold in a few countries combined with the slowing down of gold production after 1915 to justify fears that supplies would provide insufficient international reserves. These concerns persuaded the 1922 Genoa Conference to confer an official blessing on the gold exchange standard. Even supplemented by foreign exchange, 'hot money' movements in 1930 and 1931 demonstrated that official reserves were nevertheless inadequate to maintain the gold standard. International currency reserves are needed to meet temporary balance of payments deficits which do not warrant correction by exchange rate or domestic expenditure changes. The abnormal capital movements of the 1930s were a major cause of such deficits. The ratio of reserves to imports was probably higher in the late 1920s than in 1913, but the world economy had become more volatile and the official demand for reserves was much greater (Moggridge, 1972). Much gold was committed as official reserves against notes and deposits. This gold could not be used for international settlements, despite the increased need.

Weak agricultural prices in the late 1920s reduced primary good export earnings so that a number of countries lacked the foreign exchange to service their international debts. The main international debtors were exporters of agricultural goods. New Zealand and Australia with net interest payments per head of over $27 in 1928 headed the list (Ohlin, 1931). International balance of payments difficulties were compounded when the United States, a key country under the gold exchange standard, ceased to export capital in 1928, as the Federal Reserve reduced monetary growth in an attempt to control speculation on Wall Street. Germany, the principal recipient of American capital, was severely affected.

The slump in United States income in 1929 gave another twist to the contractionary forces in the international economy. New foreign lending by the United States to the non-European economies declined by $301 million in 1929, a decrease in portfolio investment, mainly to Latin America (Flessig, 1972). The balance of payments impact is revealed by a comparison with the simultaneous decline of South American export revenues of $73.8 million. By 1929 the flow of new capital from the USA to the non-European economies had fallen below the payments on past American investment of amortisation, interest and dividends. Deficits were being financed by reductions in gold and foreign exchange reserves in attempts to cushion the impact of the American depression. Reserves in many countries proved insufficient for the task.

By the beginning of 1931 seven countries (Australia, New Zealand, Uruguay, Brazil, Bolivia, Venezuela and Argentina) had been forced to allow their currencies to be quoted at discounts below their gold parities. The gold reserves moved mainly to the United States and to France. US gold reserves increased by $845 million to June 1931, but France acquired $1 billion, a sum equivalent to the gold production of the entire world. About one-third of the French increase represented the conversion of foreign assets into gold by the Bank of France (League of Nations, 1935). The gold reserves of the British Empire and the sterling bloc were relatively stable at first, with some tendency for those of the UK to increase after 1929.

The second phase of the Depression

Falling foreign investment rendered the central European economies vulnerable to the second phase of the Depression. Unquestionably, government policy, the German Chancellor Bruening's deflation, exacerbated the downturn, but could Bruening have done anything else? Relaxation of government spending by borrowing from the Reichsbank was ruled out by a law of 1924 which still commanded widespread support. It was believed (probably wrongly) that borrowing from private capital markets would crowd out private investment. And borrowing from abroad was imperilled by foreign reactions to German pressure to revise the Versailles Treaty. Moreover, Bruening's aim of renegotiating Germany's reparation payments by demonstrating insolvency required him not to undertake expansionary policies that the allies themselves believed they could not afford (Borchardt, 1991, ch. 9). Bruening's deflationary policy was so unpopular in Germany that he was obliged to rule by decree. In June 1930 the Reichstag voted to abrogate his decrees, and in the September elections the Nazis increased their seats from 12 to 107. This further disturbed international confidence in Germany, and so additionally constrained foreign willingness to lend.

International monetary contraction was precipitated in May 1931 by the failure of the Credit–Anstalt, the largest bank in Austria. Thanks to enormous holdings of industrial shares, the bank was particularly vulnerable during business downturns. At the end of May, the Bank for International Settlements (BIS) granted a $14 million loan to Austria, which quickly proved insufficient. The Austrian government guaranteed the Credit-Anstalt's liabilities, and the bank's foreign creditors agreed to stop withdrawals. However, creditors' assets lost liquidity, and confidence in Austria's ability to maintain the value of its currency began to drain away (Schubert, 1991). German banks were particularly affected, and their creditors started to withdraw deposits in anticipation of further trouble. A BIS loan of $100 million to the Reichsbank on 25 June lasted only until 5 July. Eight days later the German Danatbank failed, despite a German government guarantee for its liabilities the previous day. Germany's loss of foreign reserves was so severe that currency restrictions were instituted on 16 July. With the imposition of German controls, the pressure shifted to the United Kingdom.

The UK also lacked the international liquidity to support its position. Its reserves were higher than in 1914, but their level centred around a post-war target of £150 million that grew out of a rough guess by Lord Cunliffe. The target was later justified by the approximate correspondence with the sum of the 1913 gold holdings of the Bank of England, the clearing banks and individuals (in the form of sovereigns in their pockets) (Moggridge, 1972, pp. 18, 243). Had London been able to rely on the pre-war bank rate mechanism to attract funds, the reserves might have been adequate. But once London's changed net asset position became known, the confidence necessary to make the old system work disappeared.

As befitted the world's central banker, the Bank of England in conjunction with the BIS organised international credits to blunt the impact of the internal difficulties of the central European countries. In the UK, the Bank tried to ease the slump by bringing down bank rate on 14 May 1931 from 3 to 2.5 per cent. When figures on London's short-term liabilities were published in the Macmillan Report on 13 July, realisation of London's weakness, and of the damage done to London by the Austrian and German crises, became much more widespread.

An escape from the crisis and continued adherence to the gold standard depended on maintaining a high degree of credibility. The collapse of confidence was due, first, to acceptance of the overvaluation of the pound, especially by the Macmillan Committee, and, second, to the budget deficit, especially in August and September. Memories of the currency disorders of the early 1920s had not faded (Sayers, 1976). The Labour government could not agree to cuts in unemployment benefit to satisfy the market, and resigned on 22–3 August.

Ramsey MacDonald formed a National Government that introduced an emergency Budget on 10 September. The across-the-board pay cuts of all government employees were not well received by some naval personnel, whose objections were described in newspaper headlines as a mutiny. Britannia seemed about to cease to rule the waves, and gold, it was surmised, could be the next symbol to fall. Ministers appeared inclined to alleviate the hardship that the cuts might cause. The Bank lost increasing amounts of reserves – on 18 September, £18.75 million. Late on Sunday, 20 September, the government announced the suspension of the gold standard.

By then the great fall in raw material prices had directly affected the UK more than most industrial countries. Primary producers were its major export markets and destinations of overseas investment. British income from foreign investments was sufficient to cover 60 per cent of the balance of trade deficit in 1929. By 1931, after the two years of sinking primary prices had reduced income abroad, only 40 per cent was covered (League of Nations, 1935, p. LXIV). The international multiplier also affected the United States indirectly. Between 1929 and 1933 US imports declined by $3 billion, a fall of two-thirds, representing 13 per cent of the total decline of world imports. Imports of the non-European and non-American economies in turn fell from $10 billion in 1929 to S7.9 billion in 1930, and to $3.4 billion in 1933 (Flessig, 1972). Since a substantial proportion of these imports were US exports, American economic activity was accordingly reduced further by the international reverberations of the 1929 American downturn.

From mid-1931 to the end of 1932 most countries lost very large amounts of gold to France and other members of the gold bloc. Gold moved on a scale unparalleled in monetary history. The pound rapidly fell by 30 per cent, providing some boost to the British economy at the expense of competitors whose currencies remained linked to gold. Holders of dollars began to speculate against a similar movement of that currency. The United States lost $1 billion in gold in the second half of 1931 and the first half of 1932. In 1932 the central banks of the gold bloc countries converted the bulk of their remaining foreign assets into gold.

Primary producers began to default on their official debt from the beginning of 1931, further cutting British and other creditor income. Ratios of public debt service to exports in 1931 reached 32.9 per cent for Chile, 24.5 per cent for Bolivia and 22.5 per cent for Argentina, and continued to rise strongly (Jorgensen and Sachs, 1989). Bolivia defaulted first in January 1931. Brazil suspended debt service in 1931–2. When money was available for debt service, the Brazilian government frequently bought back bonds quoted at a large discount on the open market, thanks to default (Cardoso and Dornbusch, 1989). Ultimately, British and American bondholders lending in the 1920s recovered their principal. British lenders received rather higher returns than the Americans, but not at the time when the British balance of payments needed the support (Eichengreen and Portes, 1989).[5]

America in the Depression

American policy differed both from the German and Austrian use of controls and the British policy of devaluation; the Federal Reserve raised interest rates enough to stem the outflow of funds. This deflationary pressure coming from the international collapse squashed recovery in the United States. Prices and production fell as interest rates rose. The Federal Reserve failed to offset the contraction, even after there was no longer a conflict between the external objective of maintaining the gold standard and the internal employment and liquidity goals, because the System's decentralized organisation slowed the formation and implementation of policy. Moreover, policy-makers misinterpreted the availability of credit (Temin, 1976a; Wheelock, 1991).

A trough of the American depression was reached in the summer of 1932, but recovery was again prevented by the presidential election campaign and the long 'lame duck' period between the election and Roosevelt's inauguration. Confidence in the American banking system finally collapsed altogether during this latter period and Roosevelt was inaugurated in the midst of a banking panic. He proclaimed a Bank Holiday and began the 'New Deal': between 1929 and 1933 United States real GNP had fallen by 29 per cent. During 1932 most South American countries which had not already done so left the gold standard (League of Nations, 1935, Introduction). The United States abandoned gold on 20 April 1933, and by September the gold value of the dollar had fallen by a third, approaching the old parity with the pound. Capital flight must have depreciated the dollar; the current account continued to be favourable to the USA.

National reserves were clearly insufficient to prevent the international transmission of the Depression. The scale of the contraction, and the opportunity cost of holding sufficient reserves to neutralise the impact, mean it is implausible to expect anything else. But more thorough schemes for sharing reserves might have proved effective. Ultimately, the great size of the US economy imposes responsibility upon American economic policy. Kindleberger (1973, ch. 14) maintains that the solution would have been for America to accept the leadership of the international economy in three ways: by providing a market for the products of countries in balance of payments difficulties; by engaging in counter-cyclical overseas investment so that countries experiencing a decline in the American demand for imports would have received an offset from an increased inflow of foreign capital; and by acting as lender of last resort to countries with balance of payments problems. Britain performed these functions in the heyday of the gold standard, but was no longer able to do so in the inter-war years. Because the United States was the largest economy in the world, no effort to make national policies mutually compatible and supporting would have been possible without American participation. This was clearly shown when President Roosevelt wrecked the World Economic Conference of 1933 by announcing that he was going to ensure the restoration of equilibrium in the domestic economy before worrying about the international economy.

Whether it is historically plausible to have expected the United States to have assumed the sort of leadership required is doubtful. The American economy was largely self-sufficient by the 1920s. United States policy found little reason to be concerned about the 'Old World' and the international bankers, except to make a world safe for American foreign investment. Hence the American concern to depoliticise the German reparations problem with the 1924 Dawes Plan. Only the threat of a world dominated by an anti-capitalist ideology was to force the United States to assume the leadership of the international economy after the Second World War. In any event, the roots of the Depression lay in the inability to maintain economic activity in the American economy, a failing that augured ill for any American attempts to manage the international economy between the wars.

Exchange rates and the Depression

Another explanation for the world-wide nature of the Depression was the gold exchange standard (see, for example, Eichengreen, 1992.) In his 1925 Budget speech, announcing the United Kingdom's return to the gold standard, Winston Churchill remarked felicitously that 'All the countries related to the gold standard will move together like ships in a harbour whose gangways are joined and who rise and fall together with the tide.' (Hansard, 1925). From 1927 international economic activity was synchronised, and in 1929 nearly all the economies fell together. (Ohlin, 1931, p. 110). France remained an exception until 1930 because its massive gold reserves allowed it to defer deflationary policies for a year. Floating exchange rates probably reduced the impact of the international depression of 1974–5, and the evidence suggests that they may have done the same in 1929–32.

For small European countries, the Great Depression was an exogenous disturbance. Spain operated a floating exchange rate over the years 1929–32 and avoided much of the deflationary impact. The peseta fell by more than 50 per cent against the US dollar between 1928 and 1932. The Spanish money supply remained stable, except in 1931 when there was a contraction because of the European banking crises, which made customers withdraw bank deposits before the expected bank failures. Other small European countries on the fixed exchange rate gold standard experienced substantial monetary contractions. For Belgium, Italy, Poland and Netherlands, all on the gold standard, movements in their national outputs and price levels were closely associated with those of the United States, whereas Spain experienced no such association (Choudri and Kochin, 1980). Denmark, Finland and Norway represent intermediate cases, switching to floating rates in 1931. These economies began to recover from the Depression after 1931, as did the UK.

China reaped a similar benefit from remaining on the silver standard. At the beginning of the Great Depression, the Chinese exchange rate fell as the decline in silver prices, which began in 1920, accelerated. But although wholesale prices continued to go up, the relative rise in Chinese prices was less than that necessary to compensate for the decline in the exchange rate. Hence the effects of the Depression may have been partly offset. The value of Chinese exports in gold fell by one-third in 1930 compared with a fall in imports of one-quarter. Exports consisted mainly of raw materials for which prices had fallen more than for imports, so that it is difficult to discern whether the volume of exports fell by less than imports. Exports from Japan and the United States to China declined by the same proportion as to others, contrary to the insulation hypothesis, but exports from the United Kingdom declined by more. Indian exports were also probably more than usually affected because of the comparative ease with which domestic products could be substituted for these imports. The fortuitous benefits conferred by the silver standard ceased when the departure of the United Kingdom and the sterling area from the gold standard removed much of China's competitive advantage (League of Nations, 1935, Introduction).

Floating exchange rates would not have proved a satisfactory solution to the capital flights that troubled Germany and the UK. A stronger international net asset position would have helped. A more rapid recovery of the German economy, untrammelled by reparations and political repercussions, would have contributed to the achievement of this position, as probably would a different sterling exchange rate after 1925. Co-ordinated national reflations might have limited capital movements. If all major financial markets had been reflating, there would have been less incentive to shift capital between them. Alternatively, recycling the hot money with official loans was a possibility (Foreman-Peck *et al.*, 1992).

Substantial gains from co-operation have been calculated for these years. The model from which the conclusion was derived assumed that actual historical outcomes were non-cooperative equilibria, and that there were only two countries, the United Kingdom and America. Policy targets were assumed to be competitiveness and national output, and the money stock was the policy instrument (Broadberry, 1989).

Authorities' principal policy instrument under the gold standard was in fact the discount rate, although open market operations to influence the money supply were used as supports. However, there was a range beyond which discount rates could not be moved for political reasons, either at the negative end, or at very high levels. With falling prices, real discount rates could be stuck at a high minimum level. In the face of shocks of unprecedented magnitude, requiring great movements in discount rates, policy-makers preferred to widen their range of instruments, to include tariffs, quotas, exchange controls, debt default, restrictions on capital movements and floating exchange rates. These shifted the world economy away from co-operative to non-cooperative equilibria, with less satisfactory output and price target values.

The managed exchange rates of the 1930s

The international monetary system emerged from the Great Depression with the gold exchange standard destroyed and the few remaining countries on the gold standard under increasing pressure. By the end of the 1930s, most major countries operated a form of managed floating rate, with an exchange stabilisation fund, which often differed little from a pegged exchange rate. The pound sterling was a freely fluctuating currency only from September 1931 to the spring of 1932. The sterling area currencies were pegged to sterling. The US dollar floated freely only from April 1933 to January 1934. Having remained at its gold parity until the devaluation of September 1936, France reverted to a floating franc from 30 June 1937 to 4 May 1938, although with intervention. Some currencies of central and eastern Europe, notably the mark, were kept at the old parities with extensive exchange controls (Nurkse, 1944, pp. 122–3).

Purchasing power parity calculations (Table 11.2) suggest that by 1934 the German Reichsmark was the most overvalued currency in the world, as a result of the continued link with gold (bearing in mind the usual caveats). Where the figures are less than 100 in the table, the cost of the franc and/or sterling is less than it should be on the basis of price relations, if exchange rates were in equilibrium in 1929: in other words, the currency is overvalued.[6] The multiple exchange rate system and the bilateral clearing arrangements which were devised to maintain the overvaluation of the German mark have been described in the previous chapter. Germany adopted these policies to avoid the increase in international debt payments that would have been entailed by a depreciation of the Reichsmark.[7] At the other extreme, the yen was the most undervalued currency, but prices in Chile, India, Argentina, Australia and Spain also failed to adjust to the heavy fall in the external value of the currencies. These countries were all exporters of raw materials or foodstuffs, whose terms of trade were adversely affected by the price decline.

Table 11.2 Cost of the French franc and sterling as a percentage of purchasing power parity cost in 1934 (1929=100).

End of year	Cost of franc (1934)	Cost of sterling (1934)
Germany	74.3	62.6
Hungary	80.6	67.8
Austria	83.1	69.0
Czechoslovakia	86.0	72.8
Switzerland	87.0	71.3
Italy	98.0	82.5
Poland	98.7	83.0
S. Africa	99.2	83.0
France	—	84.2
Netherlands	101.1	85.1
Belgium	102.9	84.5
Portugal	103.3	87.0
Bulgaria	104.4	88.0
Albania	107.1	90.0
Peru	110.1	92.0
United States	114.3	90.2
Yugoslavia	115.3	97.2
Finland	115.9	97.5
Greece	116.0	97.7
Sweden	118.2	90.4
Norway	118.6	100.0
United Kingdom	118.7	—
China	110.0	100.2
Canada	122.4	103.1
Denmark	124.5	104.8
New Zealand	126.6	105.7
Estonia	127.1	107.2
Spain	137.6[a]	107.5[a]
Australia	141.6	108.0
India	145.1	122.3
Chile	150.2	126.6
Argentine	152.9	128.8
Japan	193.9	163.4

[a] August.
Source: League of Nations (1935).

Competitive depreciations

Countries that applied expansionary measures, either coupled with devaluation or floating, began recovery from the Depression earlier than those like France which maintained the old rates. Such exchange rate policies gave rise to charges of competitive depreciation, but at the time when currencies were depreciated or stabilised it was often impossible to know whether the change would entail a country 'exporting unemployment' or would restore equilibrium. The 16 per cent devaluation of the Czech crown in March 1935 was calculated on the basis of purchasing power parity with wholesale price indices, but in October 1936 another devaluation was necessary. On the other hand, the Belgian devaluation of 28 per cent in 1935 was deliberately large to allow domestic expansion. Imports were rapidly drawn in and an import surplus emerged, so this did not constitute a 'beggar-my-neighbour' policy. Even so, it did put deflationary pressures on France and the gold bloc by inducing speculative capital outflows (Haberler, 1976, p. 35).

The competitive depreciation of the yen was among the clearest and most successful examples. As Table 11.2 shows, the yen depreciated the most against the franc and sterling by 1934, and Japanese export expansion in the 1930s was very fast. But competitive depreciation was not necessarily planned or desired by the government whose currency was falling. The 1931 depreciation of the pound, in response to a capital outflow, undervalued the currency and helped economic activity increase. The depreciation of the dollar was perhaps the classic example of competitive depreciation, for there were no pressures on the current account, and gold reserves were immense. The dollar was devalued against gold on October 1933 by 59.06 per cent, by fixing a gold price of $35 per fine ounce. Part ($2,000 million) of the profits from the revaluation of the national gold stocks was used to establish an Exchange Stabilisation Fund. Gold flowed into the USA because the dollar price of gold at the franc rate was less than the price for which it could be sold to the US Treasury. The gold-buying policy was intended to raise American commodity prices and, because the United States was neither a large international supplier nor a large international buyer of most commodities, it was fairly successful (Moggridge, 1972).

The floating of the dollar in 1933 clearly reduced the undervaluation of the franc against the dollar in 1934. France and the gold bloc countries (Belgium, Netherlands and Switzerland) lost most of the advantages they had gained in the 1920s as other economies depreciated against them in the 1930s. Nevertheless they pledged themselves to stay at the existing parities at the World Economic Conference in 1933 and again at Geneva in 1934. Their commitment condemned them to depression well after recovery had begun elsewhere. Comparative success in riding out the earlier phases of the world depression and accumulating gold reserves reinforced French belief in the efficacy of the gold standard and balanced budgets (Mouré, 1991). Yet with hindsight the economy was severely weakened: between 1929 and 1937 hours worked in French industry fell by one-third. Unemployed industrial workers went back to the land, so that French agricultural employment in 1939 was higher than during the 1920s.

Sterling in the 1930s

The British float was managed by the Exchange Equalisation Account (EEA), which originated in the Bank of England's secret hoard of dollars, accumulated since 1925. By 1932 the profits and losses on exchange intervention had become too great for a privately owned bank, and the fund became officially financed, reaching £635 million in 1937 (Sayers, 1976, pp. 487–8). The rules followed by the

EEA in stabilising sterling would have been quite acceptable to the International Monetary Fund of the later 1970s (Howson, 1980). In its objective of preventing the exchange rate from being influenced by non-rational speculation, the EEA seems to have been successful. Only in December 1931, before the EEA began operations, was there evidence of severe speculative overshooting of the equilibrium value of the sterling–dollar exchange rate (Whitaker and Hudgins, 1977). The political crises of the period, such as the Belgian devaluation, the French changes of ministry in 1935 and 1936, and Hitler's march into the Rhineland in 1936, did not seem to cause untoward movements in sterling. Although the activities of the EEA were beneficial, the contribution to the stabilisation of month-to-month variations in the rate was only modest. The very existence of the EEA may have curbed short-term variability in capital flows, and this would have been valuable because controls on capital movements were incomplete and probably ineffective.

The effective exchange rate of the pound, the trade-weighted average against all other currencies, gives a rather different impression of the behaviour of the sterling exchange rate during the 1930s from the sterling–dollar rate (Redmond, 1980). The benefits of freeing the exchange, and the effects of increased competitiveness of British economic recovery, persisted in the years 1934–6, when the effective rate remained 4–5 per cent below the 1929–30 level, and 8 per cent below the level of August 1931. The effective rate shows a steady upward trend after the shock of devaluation had worked itself out by late 1932, providing a motive for the EEA to hold the pound down, as it was widely believed to be doing. Between 1931 and 1937 there was a steady capital inflow which tended to push the rate up, a tendency that the EEA may have wished to prevent in order to maintain purchasing power parities. But the EEA's resources were inadequate for that task, being merely sufficient to dampen trends.

The EEA pegged sterling closely to the dollar from late 1933 until 1938. A very large part of the trading world therefore enjoyed stability, while suffering in so far as the stabilised rates were inappropriate. In addition, pegging meant that the UK shared in the US recession of 1937–8. The pegging was in part a recognition that the United States would not tolerate a really cheap pound. It also helped maintain sterling's role in international trade, much of which was still invoiced in sterling and financed through London. Sterling balances held by other countries continued to provide the UK with a source of income. Some idea of the size of this income can be inferred from the balances of £598 million in 1938, and assuming (a) that the funds released were allowed to go into new foreign investment yielding 5 per cent; and (b) that the cost of the balances to the United Kingdom was the interest that had to be paid on Treasury bills, 0.611 per cent. These assumptions give a gross yield of £26.2 million (£598m × [0.05 − 0.00611]), a sum that amounted to 2.5 per cent of British export receipts in 1938, and 0.5 per cent of GNP (Drummond, 1981, pp. 248, 258–9).

A more widely recognised milestone on the road to the acceptance that exchange rates were matters for multilateral agreement than the pegging of sterling against the dollar was the Tripartite Agreement of 1936. Strictly speaking, there was no agreement at all, only national declarations. The governments of the United Kingdom, France and the United States issued similar statements on 25 September 1936 accepting the devaluation of the franc, agreeing to minimise the associated exchange market disturbances, and committing themselves to work for improved conditions for international trade. During the three years following the agreement, exchange rates fluctuated less (Eichengreen, 1992a, pp. 380–2). However, there was less co-operation than the agreement seemed to herald. The French government only nominally accepted the British phrase about the progressive relaxation of quotas and exchange controls. In October it suppressed one-third of its quotas, but these covered only one-tenth of the value of French imports, and the French reduced duties on the few goods without quotas. Having devalued, the French were almost as protectionist as before. For this reason, the agreement was not a bridge to the more liberal post-1945 regime. In addition, the British were already managing sterling before 1936 with an eye on American exchange rate targets, in so far as they could understand them or learn about them. The British recognised that they could not choose a dollar–sterling rate unacceptable to the United States government (Drummond, 1981, pp. 220, 227, 248). The events of the next decade were to force an even greater awareness of this balance of power.

Summary and conclusion

The First World War brought a number of political and economic conflicts to a head which, in most of the former belligerent states, generated post-war inflation and depreciating exchange rates. There is evidence to suggest that the floating exchange rate regimes of the 1920s did well their job of reflecting underlying economic conditions. They obtained their bad reputation by association with often misguided government policies, especially those that overvalued the pound and the yen, undervalued the franc and caused the collapse of the mark and the rouble.

Returning to the fixed exchange rates of the gold exchange standard during the 1920s was attractive as a symbol of 'normality'. More important, the gold link would impose restraints on governments' financial policies. The exchange standard entailed central banks holding, as reserves, currencies (usually sterling or dollars) which were convertible into gold instead of gold itself. At a time when the annual increase in gold output was small, the exchange standard was expected to enhance world supplies of liquidity. A drawback of the system was the greater mass of claims on the gold reserves of the key currencies. When confidence was delicate, these claims were likely to be cashed in. Deflationary policies in key currency countries were forced by such withdrawals during the crisis of the early 1930s. A second disadvantage was that of all fixed exchange rate regimes – that when a depression came, countries in which depression tendencies did not originate primarily in the domestic economy, such as Japan and France in 1930 and the United Kingdom in 1929, were dragged into the general collapse.

A failure of central banks under the gold exchange standard to follow 'the rules of the game' has sometimes been blamed for the poor performance of the system. France and the United States showed a persistent tendency to acquire the greater part of the gold reserves of the world without any corresponding expansion of domestic prices and incomes. Before 1914 central banks did not follow the 'rules' either. The difference between the world wars stemmed from new American monetary management and from the undervaluation of the franc. Other countries therefore lacked sufficient reserves to finance their balance of payments deficits during the early stages of the Depression, and instead were obliged to resort to deflationary policies.

Certain changes in the monetary system in the 1920s left it more vulnerable to disturbances. Prices and wages seem to have become less flexible in many countries, so that a decline in demand by the late 1920s would be met to a much greater extent than before by the contraction of output and employment, rather than by a reduction in prices. In addition, there was a much greater accumulation of debt, both private and public, between nations and within nations, than ever before. The First World War left behind a network of inter-allied war debts and reparations. Payments of these created problems of transferring the real resources in the face of the willingness of some donors to give them up, and of some recipients to receive them. The increase in debt payments due from primary producers at a time when primary prices were weakening precipitated bank collapses in agricultural areas within nations, and eventual exchange rate depreciation and deflationary policies among primary product exporters by the beginning of the 1930s.

Had there been no depression in the United States, it is unlikely that the rest of the world would have experienced anything like the slump it actually did, despite the tendencies in many primary commodity exporting countries and in Germany. Capital exports from the USA normally increased in a boom, and would have provided additional balance of payments finance for those countries in difficulty. American demand for imports, which fell off so rapidly in the Depression, would have supplied additional support. Since the United States was unable to prevent a fall of almost 30 per cent in its own national income between 1929 and 1933, it is not surprising that it failed to operate constructive policies for the rest of the world.

A much reduced volume of world trade emerged from the Depression, with the gold exchange standard gone and most nations on floating exchange rates. The managed floats of the 1930s gave the international economy rather more stability than during the 1970s, and almost as much as in the 1920s. Charges of competitive depreciation have substance only when the effect of capital flights is considered, with the exception of the United States, and capital flights came to be determined mainly by political factors. The United Kingdom continued to receive earnings from providing the reserve currency for the sterling area, and in the Tripartite

Agreement of 1936 there was a small sign of international co-operation over exchange rates. Domestic unemployment in many countries persisted partly because the inflation of the 1920s had created a fear of reflating.

Co-operative policies were generally ineffective, but were given little chance of success. Politicians or their advisers in stable international environments came to realise what foreign reactions their policies would provoke and to reformulate them accordingly. Appreciation of sequences of policy moves and reactions could lead to co-operative regimes, or at least quasi-co-operative regimes, like the gold standard. Transformed political values and objectives in Europe and in the United States of the 1920s and 1930s created a new environment where the advantages of co-operation were less clearly perceived. For most economies, the pursuit of astute domestic policies was likely to yield higher returns than positive international co-operative policies.

Notes

1. Bresciani-Turoni was a member of the Allied Reparations Commission in Germany during the period of inflation.
2. Aldcroft (1977, p. 95) also asserts that international co-operation could have produced a better exchange rate regime.
3. The forward exchange rate is the rate at which a currency can be bought or sold now for delivery or receipt at a particular future date. The forward rate therefore reflects expectations about future changes in the 'spot', or current, exchange rate.
4. The limitations of these indices must be borne in mind. Some of them are of very restricted coverage: for example, measuring only the prices of goods in the capital city, and including only a small number of commodities. In the absence of better data, however, they provide some indication of the difficulties under which the gold exchange system operated.
5. Eichengreen (1991) surveys the work on the inter-war debt crisis.
6. These observations are consistent with a tendency for the franc, sterling and dollar to move towards purchasing power parity in this period (Broadberry and Taylor, 1992). what matters for policy is the pace of adjustment to, and the distance from, the purchasing power parity equilibrium.
7. Germany also operated a successful policy of buying back its depreciated bonds so as to avoid the stigma and possible retaliation associated with default, but only paying back a fraction of what was originally borrowed (Klug, 1993).

12 The redirection of the international economy, 1939–53

In extreme contrast to the economic nationalism of the 1930s, a new international economic regime emerged from the devastation of the Second World War. National, especially United States, policy-makers chose different priorities from those of the inter-war years, and partly fortuitously, gave birth to a co-operative order. In contrast to the period after the First World War, much greater emphasis was placed on full employment, a goal which was the more readily achieved because of the greater control by governments over economic life. The United States, now the pre-eminent world power, aspired to a liberal international economic system based upon multilateral, non-discriminatory trade. It was dissuaded from withdrawing into political isolation, as it had after the Versailles Treaty, by the Soviet challenge to American values and power. European national interests also needed to be served in any new settlement. The key turned out to be a resolution of relations between the French and German economies.

The structure of the new order was greatly influenced by the perceived 'lessons of history'. The western allies believed that recovery during the 1920s had been bedevilled by war debts and reparations. They were concerned not to repeat the experience. American planners saw the 1930s as a period of economic nationalism and 'beggar-my-neighbour' policies. They intended to avoid these errors with internationalist institutions in which all countries would participate.

Those countries lacking the power to restructure international institutions sometimes adopted different interpretations of the past. Some newly independent states contended that their economic development had been stunted by trade and foreign investment imposed upon them by the imperial powers. They preferred to opt for state-controlled economic development policies, in which the only trade necessary was imports for industrialisation. Other non-industrial countries saw the depressed state of inter-war trade in primary products as demonstrating the impossibility of developing through exporting raw materials and foodstuffs, and pursued similar policies.

International economic relations during the war

In a variety of ways, the transformation of international economic relations during the Second World War influenced the structure of the system after 1945. The most important change arose from the increased economic strength, both absolutely and relatively, that the United States attained during the war. While it was still a neutral, the United States' employment and income were boosted by the massive orders for armaments placed by the British and French in 1939. Subsequently, with the entry of America into the war, the inter-allied financial transfers furthered an international division of labour in the war effort, by which American financial aid allowed Britain and the empire countries to maintain a larger proportion of their manpower in the armed forces than would otherwise have been possible (Allen, 1946). The United States could do this and yet still commit a smaller proportion of its economy to the war than the United Kingdom. In 1944, 22 per cent of the British labour force was in the armed forces and 33 per cent in war employment; for America the figures were 18.5 and 21.5 per cent respectively. Both because of the expanded American productive capacity and because a larger section of American industry was free to invest and grow, together with the absence of the direct effects of enemy action on the economy, the United States was bound to dominate the post-war world if it chose to do so. The Japanese attack on Pearl Harbor symbolised the inability of the United States to isolate itself from the rest of the world, and the necessity to exercise some influence over international relations.

Between 4 and 5 per cent of the national income of the United States was transferred to the British Empire for lend-lease goods and services. Cash payments were not required for the munitions and combat equipment (65 per cent of lend-lease) or the other goods, such as concentrated foods, sent from specially built plants in the United States to the UK. Reciprocal aid from the UK and the empire countries paid for the upkeep of US forces in allied countries. If all aid financed by the United Kingdom is included, the share of the British national income transferred to the United States was also 4 to 5 per cent. Australia and New Zealand respectively contributed nearly 7 and 10 per cent of their national incomes to the United States as reciprocal aid. After the German invasion of 1941, Russia also became a major lend-lease beneficiary. By early 1945, mutual aid supplies amounted to $5,000 million per annum. Canada operated a system similar to lend-lease in supplying allied countries and, where the UK was concerned, was equally generous after the end of the war.

In the Mutual Aid Agreement signed between the UK and America in 1942, Roosevelt was concerned that the financial costs should be divided so 'that no nation [would] grow rich from the war effort of its allies'. He wanted to avoid the problems of inter-allied war debts that had dragged on through the 1920s (Kock, 1969, p. 24; Churchill, 1951, p. 400). Churchill and Roosevelt similarly agreed that experience showed that large indemnities did not work. The settlement of December 1945 cancelled the whole of the mutual aid account between the USA and the UK, on which there was $22,000 million of lend-lease against £1,200 million of reciprocal aid. Measured in financial terms, the United Kingdom was clearly the beneficiary to the tune of at least $16,000 million. The Mutual Aid Agreement and the cancellation were tied to an agreement on commercial policy by which the UK was committed to discussions on the elimination of Commonwealth preferences established at Ottawa in 1932 (see Chapter 10). The meaning of this agreement and the objective itself remained a sensitive point in British–American relations for a number of years after the war. Churchill had secured a declaration from Roosevelt that the UK was no more committed to the abolition of Commonwealth preferences than the American government was committed to the abolition of its high protective tariff (Kock, 1969, p. 25).

American dislike of Commonwealth preferences was rooted in the belief that discrimination in international trade was a major source of political conflict and reduced welfare. Japanese expansion in the Far East, which was eventually to bring the United States into the war, owed much to Japan's exclusion from the great trade blocs of the 1930s. The Japanese concern for raw material, fuel supplies and markets led to the establishment of the Manchurian Industrial Development Corporation as the instrument of economic development in the puppet state of Manchukuo in northern China (Milward, 1977, p. 166). Korea and Taiwan for some time had been Japanese colonies, which were developed so as to complement the Japanese economy. The weakness of the European colonial powers provided Japan with another opportunity to expand its economic sphere of influence. However, in response to the Japanese occupation of the Saigon area in July 1941, the American government froze all Japanese assets and imposed an economic embargo. The Japanese decided to break the embargo by continued expansion and war. Once war with the western powers began, the difficulties of maintaining sea transport between the various states of the 'Co-prosperity Sphere' reduced the extent and effectiveness of economic relations within the Japanese Empire. By the end of the war, 88 per cent of Japan's 6.5 million ton merchant marine had been sunk.

The most important long-run effects of the 'Co-prosperity Sphere' were the political changes it precipitated. Japanese wartime expansion was pursued under the slogan of 'Asia for the Asians'. Nominally independent regimes were set up in conquered colonies: in Burma and the Philippines in 1943, and in Indo-China and Indonesia in 1945 (United Nations, 1949a, p. 32). The United Kingdom, the United States and China committed themselves in 1943 to independence for the Japanese colony of Korea. The political forces unleashed encouraged the granting of independence soon after the end of hostilities to many Asian territories, both those formerly occupied by the Japanese, and others. In 1946 the Philippines and in 1948 Burma, the Federation of Malaya and in Ceylon were granted independence. Fighting continued in Malaya and in Indo-China.[1]

India was not a battlefield and so was able to reap some economic advantage from the war. An export surplus allowed the accumulation of sterling balances, which could be spent afterwards when sterling became convertible and peacetime production began in the sterling area. With a small technological base on which to build in comparison with the western powers, the economic advantages to be gained were correspondingly small – India actually experienced a decline in coal, pig iron and steel production after 1940 (Milward, 1977, pp. 349, 353–4). Only in munitions, shipbuilding and engineering were there signs of growth adequate for industrialisation. What economic benefits there were from being on the periphery of the war were soon dissipated. In 1947 India was given independence and split into two countries, India and Pakistan, amidst fighting and forced migration. Pakistan was a producer of raw materials, especially jute and cotton, and food needed by India. India produced manufactures and coal needed by Pakistan. Because of trade barriers with Pakistan, Indian exchange reserves accumulated during the war were largely spent on importing food (United Nations, 1949a, p. 35).

One of the greatest international redirections of economic activity during the war occurred after the German invasion of France in 1940. Under the guise of occupation costs, payments extracted from France in 1943 amounted to perhaps 8–9 per cent of Germany's GNP (Milward, 1977, pp. 137–44). This figure excludes the benefits derived from French workers in Germany and confiscations. The German gains from France were much greater than those from the German-occupied eastern territories, where a combination of a 'scorched earth' policy of the retreating troops, poverty and a bestially destructive occupation policy meant that, for example, coal production was usually inadequate even for the troops on the spot. The German chemical firm I.G. Farben used the occupation in the east as an opportunity to engage in 'multinational' production with slave labour (Borkin, 1979, pp. 111–27).[2] Workers in I. G. Farben's camp could expect to live on average for three months; for the company, the supply price of this labour was virtually zero, so there was no incentive to provide more food or less brutal treatment. The absence of any mutual national gains from economic organisation within the eastern Reich rendered co-operation and even low economic returns hard to achieve (Radice, 1986, pp. 307–8).

In contrast to the Slavs of the east, the Norwegians were believed by the occupying Nazis to be 'racially pure'. Under German occupation massive investment was accordingly planned to take advantage of Norway's hydro-electric potential and for increased food production (Milward, 1972, p. 177). Although subsequent wartime shortages prevented much of this investment coming to fruition, one of the major new aluminium factories was completed, as were two of the power stations.

As with the Japanese occupations in Asia, the longer-term significance of the German-imposed international economic relations in Europe arose from their indirect results. The liberation of the eastern territories allowed Russia to pull them into the communist, centrally planned economic system, and away from their relations with western Europe of the inter-war years. In western Europe the German invasion of France for the third time within a century, instead of exacerbating mutual hostility and mistrust after the war, as on previous occasions, was eventually resolved by the formation of economic unions designed to prevent such wars ever occurring again.

The war served to ease the international economic constraints on most Latin American countries, which during the 1930s were trying to sell exports to Europe and buy imports from the United States. Enormous wartime American purchases from Latin American eliminated the dollar shortage and allowed the accumulation of dollar reserves (United Nations, 1949c, pp. 231–4). The Latin American share of world exports rose from 7.8 per cent in 1938 to 13.4 per cent in 1946. With markets and supplies outside the continent largely cut off, trade increased between Latin American countries as a proportion of their total trade. Those countries like Chile, importing mainly foodstuffs which could be supplied by other Latin American countries, maintained the value of their trade. Economies dependent on European suppliers for imports of machinery, fuel and raw materials reduced their trade markedly; Argentina cut imports to one-third of the pre-war volume and one-half of the value.

Many African economies, especially the French territories heavily dependent on Europe as a market for their primary product exports, were damaged by the war; but others, like the Belgian Congo, South Africa and British East Africa, which exported materials of strategic significance, benefited (Milward, 1977, p. 356). Both Canada and Australia made significant advances in industrialisation supplying the allied war effort. Manufacturing output in Canada at the end of the war was 2.5 times the average between 1935 and 1939, which had been, however, considerably less than full employment output. Overall the war did not produce import substitution on the same scale as had the First World War for those not heavily involved in the fighting. Although the Second World War similarly provided protection from foreign competition, this alone was inadequate for the mass production (outside textiles) by then necessary for industrial development.

The post-war international system

As already noted, a preliminary agreement between the United Kingdom and America about the ultimate shape of the post-war international trading system had been reached in 1942. Planning for the new order in both countries began the same year. Not surprisingly, the planners failed to foresee and allow for the magnitude of the post-war dollar gap, stemming from the reconstruction needs of the devastated areas. What is perhaps surprising is that the intentions behind the system were eventually put into practice despite the difficulties of the late 1940s. In 1943 preparatory negotiations between British and American Treasury experts about the post-war monetary system were based on their plans, prepared respectively by J. M. Keynes and Harry Dexter White (Block, 1977, ch. 3; Gardner, chs. 5, 7, 13; Kock, 1969, ch. 1; Horsefield, 1969, vol. 1, chs. 1–4). The common ground of the two plans lay in their opposition to floating exchange rates and to competitive trade restrictions. They also both favoured the national right to control short-term capital movements. This consensus derived from a shared interpretation of the inter-war years, which owed much to the analysts of the League of Nations. But hostility to floating rates can also be seen as means of committing policy-makers to consistent, low-inflation policies. Politicians would be unable to engineer a burst of inflation to enhance their popularity, while maintaining an exchange rate pegged to other currencies and, indirectly, to gold (Nurkse, 1944; Bordo, 1993).

The United States differed from the United Kingdom in its concern with the incidence and application of barriers to trade, rather than with tariffs themselves. With a continent-sized domestic market, the United States could afford to assert the need for non-discrimination, and for the elimination of the British Commonwealth preference, as it had in the inter-war years. A second difference arose from the distrust by American officials of banking and 'big money' interests – the same attitude that had introduced the decentralised American central banking system with twelve Federal Reserve Banks. Related to this attitude was a belief in the necessity for a written constitution for any new institutions. In addition, the arrangements had to be acceptable to Congress, which could nullify any agreement (Scammel, 1980).

To the United Kingdom, the national circumstances made it more obvious than to the United States that a relaxation of exchange controls and restrictions on trade would be feasible only if American aid and credit were very liberal. The British were especially concerned to maintain full employment and to be allowed to choose the exchange rate necessary to achieve that end.

The economic dominance of the United States ensured that the Bretton Woods Agreement of 1944, which set up the two new international monetary institutions of the post-war world, more closely resembled the American plan than the British. This was despite the view expressed in a verse allegedly salvaged from the first Anglo-American discussions during the Second World War:

> In Washington Lord Halifax
> Once whispered to Lord Keynes:
> 'It's true they have the money bags
> But we have all the brains.'

The two institutions were the International Monetary Fund (IMF), for the maintenance of exchange stability and to address balance of payments problems, and the International Bank of Reconstruction and Development (IBRD) to encourage long-term international investment. The IMF was to achieve its objectives by insisting that member countries establish a par value for their currencies in terms of gold or the US dollar (Argy, 1981, ch. 2; Horsefield, 1969, pp. 110–13). These par values could be changed only to correct a 'fundamental disequilibrium' in the balance of payments. Reserves for the support of the fixed exchange rate could be supplemented by the Fund's resources. The resources were obtained from the quotas assigned to member countries. Quotas were assigned according to a country's national income, trade and international reserves. One-quarter of the quotas had to be paid in gold or US dollars, and the rest in the member's own currency. A member country could borrow up to the point where the Fund was holding currency equal to 200 per cent of the member's quota. The member supplied its own currency in exchange for the currency of a member country whose reserve position was relatively stronger. A member enjoyed automatic access to borrowings up to 25 per cent of its quota. Additional drawing rights were subject to increasingly restrictive conditions generally supporting a programme that would establish or maintain the stability of the member's currency at a 'realistic' rate of exchange.

The IMF thus acted as a bank making loans and receiving deposits, and could increase the stock of international reserves. So, for example, if a borrower drew on its first credit tranche and received US dollars, its reserve position (the country's quota less the Fund's holdings of the country's currency) increased by that amount. At the same time, the United States' reserve position increased by the same amount, so that world reserves increased by double the amount originally drawn. However, no explicit allowance was made for a long-term growth in the stock of world international reserves to accommodate the growth in world trade.

The chaotic state of so many economies immediately after 1945 meant in fact that the IMF had little to do, and its rules were largely ignored after the disastrous experience of the United Kingdom's attempt at restoring currency convertibility in 1947.

The International Bank for Reconstruction and Development, later better known as the World Bank, had a capitalisation of 2 per cent to be paid in gold or dollars, 18 per cent in the currencies of the member countries and usable for lending purposes only with the consent of the contributing country, and 80 per cent as a guarantee fund (Mason and Asher, 1973, ch. 5). This capital structure sought to ensure that every country, rich and poor, would participate in providing the capital (the 2 per cent portion). It recognised that countries temporarily impoverished and in balance of payments difficulties would be able to contribute usable capital later (the 18 per cent), and it embodied the expectation that the mobilisation of capital from other sources would be much more important than the use of the Bank's own assets (the 80 per cent). Unlike the IMF, the World Bank did make a significant though small contribution to the international economy by the end of the period of redirection, by lending first to war-torn Europe, and then to poor countries.

The plans for the new liberal order in international trade went even more awry than did those for international monetary arrangements. But the institution that emerged as a by-product to liberalise international trade was at first unexpectedly successful. In December 1945 the American government published proposals for an international trade organisation, and invited various countries including the USSR to take part in negotiations for reductions in barriers to trade (Kock, 1969, p. 62). Out of the international deliberations in London, Geneva and Havana between 1946 and 1948 came the Havana Charter for creating the International Trade Organisation (subject to ratification, which never took place).

The discussions were marked by confrontation between the United Kingdom and America in which the British and then other countries successfully pressed for changes in the Charter, allowing trade discrimination and quantitative restrictions under various circumstances. As these loopholes widened and increased, the Charter became less and less attractive to American multilateralists (Block, 1977, pp. 75–6, 84; Gardner, 1980, chs. 14 and 17). At the same time, the success of the first GATT session in 1947 did much to reduce the urgency for an International Trade Organisation.

The General Agreement on Tariffs and Trade (GATT) originated in the recommendation of the Preparatory Committee for the Havana Conference at the London meeting of 1946 that negotiations for the reduction of trade barriers should be held under the sponsorship of the Committee (Kock, 1969, p. 62). The code of conduct incorporated in GATT involved two major principles: first, a multilateral and non-discriminatory approach to international trade; and second, condemnation of quantitative trade restrictions. The first of these principles was implemented through the inclusion in the code of the 'most favoured nation clause'. Under this clause, GATT prohibited any preferential trading agreement designed to favour one nation over another. Before negotiations started, each member was to transmit to all others a preliminary list of tariff reduction concessions which it proposed to request. When negotiations began, they were to present a corresponding list of concessions they were prepared to grant. Negotiations would then take place between two or more countries.

The procedure allowed participating countries to assess the value of concessions granted by other countries over and above the direct concessions negotiated. Negotiations began in April 1947. Twenty-three countries signed GATT on 30 October 1947. The first round of negotiations in 1947 between these countries resulted in 123 agreements and twenty schedules covering about 45,000 tariff items relating to about one-half of world trade. In this and subsequent GATT rounds, negotiations were threatened by a breakdown due to different interpretations in America and the United Kingdom of the rule concerning the elimination of preferences. The Commonwealth preference system continued to irritate the United States at later GATT meetings and retarded tariff reduction.

In January 1952 the thirty-four contracting countries accounted for more than 80 per cent of world trade. By the mid-1950s it was estimated that a net reduction in United States duties of 50 per cent had been achieved since 1934 by tariff concessions alone, the greater part of which had been accomplished in the period after 1945.

The dollar gap

The United States made the products that the rest of the world wanted to buy after the war, both for present consumption and for reconstruction, but the devastated areas did not have suitable goods and services to exchange. Only if they could borrow or run down their foreign exchange reserves could these countries import anything like the quantities of goods they needed. The alternative of exchange depreciation does not seem to have been considered a serious possibility until 1948.

The United States produced nearly one-half of the world's manufactured goods in the late 1940s. Wartime increases in American productive capacity meant that, for the world as a whole, industrial production by 1947 was 42 per cent above the 1938 level (United Nations, 1949b, p. 149). The lesser importance of the United States in world agricultural supplies, and the devastation in Europe and the Far East, explains why world agricultural production in 1947/8 was 4 per cent below 1934–8 average levels. Practically all of Europe and Asia suffered a decline in food consumption measured in calories, and a deterioration in the quality of their diet (United Nations, 1948a). In 1946/7 there was an average shortfall of 18 per cent in average calorie consumption per head in comparison to pre-war diets in deficiency areas (excluding China and the USSR). Poor harvests cut the availability of food supplies to Europe in 1947/8. Similarly, there were deficiencies in coal supplies: European coal consumption fell by 12 per cent between 1937 and 1947. Even allowing for compensation from other energy sources, coal supplies in 1947 were inadequate in most European importing countries to meet requirements in industry, transport and domestic use.

In 1947 merchandise imports of the devastated areas were $20.2 billion, while exports and other current receipts amounted to $13.1 billion, leaving a deficit of $7.1 billion (United Nations, 1948b). Because the currencies in which exports were valued were inconvertible, they could not all be used to buy the desired imports, so the problem was larger than the dollar value of the deficit indicated. The net deficit had been reduced to $1.6 billion by capital account transactions and by the United Nations Relief and Rehabilitation Administration (UNRRA), but UNRRA operations in Europe ended on 30 June 1947 and therefore no more finance could be hoped for from that direction.

Sterling convertibility from 15 July 1947 reduced the United Kingdom's ability to import by depleting its dollar reserves. As part of the American Loan Agreement at the end of lend-lease, the UK was required to make sterling freely convertible into other currencies within two years (Block, 1977, ch. 3; Gardner, 1980, ch. 16). Lend-lease terminated on VJ Day, at 12.01 a.m. on 2 September 1945, but the need for transfers from the United States did not simultaneously end. The UK did not expect the war against Japan to end so soon and planned on receiving aid until mid-1946. It ran a large trading deficit with America and many countries had accumulated sterling balances during the war because they had supplied the UK with goods without taking any goods or services in exchange. These countries wanted American products and converted their sterling into dollars, which the British authorities were obliged to supply at the fixed exchange rate of $4.03 = £1. By 20 August 1947 dollar reserves were nearly exhausted and sterling returned to inconvertibility.

These difficulties reduced European foreign trade in 1948 to a level lower than it had been ten years earlier. Had it not been for the remarkable British export performance (by the last quarter of 1948 export volume was 147 per cent of the pre-war level), the European export trade would have been even lower than the 18 per cent below the 1938 level that it was (United Nations, 1949b, pp. 35–6). The reduction of European imports by 14 per cent over the same period was due mainly to reduced British and German imports. These were lower for different reasons. British production was higher than the 1938 level because some import substitution had been induced by controls, whereas German imports were low because of the failure of the economy to recover from the war; German production was still only 64 per cent of the pre-war level.

Apart from the problem of the European deficit with the United States, and of Germany, European recovery proceeded well. Unlike the period after the First World War, there was no mass unemployment. Improved supplies of raw materials imported from abroad, and the restoration of incentives following monetary reforms, pushed output for fifteen European countries approximately back to pre-war levels by 1948, when Marshall Aid began flowing (see below). Price and wage inflation was reduced to the 2–5 per cent range in 1948 as against a range of 10–20 per cent the previous year, with the exceptions of France and Greece.

In eastern Europe (excluding USSR and East Germany) trade with western Europe reached the pre-war level in real terms by 1948 (Holzman, 1976, ch. 3). Before the war the USSR had almost no economic links with the nations of eastern Europe; Germany was the main focus of east European trade before 1945. The collapse of Germany gave these countries little choice but to trade with each other and the Soviet Union if they wished to trade at all. The former enemy countries of East Germany, Hungary, Romania and Bulgaria were in any case occupied by the Soviet Union, collecting maintenance payments for its troops, removing German and Italian plant and equipment, and establishing and running jointly owned enterprises. Unlike the British and Americans, the Soviets did extract substantial reparations from their defeated enemies. Nominal reparations of $500 million were imposed on Romania and Hungary. Moreover, these payments were valued in goods at pre-war prices fixed arbitrarily in favour of the USSR. Reparations therefore served as a means of appropriating a substantial portion of the satellite country's output (Zauberman, 1955, p. 11).

Transfers from Hungary, according to an American estimate, amounted to 35 per cent of Hungary's national income. A six-year contract for Romania to supply 1.7 million tons of petrol a year valued the petrol at one-half of the world price. Even liberated countries suffered: coal exported from Poland to Russia was valued at 10 per cent of the world market price, and the tariff of the joint Yugoslav–Soviet navigation company provided that the rate for Yugoslav use should be double the Soviet rate. Goods and services were transferred more directly from East Germany to the Soviet Union. By 1946 one-quarter of total industrial production was run by Soviet-controlled companies, including the whole of the iron and steel and motor vehicle production.

Japan's economic plight was the most serious, with two-thirds of its large cotton textile capacity destroyed, and with food and raw materials supplied only through the allied occupation forces, financed mainly by the United States (United Nations, 1949a, pp. 27, 34). About 6 million Japanese were returned to Japan from overseas territories, only partly offset by the emigration of 1 million of other nationalities, especially Koreans, from Japan. Between October 1945 and October 1948, 2.2 million Koreans moved into South Korea. Wartime devastation, as in Europe, reduced the ability of the Asian economies to export. Asia's trade with the United States, which had been in balance before the war, was now in deficit (United Nations, 1949a, ch. 12). Because of quotas and exchange controls, so too was the trade balance with the rest of the world, which had formerly been in surplus.

In contrast to the devastated areas, Latin America did not suffer from a dollar shortage in the immediate post-war years, but a return to the pre-war pattern of trade with Europe was difficult because of the scarcity of convertible currencies. Consequently, Bolivian tin exports went to the United States instead of to Europe (United Nations, 1949c, pp. 209, 247). American policy was in any case directed to replacing Asia as a source of strategic materials such as rubber and tin, and Latin America was an obvious substitute supplier. Argentina and Brazil believed in the continuance of the pre-war importance of Europe in Latin American trade, and so extended credit to their traditional buyers. Although this type of policy was necessary to solve the dollar gap of the post-war world, the scale of the credits needs to reduce the problem significantly was so great that only the United States could provide them.

The Marshall Plan and west European reconstruction

American planners, meeting in a high-level State–Navy–War Co-ordinating Committee in early 1947, predicted that the world would be unable to continue to buy US exports at the 1946–7 rate for more than another twelve to eighteen months (Block, 1977, p. 82). Foreign currency reserves of the dollar gap countries would soon be exhausted, there was little international credit and UNRRA relief was ending. The Committee warned that the substantial decline in the US export surplus would have a depressing effect on business activity and employment in the USA. A major US aid programme to finance a continued high level of American exports was the solution proposed by the Committee. This was the embryonic form of the plan revealed in Secretary of State Marshall's Harvard speech in June 1947. The Marshall Plan also neutralised the forces moving western Europe permanently away from multilateral trade: the strength of the European left-wing parties, the relative weaknesses of the European economies, and the pull of the

Soviet Union. Without the intensification of the Cold War, it would have been impossible even to contemplate sending such a massive aid programme to Congress. The ground was prepared by the enunciation of the Truman Doctrine in March 1947, pledging the United States to supply economic aid to uncommitted countries.

From 1948 to 1952 about $13,150 million was made available under the Marshall Plan to Europe, where the major recipients were the United Kingdom ($3,176m), France ($2,706m), Italy ($1,474m), West Germany ($1,389m), the Netherlands ($1,079m), and Greece and Austria (each $700m) (Price, 1955, pp. 88–90). This distribution of aid shows that the fear of communism was not the sole inspiration of the programme (although a necessary one as far as Congress was concerned). The UK, the recipient of the largest sum, was in little danger of a communist takeover. So far as there was a principle governing the allocation between nations, it was the volume of national foreign trade. After June 1951 Europe received a further $2,600 million in Marshall Aid mainly in the period up to mid-1953. The aid offered under the programme took the form of grants for commodities produced predominantly in the United States. Inevitably there were pressures from interested American parties influencing the programme. More agricultural products were offered than requested, and instead of scrap and semi-finished iron and steel for which Europe had asked, more finished iron and steel than was wanted was sent (Harris, 1948, p. 12).

Compared with European reconstruction after the First World War, economic recovery during the Marshall Plan period was rapid (Hartman, 1968, p. 63; Wexler, 1983). The most enthusiastic advocates of the programme thought that it would act as a catalyst, raising income by possibly five times the amount of the aid. The plan would achieve this boost by removing serious bottlenecks, by making dollars available in order to prevent a collapse of international trade, by providing a weapon for making the European countries pursue responsible fiscal and monetary policies, and by encouraging them to co-operate with each other. On the other hand, European recovery, except in Germany, was apparently already strongly under way (Milward, 1984). Balance of payments difficulties were caused by the strength of the upswing rather than by poverty. Moreover, the contributions of Marshall Aid to European national income, were generally too small to make a great difference to Europe as a whole. The Netherlands (Table 12.1), and probably Austria, proportionately benefitted the most by far. France, the United Kingdom and Italy, gained from Marshall Aid roughly the equivalent of two years' economic growth, which was anyway high at the time. About two-thirds of all imports into the western occupied zones of Germany were financed by American aid between 1945 and 1948, but the European Recovery Programme (ERP) was only a contributor to this programme. For France imports covered by Marshall Aid payment accounted for one-fifth of the 1949 total, whereas for Britain they were little more than one-tenth. British ERP imports were mainly food, but this was not true elsewhere.

The impact of Marshall Aid turns on the alternative policy scenario. If Europeans were to have eaten as well as they actually did in 1949, they would

Table 12.1 Total net ERP aid after utilisation of drawing rights, as a percentage of 1949 GNP.

	At pre-September 1949 exchange rates	At post-September 1949 exchange rates
France	9.9	11.5
Italy	8.8	9.6
Netherlands	16.1	23.1
United Kingdom[a]	5.2	7.5
West Germany[b]	4.7	5.9

[a] GDP.
[b] 1950.
Source: Milward (1984).

have been unable to import the same volume of capital goods. European economies were constrained by shortages of imports. Once the constraint was removed by aid, income could rise rapidly (Kindleberger, 1987). In the limiting case where capital imports rose by the amount of aid, investment would also have been higher by that amount, and European output would have increased correspondingly. There is statistical evidence that Marshall Aid did increase investment quite markedly, but allowing for that, together with fiscal and foreign exchange impacts, does not in total yield a substantial contribution. Quite independently of these channels, Marshall Aid raised European GDPs. The Marshall Plan may have been so effective because it resolved the marketing crisis. Prices were held at unsustainably low levels encouraging hoarding. Controls and rationing were universal. Marshall Aid helped break the political deadlock as to how to deal with open and repressed inflation. It reduced the sacrifices necessary to rebalance the sum of notional demands with aggregate supply (Eichengreen and Uzan, 1992).

Part of the contribution was the American ERP administration adoption of a positive role in restructuring European economic relations to encourage recovery. The Organisation for European Economic Co-operation (OEEC), as a committee, was formed to draft the request for American aid in the first place. In order to stimulate intra-European trade, the OEEC member governments in 1949 began to liberalise their trade, and in 1950 they accepted a Code of Liberalisation, the aim of which was the gradual freeing of up to 75 per cent (later 90 per cent) of their mutual trade from the network of quantitative restrictions imposed in response to balance of payments difficulties after the war (Kock, 1969, p. 115). Liberalisation was not automatically extended to GATT members outside the OEEC. The code was thus a violation of the general GATT rule of non-discrimination in the application of quantitative restrictions. However, the United States and others felt that the code would speed up the return to more normal conditions in trade and payments, and for that reason did not object.

A second mechanism by which American administrators planned the economic integration of Europe was the European Payments Union (EPU). The EPU was to revive intra-European trade by creating a multilateral payments mechanism within Europe. A network of more than 200 bilateral trade agreements had been negotiated by the European countries by 1947 (Patterson, 1966, pp. 75–119). They temporarily removed some of the pressure for strict barter in the absence of convertible currencies, but since the credit margins in the trade agreements were small, their effects were short-lived. The bilateral agreements tended towards closed bilateral balancing, and were both restrictive and distorting.

Nine months after the 1949 devaluations (see below) the new EPU was agreed. Once a month the central bank of each Union member, including its associated monetary area if any, reported to the Bank for International Settlements, the EPU agent, the net surpluses or deficits of its current transactions with each of the other members. The BIS then offset each country's total net surpluses and total net deficits, and arrived at a single net figure for each member *vis-à-vis* all other members. The net balance for each member was accumulated month by month. This was then set off against a quota assigned to each country at approximately 15 per cent of each member's total intra-European visible and invisible trade in 1949. Accumulated balances were settled partly in gold and partly in credit on a sliding scale. The Code of Trade Liberalisation, effective after the EPU was signed, attempted to reduce quotas, which GATT did not touch. The EPU proved temporary, as intended, rather than permanent because the participants avoided inflationary policies and because those countries which had to pay the highest price for the continuation of the scheme, the United Kingdom, Germany and the Benelux countries, were among the more economically and politically powerful.

In the long term, one of the most significant American contributions to European economic relations was the support for the formation of the European Coal and Steel Community, despite its discriminatory implications and its violation of the principles of GATT. The common market in iron and steel began with a speech by the French foreign minister Robert Schumann in May 1950, proposing that the entire French–German production of coal and steel be placed under a common authority in an organization which other European countries would be free to join (Swann, 1972, pp. 19–21). The motivation was to draw France and West Germany together, making a future war impossible. The treaty was signed in 1951 and included the Benelux countries and Italy. The UK declined the invitation to take part in preparatory work when France indicated that participation meant acceptance of the goals of a supranational authority and, ultimately, of political unity. The treaty went further than a customs union in aiming at an integration of production, but less far in covering only a proportion of trade.

The Cold War and the reconstruction of eastern Europe

The war inevitably reduced economic relations between eastern Europe and the West, and the Soviet Union in 1945 was necessarily the major market and source of supply. But in the immediate post-war years, helped by the payment of $1.1

Table 12.2 East-West trade as percentages of the trade of the developed West and of the East, 1938–53.

Year	West	East
1938	9.5	73.8
1948	4.1	41.6
1953	2.1	14.0

Source: Wilczynski (1969, p.54).

million by UNRRA to the Soviet satellite states, trade with the West increased and the importance of the Soviet Union declined (Zauberman, 1955, pp. 21–2). In 1946, the peak year for UNRRA deliveries, goods delivered were 5 per cent of Czech national income, and 60 per cent of Polish merchandise imports (Brus, 1986b, p. 577). Between 1945 and 1948 Polish imports from the Soviet Union fell from 90 per cent of the total to less than 25 per cent. The UK and the USSR had concluded a trade agreement which foresaw expanding trade (Wilczynski, 1969, pp. 48–9; Holzman, 1976, p. 40). The Marshall Plan promised to break the East–West tension caused by Soviet aggressive tactics in Romania, Bulgaria and Poland. The USSR was included in the offer of financial assistance for the reconstruction of Europe. It accepted an invitation from France and Britain to discuss procedures in Paris in June 1947. The conference failed because the Soviets would only accept unconditional aid, and the Americans were concerned that this would be used to strengthen communist governments. When the Soviets withdrew, they forced the Czechs to withdraw their acceptance and prevented the participation of the other east European nations. During 1947 Czechoslovakia, Poland and Yugoslavia ran large deficits which could have been financed only with foreign assistance. The communist coup in Prague occurred in February 1948, and a few months later the Soviets sealed off Berlin. The USA began constructing its economic warfare apparatus against the communists, culminating in the Export Control Act of 1949. Nato was formed in 1949 and hostilities in Korea began in June 1950, resulting in a complete US embargo on trade with China.

Table 12.2 shows that East–West trade both before and after the war had been much less important to the West than to the East, and that, although the war and occupation had reduced the trade's significance, the additional fall between 1948 and 1953 as a result of the Cold War was very severe. Czechoslovakia, the most open economy of eastern Europe, was able to compensate in part for the loss of the German goods by exporting its own manufactures.

The greater part of the reduction in trade was due to western, rather than eastern, controls. The USSR continued to ship chrome and manganese ores to the United States throughout the period of most intense economic warfare. Tension was eventually reduced by the end of the Korean War and the death of Stalin in 1953, together with the armistice in Indo-China the following year.

The Marshall Plan foiled the influence of the communist parties in western Europe. The USSR therefore tightened its control on eastern Europe and began to develop east European resources by industrialisation. Reparations were reduced and eliminated completely for Hungary, Romania and East Germany in 1952, 1953 and 1954, respectively. However, together with troop maintenance payments, reparations remained sufficiently heavy to require Soviet tanks to quell the East German riots of 1953. Equipment removed from East Germany and Romania to the USSR was brought back. The first international organisation of the eastern nations, the Council for Mutual Economic Assistance (CMEA), was established on 25 January 1949. The CMEA was in part a rebuff to Yugoslavia, which had broken with the USSR the previous year, and in part a response to the Marshall Plan and the formation of the OEEC. The Council did little to facilitate trade in the early years, but intra-bloc trade rapidly came to dominate the trade of members, accounting for 80 per cent of the total trade by 1953 (Holzman, 1976, p. 40; Wilczynski, 1969, pp. 48–9). Before the formation of the CMEA, intra-member trade was about 15 per cent of total member trade. The great shift in trading patterns undoubtedly imposed large economic losses from trade diversion on all participants including the USSR. All nations would have been better off if they could have continued or begun trading with the West. The losses were greater for the smaller countries, which were more dependent on trade. These sacrifices emphasise the political nature of CMEA, although once the Cold War had begun in earnest there was little alternative.

The industrialisation of the dependent countries of eastern Europe solved one of the Soviet planners' problems: how to gain foreign currency to buy goods, especially copper, rubber, tin and wool in which the Soviet bloc was deficient. So long as they maintained a surplus in their trade with western Europe, the satellite countries supplying Romanian oil, the Czech and Polish timber, could earn this currency. The ability to earn was enhanced by economic integration, which had been prevented by the nationalism of the inter-war years. The transmission of electricity and the supply of steel-rolled products across national borders were among the most common of the transnational projects, inevitably supervised and standardised by Soviet administration (Zauberman, 1955, pp. 39–54).

The industrialisation effort of the Soviet Union was concentrated in five designated zones, accounting for 7 per cent of the land area of the USSR. The dependent countries formed a sixth area complementary to the western regions of the Soviet Union. Iron ore was abundant in these western regions, but most of the increase in Soviet coal output since 1940 had occurred in the Urals or further east. The dependent countries had substantial coal-mining sectors to which the iron ore from the western USSR could be shipped (Zauberman, 1964, pp. 156, 217).

Yugoslavia was the one communist European country not to become integrated into the Soviet system, and it suffered accordingly. Between 1948 and 1953 the Soviets gradually escalated towards a total economic blockade of Yugoslavia because it was unwilling to become an obedient satellite. Western assistance reduced the damage, and although the blockade harmed the Yugoslav economy, it did not achieve the desired political effect.

The pattern of trade and finance

Government aid capital flows played little role in eastern Europe (mainly going to Poland, $251 million and Czechoslovakia, $168 million, between 1945 and early 1947), but were crucial for financing the west European balances of payments in the recovery period. By the end of the 1940s with diminished volume they became a major support for the international payments of many less developed countries, assuming some of the role of private investment during the period before 1914. These non-industrial areas received $2,000 million per annum in official grants and loans on average between 1953 and 1957 (GATT, 1958, pp. 32–43). The greater part was government-to-government aid, but the World Bank also made a significant contribution. By 30 June 1952 the Bank had made sixty-eight loans totalling the equivalent of $1.4 billion (Mason and Asher, 1973, ch. 6). More than 35 per cent of the loan total was accounted for by the four European reconstruction loans of 1947, for France, Netherlands, Denmark and Luxembourg.

Under its Articles of Agreement, the Bank was expected to finance only those productive projects for which other financing was not available on reasonable terms. The management thought private capital would be most readily available to the low-income countries for the development of export products, and was opposed to financing government-owned industry. The Bank also eschewed financing sanitation, education and health facilities because of their less measurable contribution to production and the associated problem of Wall Street's view of the soundness of the Bank's management. (Initially, the Bank's securities were sold mainly on the American market.) The Bank came to concentrate on investment in power plants, railways, roads and similar physical facilities in part because such projects were large enough to justify review and appraisal by a global agency. The selection was the result of circumstances, but it led the Bank to argue for public utility investment, financial stability and the encouragement of private investment as the best means of economic development.

Official funds often tended to flow to areas and industries neglected by past private investors, and they increased the volume of capital flowing to non-industrial countries over the previous peak of the late 1920s. Net private investment of all industrial countries in the non-industrial areas in the years 1952–7 returned to the classic nineteenth-century pattern where most funds went to Latin America or the regions recently settled by Europeans. The one change was that India ceased attracting long-term private funds on balance as a result of the new economic policies pursued after independence. The widespread governmental restrictions on international capital movements in the early 1950s were reflected in the altered relation between world exports and foreign investment: world exports had tripled since 1928, but private foreign investment had only attained its 1928 level.

The non-industrial areas remained dependent on the industrial economies (GATT, 1958, pp. 32–43) not only for capital imports, but also as a source of demand for their exports. The United States recession of 1948/9 caused primary product prices to fall by about 10 per cent (GATT, 1958, p. 17). Only the uninterrupted prosperity of western Europe offset similar repercussions from the American recession of 1953/4. Although the volume of trade quickly recovered from the war and exceeded the levels of the 1930s, the growth of production was still more rapid, so that economies were proportionately less open in the mid-1950s than they had been in the late 1920s. The composition of trade had also changed. Exports of manufactures from major suppliers rose by one-quarter between 1937 and 1950, and by three-quarters between 1937 and 1955 (Maizels, 1963, ch. 8).

Under the stimulus of a government export drive to earn the currency to repay foreign loans, benefiting from the collapse of the Japanese and German economies, and helped by the devaluation in 1949 (see below), the United Kingdom increased competitiveness in international markets for manufactures between 1937 and 1950 for the only period in the twentieth century (Table 12.3). Half of the rise in world exports of manufactures came from the UK and the other half from the United States between 1937 and 1950. The size of this increase corresponded with the decline in Japanese and German exports of manufactures over the same period.

Primary product trade increased much less strongly. Exports from non-industrial countries in the period 1937–55 rose by little more than one-quarter (Table 12.4). Greater agricultural protectionism in industrial countries cut trade in non-tropical foodstuffs. Oil exporters benefited most from prevailing technological trends and income effects on demand, which were a continuation of those of the inter-war years. The volume and value of oil and oil products rose from 10 per cent of exports of non-industrial areas in 1937/8 to 20 per cent in 1955 (GATT, 1958, p. 21). Trade in manufactures was affected by the same trends: machinery and transport equipment for recovery, industrialisation and consumption became more important, and the share of textiles in manufactures trade declined as former importing areas developed their own textile industries (Svennilson, 1954, ch. 9; Maizels, 1963, ch. 8). The prices at which goods were traded were considerably higher than before the war because of wartime and post-war inflation, a response to political and social pressures.

The Korean War boom of 1951–2 radically altered relative prices. Before then foodstuffs were expensive because of the disruption of agriculture, and raw materials were relatively cheap because the industrial demand for them had not recovered. Thus Chile as an exporter of minerals and an importer of food in the immediate post-war period suffered a deterioration in the terms of trade, whereas Brazil, as an exporter of coffee and an importer of raw materials and machinery, experienced a favourable shift in the terms of trade. With the emergency stockpiling and direct demands of the war, raw material prices increased massively: the price of rubber in the London market quadrupled in 1950 and wool prices tripled (United Nations, 1953, p. 7). Fears of a new Malthusian crisis emerged in which world manufacturing capacity would outstrip the supply of raw materials, threatening mass unemployment (Royal Institute of International

Table 12.3 Changes in the volume of exports of manufactures, 1937–50 ($bn at 1955 constant prices)

	United Kingdom	France	Germany	Other Western Europe	Canada	United States	India	Japan	Total
1937 exports	4.03	1.15	3.24	3.87	1.16	4.08	0.48	1.89	19.90
Change to 1950 attributable to:									
market growth	+1.00	+0.29	+0.80	+0.96	+0.29	+1.00	+0.12	+0.47	+4.92
diversification	−0.09	−0.08	+0.16	−0.26	0	+0.91	−0.20	−0.43	–
competitiveness	+1.58	+1.11	−2.39	−0.12	+0.04	+0.74	+0.15	−1.12	–
Total	+2.49	+1.32	−1.43	+0.58	+0.32	+2.65	+0.06	−1.07	+4.92

Source: Maizels (1963).

Table 12.4 World trade, 1937–55 (volume and unit values, 1928 = 100) (value, $000 f.o.b.)

	Exports from non-industrial countries			Exports from industrial countries			World exports		
	Value	Vol.	Unit value	Value	Vol.	Unit value	Value	Vol.	Unit value
1937–8	7.80	108.5	69	15.11	85	83	22.91	93	78
1955	22.22	138	197	53.44	139	180	81.66	139	185

Source: GATT (1958).

Affairs, 1953, pp. 1, 13, 16–19). As the world's largest consumer of practically all raw materials, the United States was at the centre of these fears. The President appointed a Materials Policy Commission, and in January 1951 an International Materials Conference was convened to introduce rationing schemes. By 1952 most of the more far-fetched Malthusian concerns had disappeared, although raw material prices remained above their pre-Korean War levels. Large-scale investment in primary production, having been neglected throughout the depressed 1930s, began after the outbreak of the Korean War. The American government at the same time gave lavish financial assistance for the exploitation of new deposits and the expansion of existing sources of supply within its territory. These stimuli soon achieved the intended result.

Despite the hopes of the American planners, discrimination remained an important determinant of the pattern of trade and finance throughout the period of redirection, and afterwards. Countries belonging to the overseas sterling area and the associated territories of western Europe, together with Cuba, increased their exports by 227 per cent between 1928 and 1955, whereas 'unsheltered' countries, with no privileged access to markets, increased their exports by half that proportion over the same period (GATT, 1958, p. 26; Bell, 1956). The sterling area had only become legally discriminatory from 1939, when sterling was made inconvertible outside the group of countries that used sterling as a normal means of international settlement. This meant that member governments strictly controlled payments out of the area, while payments within the area were relatively unrestricted (Patterson, 1966, pp. 67–75). Members sold their hard foreign currency earnings, especially dollars and gold, to the United Kingdom Treasury in exchange for sterling, and agreed to limit their drawings on this 'dollar pool' to amounts needed for certain purposes – generally to buy items unattainable or very expensive within the area. Thus non-sterling area less developed or industrial economies had difficulties selling to sterling area countries until the area was abolished. In 1952 the UK and other members of the area began a new drive to make the pound convertible, not altogether successfully, but not as disastrously as the 1947 attempt. Not until 1961 did the UK assume full responsibility for the external convertibility of sterling as the IMF agreement required.

Not only did the sterling area violate IMF principles, but the greater part of the international economy ignored them. Despite the considerable international payment difficulties for some time after 1947, the Fund's currency transactions over the first five years of its operations amounted to only $851 million, of which $606 million were drawn in the first year of the Fund's existence (Horsefield, 1969, vol. 2, p. 397). The Fund's activities were limited partly because members were expected to rely largely on exchange controls during the transitional period. In addition, when Marshall Aid began, the Fund adopted the policy of refusing recipient countries access to its resources, so that in 1950 there were no drawings at all (Horie, 1964, pp. 138–9). (Under Marshall Aid the country receiving assistance from the IMF had aid reduced by the amount of the assistance.)

The one area in which the IMF might be held to have had some discreet success was over the 1949 devaluations. In 1948 the IMF thought that the dollar gap required changes in European exchange rates, and discussed the matter confidentially with member governments (Horsefield, 1969, vol. 1, pp. 234–42). The Fund's Managing Director was privately informed in advance the following year of the United Kingdom's intention to devalue the pound by 30.5 per cent, with a view to ending the growing loss of gold reserves. British officials also notified Governors of the other sterling area countries to arrange a simultaneous devaluation. The board of the IMF provided a means of keeping governments informed of the views and plans of other members. Each country's proposal was submitted to the criticism of other directors so that the decisions reached were more acceptable to world opinion than those taken after 1931, and did not provoke retaliatory action. Although some non-sterling area countries also devalued at the same time, they did so by a smaller proportion – West Germany by 20.7 per cent, France by 5.6–22.2 per cent (because of the multiple exchange rates which had earlier incurred IMF disapproval), Belgium and Luxembourg by 12.3 per cent, Canada by 9.3 per cent and Italy by 8.1 per cent (Flanders, 1963). One of the two countries which rapidly bit into the UK's export market share after 1950, Japan, did not devalue at all, while the other, Germany, also owed its export success more to recovery than to altered price relations.

As the IMF hoped, the general devaluation against the dollar slowed US export growth in 1950 and US imports rose. The trade-weighted devaluation of sterling was 9 per cent, larger than in 1931. British competitiveness increased and was at least maintained for more than five years. In 1950 the proportion of British exports sold in dollar markets rose abruptly, and the share of imports from the dollar area fell correspondingly. Comparison of British and US price increases by 1954 showed only a 3 per cent difference, indicating a long-term improvement in competitiveness against the dollar (Cairncross and Eichengreen, 1983, pp. 151–4). British gold and dollar reserves were 70 per cent higher in the first nine months after devaluation and continued to strengthen.

Summary and conclusion

International economic relations by 1953 had not adopted the pattern proposed by the American wartime planners. Yet trade was becoming freer and there were some relaxations of currency and capital controls. The trend was facilitated by rapid expansion of national incomes in western Europe, which itself owed much to the easing of trade restrictions. In some respects, the new order towards which much of the international economy was changing resembled that of the classic free-trade era of 1850–75 in its internationalist orientation. The major difference, however, was the new blend of liberal, nationalist and socialist economic policies. These gave a much greater role to governments, even in the market economies, and committed them to maintain full employment.

In honouring this commitment governments were successful, especially by contrast with the period after the First World War. Their success was due in part to the conditions that allowed relatively stable and expansionary fiscal and monetary policies. Most countries would not again allow the subordination of domestic policies to international considerations, which is why the dismantling of controls took so long. Military spending played an important part in reaching and maintaining full employment in the United States. The causes of this spending, first the Second World War and then the Cold War, accounted for one of the major structural changes of the international economy: the dragging of eastern Europe into the economic orbit of the Soviet Union, away from the central European bloc of the 1930s centred on Germany. China also became a major Soviet economic satellite, closed to economic relations with the non-communist world.

The Second World War triggered nationalist movements in the colonial territories of Asia, encouraged by the Japanese invasions. These contributed to the eventual political independence of some territories. Independent trade policies then followed. India's independence, which owed little to the war, introduced pervasive and influential state planning for economic development.

Imports were limited to protect infant industries, and private foreign capital was rigorously controlled. In the early GATT negotiations, the Indian representative argued for exemptions from multilateralist principles for developing countries. Their demands were the origin of the third system of post-1945 international economic relations: in addition to the western liberal and the Soviet systems, the state-controlled developing country system.

These less developed countries gained from a prominent innovation in this period: the rise of official international capital flows. At first, these movements were between wartime allies (lend-lease and mutual aid) or to finance the dollar gap created by the devastation of war (Marshall Aid). Subsequently, official aid, although diminished in volume, was rerouted to less developed countries. Official capital inflows allowed them to run an import surplus, in contrast to the pre-war period when they needed to balance their trade. Private foreign investment returned to peak inter-war levels, but remained low in relation to production.

A further benefit to some poorer countries was the Korean War boom in raw material prices. The lack of investment in supplying these commodities throughout the 1930s and 1940s, combined with the stockpiling of the Korean War and the recovery of industrial Europe, rapidly pushed up material prices, and the shortages encouraged fears of a new Malthusian crisis. As investment recovered and supplies increased, prices fell relative to other goods during the 1950s and 1960s; the pattern became reminiscent of the nineteenth-century cycle in investment.

Manufacturing trade volumes expanded more rapidly than trade in primary products, thanks to the recovery of industrial Europe and the greater scope for technical progress in manufactures production. The much higher production and national incomes of the early 1950s compared with 1937/8 were not matched by international trade, which continued to be subject to a large number of restrictions. Some of these the American planners of the post-war economic system had hoped to eliminate: inconvertible currencies, multiple exchange rates and floating exchange rates were outlawed by the International Monetary Fund; discriminatory tariffs and quantitative restrictions were condemned by the General Agreement on Tariffs and Trade. The sterling area and Commonwealth preferences were therefore a major source of friction between America and the United Kingdom. By tying the post-war British loan in 1945 to sterling convertibility within two years, the United States hoped to eliminate them. The attitude of the United States changed somewhat with the intensification of the Cold War in the late 1940s. If western Europe was to remain non-communist, then rapid recovery and prosperity was required. Discriminatory European trade practices, in the form of the Payments Union and the European Coal and Steel Community, were accepted as the price.

Throughout this process, the IMF remained ineffective, except perhaps in the devaluations of 1949, when it may have helped to avoid retaliatory action that could have reduced world trade and employment. The other institution of the Bretton Woods Agreement, the World Bank, was more significant, although it provided only a small proportion of the capital flowing to less developed countries. The Bretton Woods 'system' seemed to have been still-born.

Notes

1. In 1933 the US Congress had voted that the Philippines should become independent in ten years.
2. Primo Levi survived his experience in a camp that failed to produce artificial rubber to describe it in *If This is A Man* (1987).

13 The new liberal trade order

Between the end of the Korean War and the oil crises of the 1970s the world economy boomed. Never before or since have so many economies shared such extraordinary and sustained expansion. Restrictions on cross-border transactions and travel were progressively relaxed. Economic integration surpassed the levels attained under the pre-1914 gold standard. At first, most economies grew by catching up on the backlog of investment opportunities accumulated during the troubled years after 1914. But as countries became more and more open to world trade and investment, new technologies were generated and diffused more rapidly. The institutional framework of the new order supported a more interventionist form of economic liberalism that proved remarkably effective at maintaining prosperity in the two decades after 1953. The General Agreement on Tariffs and Trade, and Cold War competition, in particular, lowered barriers to trade in manufactures between industrialised countries, although there was much less success with primary product trade. Governments maintained high levels of aggregate demand, which induced investment and massive outlays by firms on research and development. These expenditures spawned numerous inventions and innovations that fuelled the engine of international economic growth.

New technologies shaped the world economy both by the channels through which they were transmitted between economies and by their impacts upon the receiving society. In richer countries, the main effect was to accelerate economic growth. In poorer regions, although standards of living also usually rose, the new ideas and techniques indirectly triggered a population explosion. Measures such as chemical spraying to eliminate malaria-carrying mosquitoes suddenly and radically reduced death rates, while fertility remained high (Preston, 1980).[1] From an annual rate of population increase in the second quarter of the century of 1.5 per cent, the southern group of countries in Latin America, Africa, southern Asia and Oceania raised their population growth to 2.4 per cent per annum between 1950 and 1975 (United Nations, 1974, p. 65). Prosperity boosted birth rates in the northern group – North America, Europe, the USSR and eastern Asia – but their population growth increased from 0.6 per cent to only 1.3 per cent over the same period.

In view of the fundamental nature of the advances in industrial knowledge, the first section of the chapter describes the generation and international transfer of technology, going on to consider their influence on economic growth and the pattern of trade. Trade directions were radically changed by the commercial policies of both industrial market economies and non-industrial countries, which form the subject of the following two sections. Less obvious forces also need to be invoked if actual trade flows are to be explained, and we therefore go on to consider the role of spheres of influence. Policy in non-industrial countries was particularly concerned to avoid being trapped by declining relative prices of primary goods. For this reason, the terms of trade between manufactures and primary products are discussed next. Operating a far more restrictive foreign trade regime, the centrally planned economies described in the sixth section were less radically affected by the international price explosion of the early 1970s, but they experienced different problems. In contrast to the less developed countries and the industrial market economies, their productivity growth decelerated in the 1960s. By the 1970s eastern Europe was urgently trying to import western technology to support flagging growth rates (Holzman, 1976; Wilczynski, 1974, ch. 1). Yet another difference between centrally planned and market-based international economic relations arose in migration policy, explored in the next section. How much international relations contributed to the overall growth performance of the world economy, together with a discussion of the value of that growth, concludes the subject matter of the chapter.

The new technology: generation and international transfer

The most advanced western industrial countries accounted for the bulk of major product and process innovations. Of one hundred major innovations between 1945 and the late 1960s, about 60 per cent were introduced by US companies, 14 per cent by UK companies and 11 per cent by West German companies (OECD, 1970). Such clear American technological leadership until the early 1960s was based on a higher ratio of research and development (R & D) expenditures to gross domestic product than other countries. In addition, a much larger GDP enabled the United States to undertake by far the greatest volume of R & D. As the highest-income country in the world with the most productive industry, the United States could rarely copy technology from other countries, and therefore needed to invest more in achieving a given rate of technical progress. Second in both total R & D and the ratio of R & D to GDP was the United Kingdom. High government spending on defence, nuclear and space research and development accounted for much of the higher than average R & D ratios of both countries in the early 1960s. During that decade the gap in non-defence R & D ratios between OECD countries narrowed rapidly (OECD, 1980). By 1975 the non-defence ratios in Germany, Japan and the Netherlands exceeded those of the USA and the UK.

Typically, about 60 per cent of the R & D of OECD countries, particularly France, Germany, Sweden, the UK and America, government-financed programmes accounted for a considerable part of total industrial R & D. About three-quarters of the private industry-financed R & D was concentrated in electricals and electronics, chemicals, machinery and other transport (excluding aerospace). State-financed expenditures were directed to the same sectors, except that governments excluded other transport and included aerospace. The huge sums spent by the US government on developing a computer for ballistics calculations conferred an advantage on US companies when computers became commercial propositions around 1960. Once Bell Telephone Laboratories had demonstrated the potential of the semi-conductor, the armed services and later NASA financed development and indicated that they were a large and stable market. Military design experience in transistors carried over to civilian work (Nelson and Wright, 1992; OECD, 1980, p. 30).

Research and development clearly raised economic growth. Rapid productivity growth industries or sectors either were high R & D spenders, such as chemicals, or bought equipment from big R & D spending sectors, such as air transport. Even in such a traditional industry as textiles, research and development began to make a major impact. In the years after 1960, textile production in OECD countries was transformed from a labour-intensive to a capital-intensive industry (OECD, 1980, p. 51). Textile machinery incorporated progress in materials science, fibre technology, hydrodynamics, aerodynamics and, later, electronics. From 1970 numerical control spread rapidly in the machine-tool industry. Most dramatic of all the industrial transformations was perhaps the 'micro-electronic revolution' beginning in the second half of the 1970s. The capabilities of one of the first electronic computers, ENIAC, built in the 1940s for several million dollars, could be reproduced in 1978 for less than $100 in a microcomputer which calculated twenty times faster, was 10,000 times more reliable, and required 56,000 times less power and 300,000 times less space. Such radical innovations were bound to have pervasive effects in many sectors. On the other hand, innovations in some other science-based sectors slowed down in the 1970s, especially in pharmaceuticals and in pesticides, because of more stringent safety and environmental standards.

The new technology was transferred from the producing countries in a variety of ways. In the non-communist world, multinational companies played a major role: a large proportion of international payments for technology were between parent and subsidiary companies. It was to encourage such flows that Japan slightly eased restrictions on foreign direct investment in the 1960s (Hanson, 1981). Countries such as the USSR that excluded this investment handicapped themselves, although there may have been offsetting political advantages.

Official aid flows contributed to transferring technologies from the advanced industrial countries to the less developed economies. The appropriateness of some of the technologies imported by LDCs through both official channels and multinationals has, however, been questionable. The costs of redesigning equipment to use efficiently the resource endowments of less developed countries may have been excessive, or the decision to buy particular technologies may have been economically irrational. A typical criticism was that imported technologies failed to utilise abundant labour, but made exorbitant demands on scarce capital. There is evidence of some sensitivity to factor endowments, but also that techniques were often transferred unmodified. In nine out of eleven industries in which 1,484 US multinational enterprises operated in Europe and in less developed countries during 1970, the production process was run more labour intensively in the less developed countries. However, there was no significant difference between the installed techniques of developed and less developed country affiliates in five of the eleven industries (Courtney and Leipziger, 1975; Jenkins, 1977, ch. 5; Odell, 1975).

At least some of the production methods adopted in LDCs were responses to government regulations. The transportation industry employed identical methods in the two groups of countries, but 40 per cent more capital per person was used in less developed countries than in developed countries. In Latin America, at least, this was attributable to regulations requiring specific proportions of motor vehicles to be manufactured domestically, regardless of the small size of the market. The emphasis on technologies using oil, instead of traditional indigenous energy sources, or using less energy-intensive techniques, raised the total demand for energy in Latin America almost fivefold between 1939 and 1973. Motor transport and urban motorways were energy-intensive choices compared with mass transit facilities. Rural electrification displaced wood as an energy source. For Brazil and Argentina by the 1970s, oil was accounting for 25 per cent of available foreign exchange, even though oil prices tended to fall until 1970.

Technology imports and economic performance

Despite the possible transfer of inappropriate technology in some instances, technology imports boosted the growth rates of most of the recipient countries. The greater willingness and ability of all industrial countries to take advantage of new techniques compared with the inter-war years was based on the strength of demand. High demand pressure reduced the risks inherent in utilising new techniques by lowering the probability that the new products or processes would be unprofitable (see, for example, Cornwall, 1977). The strength of the reconstruction boom, fuelled by liberalisation and stabilised by a large government sector, was mainly responsible for the great strength of demand in the two decades after 1953. Expectations of continued price and wage stability in the face of fiscal and monetary expansion were self-fulfilling for many years. Instead output and employment rose. Only towards the end of the 1960s and in the early 1970s when these policies, especially in the United States, became more strongly expansionary did expectations of future inflation begin to rise. Wage-earners anticipated higher prices in their money wage demands and in so doing raised unemployment by pricing the marginal workers out of the market.[2]

Demand pressure, however, cannot adequately explain differences in growth rates between industrial countries. Rather the major determinants were the extent of the technological lag and the ratio of investment to GNP, reflecting entrepreneurial vigour in innovating and borrowing technology, and flexibility of the labour force in accepting new production tasks, work rules and equipment, and being willing to move to new jobs and areas. Surplus labour from the agricultural sector, where the growth of labour productivity usually outstripped demand for agricultural produce, fuelled the growth of many economies, but it was not an essential contribution.[3] Labour could be drawn from other sectors if required.

Growth rates decelerated in the 1970s not because of the exhaustion of labour supplies, but because other OECD countries had largely caught up technologically with the United States. Changes in the composition of output also helped. The manufacturing sector had the greatest scope for productivity improvement but with increasing affluence more services, rather than manufactures, were demanded. Oil price hikes in 1973/4 and 1979, which reduced demand in industrial countries, combined with restrictive monetary policies in the face of rising inflation, also contributed to declining growth rates. 1974 was the first year since 1945 in which the combined growth rates of the OECD countries failed to increase. 1975 was the first year in which aggregate income actually declined.

By contrast with many western industrial countries, the communist bloc countries required increasing imports of western technology in order to prevent growth rates declining. Soviet total factor productivity growth decelerated from the high levels of the early 1950s, becoming negative between 1967 and 1973. The central planning system seemed to be inadequate for the development and application of new technology, except in the high-priority areas of the military and aerospace industries. Consequently, in an effort to remedy this deficiency, Soviet imports of western machinery rose rapidly between 1955 and 1978, as a proportion of Soviet equipment investment from around 2 per cent to about 5.5 per cent (Hanson, 1981, chs. 8 and 9). Even so, technology imports remained small by comparison with those of the Japanese. Soviet licence purchases in 1970 were about one-eighth of the value of Japan's (although this statistic conceals the high import dependence of some Soviet sectors such as chemicals, shipping in the late 1950s and early 1960s and, by the 1970s, the motor industry).

The overall effect of imported western technology was probably to raise Soviet industrial growth by between 0.2 and 0.4 per cent per annum. In Poland, where the share of western machinery in industrial equipment investment reached as high as 30 per cent in the mid-1970s, the impact was considerably greater, at least temporarily, although the subsequent crisis suggested an inability of the central planning system to absorb new technology on this scale.

Trade patterns and new technology

Rapid technological change inevitably affected the pattern of international trade. In the three decades after the Korean War, trade in manufactures between industrial countries grew remarkably. Mainly for this reason, by far the largest

Trade 13.1 The network of world trade, 1963–78 ($m, f.o.b.).[a]

Origin	Year	Destination		
		Industrial areas	Developing areas	Eastern Trading area
Industrial areas	1963	69,285	21,900	3,495
	1973	288,915	68,740	18,160
	1978	578,760	199,570	41,955
Developing areas	1963	22,140	6,685	1,670
	1973	79,475	22,540	5,290
	1978	215,120	68,765	12,855
Eastern Trading Area	1963	3,505	2,465	12,375
	1973	15,370	6,545	32,390
	1978	33,445	18,410	69,400

[a] Australia, New Zealand and South Africa are excluded.
Source: GATT (1982).

market for the industrial area's exports was the industrial world itself (Table 13.1). Communist bloc countries were the most rapidly growing market between 1963 and 1978, but remained a small proportion of total trade, in the last year amounting to only 7 per cent of the industrial area's exports to itself. The developing areas increased their importance as markets for the industrial countries, but in 1978 were only one-third as important as the industrial economies for themselves.

Between 1963 and 1978 exports from developing countries to the industrial countries rose by almost ten times, and within the developing area they increased proportionately even more. Throughout the period, the industrial countries remained the main market for developing country exports, taking three times as much as the developing countries themselves. Communist bloc economies were considerably less dependent on trade outside their area, but this dependency increased over the period. Exports to industrial areas were the most dynamic component of world trade, by 1978 accounting for almost half of intra-bloc exports. Exports to less developed countries also rose markedly, amounting to almost one-quarter of intra-bloc exports.

The product composition of the trade of industrial and oil-importing developing countries (see Figure 13.1) shows how wrong it is to identify industrial countries solely as exporters of manufactures and as primary product importers. During the 1970s, the proportion of non-fuel primary products in total imports of the two groups was fairly similar. Although non-fuel primary products were a smaller proportion of industrial than of developing country exports, the value of these goods exported by industrial countries was much greater than those exported by developing countries.

Manufactures increased their share of the exports of both groups, but the rise was much more rapid for the developing countries. The composition of the latter

264 *A History of the World Economy*

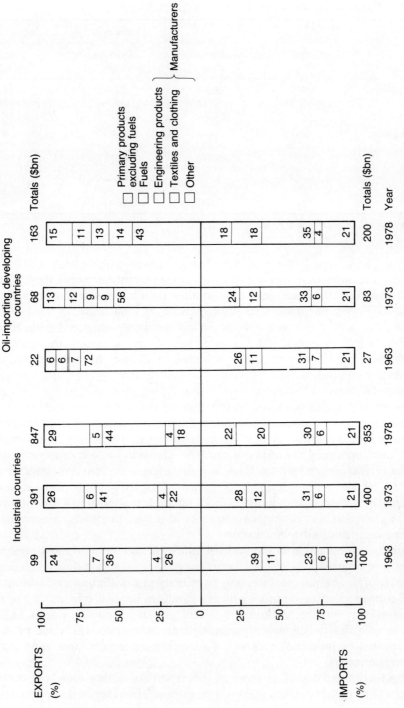

Figure 13.1 Product composition of the trade of industrial (GATT, 1979) and oil-importing developing countries

group of manufactures exports was more heavily weighted towards textiles and clothing, whereas industrial countries' exports were biased more towards engineering products (although the industrial country proportion of imports accounted for by engineering products was much lower than for developing countries). This emphasis on textile exports bore important consequences for the average tariff faced by developing country exports, discussed below.

How can technological change account for these patterns? A Hecksher–Ohlin theory in which the key factors of production are land and labour, or capital and labour, is of little use because the theory predicts that capital-abundant or land-scarce industrial economies should export manufactures and import primary products. Whereas trade in the mid-nineteenth century seems to have been based mainly on natural resource endowments (see Chapter 3), the rate of technical progress was then considerably slower than in the third quarter of the twentieth century (to judge by the growth rates of income per head). In the later period, technologies were more likely to differ markedly between countries because of lags in diffusions. Trade in manufactures between industrial countries may therefore have arisen from this technology gap giving rise to comparative advantages. Alternatively or additionally, the new trade pattern may have stemmed from increasing product variety. Trade is driven in the first case by process innovation and in the second by product innovation.

A very substantial portion of the trade in manufactures between industrial countries entailed the exchange of goods of similar industries. For ten countries accounting for 58 per cent of world exports in 1967, the average level of this intra-industry trade was almost two-thirds of their total trade (Grubel and Lloyd, 1975, p. 36, ch. 6). Some trade was the exporting and importing of differentiated products that were close substitutes, and some involved semi-processed materials and components passed between vertically integrated subsidiaries of multinational companies – intra-firm trade. Intra-industry trade was highest for the engineering and transportation industries in 1976 (United Nations, 1980).[4] Economies of scale therefore probably played a major role in generating the trade, for these industries tended to produce under such conditions. The disappearance of many small car manufacturers in Europe since 1958 with the creation of the EEC is consistent with this view. Several types of differentiated product in France and Germany turned out to be very close substitutes for each other once tariffs were reduced. Whichever firm produced one of these competing models with the largest runs or lowest costs (or both) increased its sales and achieved still larger cost savings. The other competing models – Borgward, Lloyd, NSU, Simca and Citroën – either went bankrupt or were absorbed by larger firms.

Rising exports of manufactures by LDCs may be explained in part by the product cycle theory.[5] New products will tend to be manufactured first in a country with a high income and mass market, because the opportunities for profitable sales are greater and because entrepreneurs need to be close to the consumers to judge their wants. When the product has been successfully introduced for long enough, it becomes standardised and the production can be

Table 13.2 US net exports[a] of manufactures by production characteristics, 1960–70 ($m, f.o.b.).

	1960	1970
R & D intensive, high wages	1,800	4,096
R & D intensive, high wages, capital intensive	1,537	4,392
Capital intensive, high wages	716	−1,334
Capital intensive	285	−424
Residual (mainly labour intensive)	1,170	−2,145
All manufactures	5,508	4,585

[a] Net exports are total exports minus total imports of that category.
Source: GATT (1972, p.24).

relocated to areas that have lower labour costs, but which still provide a substantial market, so that advantages can be taken of scale economies. The high-income country exports the new products and the lower-income country exports the standardised products.

If a country maintains a comparative advantage in innovating and continually develops new products and processes, then the Hecksher–Ohlin theory can be modified to take this into account. The United States in the post-war years, at least until 1971, maintained and increased an advantage in innovating: that is, an advantage in research-intensive products (Balassa, 1977). A study of US trade in 1947 showed, apparently contrary to the Hecksher–Ohlin theory, that America exported labour-intensive products and imported capital-intensive products (Leontief, 1969), exactly the opposite of the result expected for a relatively capital-abundant economy. However, in this and other instances of the 'Leontieff paradox' discovered for other countries, the introduction of human capital as a factor of production, in the form of skilled labour, resolves the paradox. Rich countries tend to be relatively abundantly endowed with skilled labour, and therefore expect those products intensive in this factor. Thus in the period of rapid US export growth, between 1960 and 1970, US comparative advantage shifted further towards commodities intensive in research and development (Table 13.2), which were not as subject to import substitution as the fairly standard manufactures mainly exported between 1953 and 1960.

As a technological leader, the United States' exports reflected the continuous generation of new products and processes. Japan was a technological follower economy during the 1950s and 1960s. Japanese payments for foreign licences and patents rose from $20 million in 1955 to nearly $350 million in 1969. These were payments that allowed Japan to utilise foreign R & D, although Japanese research and development expenditure grew at the same time, as Japan diminished the technological gap (GATT, 1971, p. 8).

A constant market share analysis of trade in manufactures (Table 13.3) between 1963 and 1967 shows the massive increase in Japanese competitiveness.

Table 13.3 Changes in the volume of exports of manufactures attributable to changes in the size of the world market, the pattern of world trade, and market shares, 1963–7 ($m, 1963 prices).

	Base year exports	World market growth	Commodity pattern	Area pattern	Market share (current weighted)	Actual change
United States	13,503	5,500	127	606	710	5,523
United Kingdom	9,984	4,578	71	−217	−3,299	1,133
Japan	4,949	2,275	−149	253	2,692	5,071
EEC Six	29,168	13,554	−58	−462	712	13,746
France	5,841	2,637	8	−278	−382	1,985
Germany	12,910	6,029	193	39	−197	5,986
Italy	3,877	1,870	−71	−64	1,253	2,988
Belgium–Luxembourg	3,830	1,769	−180	−58	−96	1,435
Netherlands	2,710	1,249	−8	−23	134	1,325
Canada	2,830	1,265	−100	607	946	2,718
Sweden	2,195	1,026	−22	16	−56	964
Switzerland	2,196	1,030	−13	14	−289	742

Source: Batchelor *et al.* (1980).

Germany as well as the United States by now had overtaken the United Kingdom in the size of their manufactures exports, but they both also lost market shares. The much greater decline in British competitiveness was temporarily corrected by the 1967 devaluation. Thereafter the deterioration continued at a slower rate. Italy's arrival as a major exporter and the increase in Italian competitiveness were new features of the post-war era.

The commodity composition of trade adversely affected Japan, Belgium–Luxembourg and Canada. Differential rates of growth by market area were more important. British trade continued to suffer from the high proportion of exports still sold to markets outside the fast-growing industrial areas, especially the semi-industrial countries. Unlike Japan, the UK gained no compensation in the form of an unusually high dependence on the US market with its rising import propensity.

The country and commodity composition of trade reflected commercial policies of governments of both industrial and developing groups which resisted the market forces tending towards changing and increased international specialisation. This was one reason why trade in manufactures greatly exceeded that in non-fuel primary products, a break with the experience of the preceding century. Much of the expansion was explicable by the easing of restrictions on manufactures trade during the 1960s, together with the growth of national incomes. Agricultural trade remained tightly regulated, but with far lower transport costs than ever before, and economic integration reached new peaks. By the 1970s world trade/GNP ratios were surpassing 1913 levels (Grassman, 1980).

Trade policy of the industrial world

After the early tariff reduction success of GATT, the item-by-item approach to negotiation began to flag. GATT followed the path of least resistance, allowing each negotiator to select those items on which concessions might be given. Countries with low tariffs could offer little in exchange for a reduction in their partners' high tariffs, and the US Reciprocal Trade Agreement Act only gave the US administration power to cut tariffs by 5 per cent per annum. The inability of GATT to deal with the European tariff problems encouraged a move to complete trade integration among the six countries that eventually formed the European Economic Community with a common external tariff (Curzon and Curzon, 1969; Curzon, 1965, pp. 87–107). They signed the Treaty of Rome, laying down the rules for the Common Market in 1957. The Americans and the British believed that the new grouping of major export markets would raise discriminatory tariffs against them. The UK proposed as an alternative a European-wide free-trade area within the OEEC. French industry and agriculture objected strongly to the international competition implied by this scheme, and the French government thus apparently unilaterally announced in November 1958 that the free-trade area was not possible (Camps, 1959).

As a temporary measure, the United Kingdom created the European Free Trade Association for those countries remaining outside the EEC. The Americans set about obtaining a reduction in the proposed common external tariff of the EEC. Their efforts culminated in the Dillon Round of multilateral negotiations within the GATT framework. Although this was clearly inadequate for what was intended of it, the EEC used the round to help the US administration obtain more negotiating freedom from Congress. A linear cut of 20 per cent in the Community's external tariff was to become operative if and when Congress gave the US administration power to reciprocate. The new administration of President Kennedy took the opportunity to promote a 'Grand Design' for Atlantic partnership with the US Trade Expansion Act of 1962, intended to push Britain into the Common Market (Schlesinger, 1965, pp. 842–4). The Act proposed a 50 per cent linear tariff cut and the complete elimination of duties on those products in which the EEC and the USA accounted for more than 80 per cent of free-world exports.

General de Gaulle was unwilling to allow the United Kingdom to join the EEC. To avoid any possible challenge to French political influence in the Community, he vetoed the UK's application in 1963. This failure to enlarge the EEC blocked the proposal to eliminate duties on goods mainly supplied by the United States and the EEC, because this category was now small. Nevertheless the Act was even more successful than the Anglo-French treaty of 1860, which had been a major step in spreading free trade in the first liberal era (Chapter 3). The industrialised countries participating in the Kennedy Round made tariff reductions on 70 per cent of their dutiable imports (excluding cereals, meat and dairy products). Moreover, two-thirds of these cuts were of 50 per cent or more. Trade in temperate zone agricultural products, however, was relatively untouched. Other achievements of the Kennedy Round included the discussion of non-tariff barriers to trade, and a protocol by which a planned increase of imports in communist countries was to be considered equivalent to a tariff concession on the part of a market economy. The Tokyo Round, ending in 1979, was distinguished by a less open-handed US policy. Codes of practice on subsidies and government buying were agreed and the impact of tariff cuts conceded was felt in the 1980s (Chapter 15).

Almost all the major reductions in tariffs took place between 1959 and 1971. The most important cuts, apart from the Kennedy Round, were those of the Dillon Round, and Canadian–United States Automotive Agreement and those associated with the formation of the European Economic Community and the European Free Trade Area (EFTA). They were reflected in the area and commodity pattern of trade. More than half the trade increase between the United States and Canada was accounted for by transport equipment resulting from the 1965 free-trade agreement, and over the period 1959–72 total manufactures trade between the two countries increased at 12.5 per cent per annum (Batchelor *et al.*, 1980, ch. 4). The growth in EFTA manufactures trade was slightly higher at 13 per cent per annum, but because of slow British economic growth, the rate was well below the expansion of EEC intra-trade in manufactures (of nearly 18 per cent per annum in dollar value).

Some of this increase in EEC trade was diversion by the common external tariff from non-EEC suppliers with lower opportunity costs to EEC producers with higher opportunity costs.[6] But the trade creation effect, stemming from improved resource use as internal tariffs were eliminated, is generally thought to have outweighed the diversion, and was probably about 25–35 per cent of trade in the absence of the EEC. The EFTA effect was smaller: EFTA imports were probably about 10–15 per cent higher because of trade creation. A lower-bound estimate of the effects of the US–Canada Agreement is that it accounted for two-thirds of the 'additional' Canadian imports in 1967, and around 30 per cent of the 'additional' US imports. Perhaps 10–15 per cent more imports were drawn into the United States by the Dillon Round reductions. In the Kennedy Round, the biggest cuts in duties of 50 per cent fell on machinery and vehicles. It is likely that American and British imports were affected most strongly by these reductions; in the British case the upper-bound effects were probably to raise imports by 15 per cent, and the United States' imports were increased by more. These estimates of the effects of tariff cuts, together with changes in price competitiveness, explain most industrial countries' rising propensity to import. But for the United States, the United Kingdom and Germany between one-third and a half of 'additional' imports remain unaccounted for; apparently in all three countries import propensities were rising autonomously. The explanation was most probably an increasing consumer demand for variety that could only be supplied efficiently from a wider market area, transcending national boundaries.

Japanese trade policy by 1960 was under considerable international pressure to liberalise. Until then control by the Ministry for International Trade and Industry (MITI) had been very tight. During the 1950s MITI set up a rationing cartel for imported scrap iron when there was a glut, and rationed raw cotton as a means of controlling cotton-spinning production. In the following decade, moral suasion was more commonly employed than rationing. But Japan's low propensity to import manufactures was supported by a host of regulations and standards that increasingly irritated European and American governments as the Japanese trade surplus grew inexorably.

A common response to such competitiveness, which threatened jobs in industries with political influence, was the Voluntary Export Restraint or Voluntary Restraint Agreement. Voluntariness was actually the exercise of bilateral bargaining power. Low-cost textile exporters would not have agreed to limit their sales unless they believed the failure to do so would bring worse consequences. Perhaps the best known agreement was the periodically renegotiated Multi-Fibre Agreement (MFA). This agreement, which began in 1974 and expired in January 1982, promised a 6 per cent annual growth rate in the textile trade. Hong Kong, South Korea, Taiwan and India all suffered export growth restriction over this period from unilateral cuts under the MFA's 'market disruption' clause. If imports caused or threatened serious economic disruption, the importing country could take temporary action that did not discriminate against imports by country of origin. What in fact happened was that some exporters were bullied into cutting back because they were more vulnerable to economic retaliation, while others were allowed to maintain their sales. The market disruption clause was invoked almost 800 times during the life of this agreement, usually with little or none of the prior consultation required by the MFA.

Primary product trade policy

Developed countries became more protective of their agricultural sectors, rather than less, just as the developing countries increasingly feather-bedded their industrial sectors. Trade in temperate zone agricultural products therefore showed nothing like the same expansion as trade in manufactures. The fundamental difference was that the developed countries were sufficiently rich to use resources in this fashion without dire consequences for themselves. The industrialised or high-income countries were located in temperate zones and were the main consuming countries for temperate zone products (grains, meat, dairy products, sugar beet). This meant that production could be, and was, expanded in countries which had been traditional importers. Temperate zone products at the beginning of the 1970s accounted for about 30 per cent of world agricultural exports, or about 10 per cent of total world trade (Grogan, 1972; Nagle, 1976).

Although they were apparently contrary to the General Agreement on Tariffs and Trade, the United States maintained import controls on cotton, wheat and wheat flour under the Agricultural Adjustment Act (Kock, 1969, ch. 7). After 1951 American butter imports were virtually eliminated by controls. On the other hand, because exports of agricultural products were the single most important export trade of the United States, America was interested in expanding world markets for temperate zone primary produce. In 1956, 60 per cent of total US wheat shipments abroad and 80 per cent of cotton were non-commercial (Curzon, 1965). Disaster relief, economic aid and surplus disposal agreements were the main means of disposing of US agricultural produce that would otherwise force down domestic prices. In the late 1960s, about one-quarter of total US agricultural exports were dealt with in this fashion.

With the formation of the European Common Market by the Treaty of Rome in 1957, a common agricultural policy among the six members began to emerge. France, the Netherlands and Italy would not open their markets to industrial goods if Germany in particular did not admit their agricultural exports. National policies therefore needed to be replaced by a common agricultural regime. Resolution of these interests with the EEC Common Agricultural Policy (CAP) in 1962 greatly increased agricultural protectionism. A variable levy raised foreign prices to domestic price levels, similar to Britain's Corn Laws which had been abolished in 1846. In 1962 Germany's 'threshold price' of wheat set by the EEC was 106 per cent above the landed price. A refund system to EEC exporters of cereals of the same size as the levy allowed them to sell below world prices (Conybeare, 1987, ch. 7; Curzon, 1965, p. 202).

Immediately the CAP provoked a confrontation with the United States, and continued to do so sporadically for at least the next thirty years. US poultry meat exports, especially to West Germany, were among the first victims of the new policy. French and Dutch producers gained. A series of negotiations and threats culminated in an agreement to take the dispute to arbitration by GATT in September 1963. GATT found in favour of the USA and damages of $26 million were awarded, which the USA recouped by imposing duties on selected EC exports to the United States. The EC was unable to back down because of the dependence of the French government on the agricultural vote, and the pre-eminent position France held in the Six. Renegotiation was difficult because of the extremely cumbersome machinery by which CAP policy was formed, and the Six's perception that the CAP was central to the Community. GATT arbitration both prevented an escalation of the dispute and diminished incentives to find more co-operative and beneficial solutions.

In addition to the potential for starting trade wars, the cost of agricultural protection to the consumer or taxpayer in the six EEC countries and the UK in 1961 has been estimated at $3.75 billion (Grogan, 1972). Traditional justifications for such enormous costs include the benefits of strengthening and maintaining the peasantry. Where the peasantry was an important electoral force, these gains were understandably likely to be reflected in policy, but for a country such as the UK from which the peasantry had long disappeared such an explanation will not do. Here the effects of wartime expansion of domestic food supplies played an important role. Unfortunately, continental European farmers remained discontented with the CAP, for their incomes did not rise as fast as in other sectors and a substantial difference persisted.

Agriculture in industrialised economies faced a limited demand for its products, even in a protected market. The high income levels, and high calorie consumption, of the developed countries implied that market growth would be driven mainly by now slow population growth. As prosperity spread in the West, so the composition of demand changed: cereal consumption declined, and meat, milk and sugar consumption all rose after the war. On the other side of the market, technological advance in some countries was extremely fast. Workers were forced to leave the land in large numbers, if market forces were not restrained. Gross output per man-hour in US agriculture more than tripled between 1950–2 and the mid-1960s. World production of temperate zone products after the war increased by very much less: about 50 per cent for feed grains, mutton and lamb, and about 100 per cent for sugar beet.

Protection for agriculture in the industrialised economies reduced the demand for produce from regions specialising in agriculture. The level of protection for a range of commodities and countries varied from 20 to 40 per cent.[7] The exports of LDCs may have been cut by more than $3.5–4.5 billion by this protection (Johnson, 1964).[8] Some economies were especially vulnerable to protection policies for particular commodities. Exports of particular products such as Canadian or Australian wheat, or New Zealand meat and dairy products, sometimes exceeded 75 per cent of national production. Canada, Australia and New Zealand increased their exports only moderately in contrast to the major proportionate and absolute rise in exports by the protected agricultural sectors of the USA and the EEC.

The difficulties caused by the increased temperate zone agricultural protectionism after the war were felt most acutely perhaps by New Zealand. Agriculture dominated New Zealand's GNP and accounted for 92 per cent of its exports (Grogan, 1972, p. 59). Moreover, industrial growth was limited both by the country's resource base and its small population (2.6 million). The economy was therefore extremely vulnerable to external influences on its agricultural trade. Traditionally, New Zealand had been dependent on the British market for about half of its exports. Increasing British and other countries' self-sufficiency, culminating in the UK's entry into the Common Market which limited New Zealand agricultural exports, accentuated New Zealand's balance of payments problems and required the continuation of the import licensing first adopted in 1938. Efforts to develop alternative markets in Japan and the USA met with considerable success, but both markets pursued protectionist policies. New Zealand therefore had to try to diversify its economy into secondary industries including textiles, forest products, and basic iron, steel and aluminium smelting, a policy apparently similar to the import-substituting industrialisation of LDCs (see below).

The industrial countries were less unwilling to accept the changing pattern of comparative advantage when adjustment was required in their industrial structures, although even here the market was not allowed free play. Japan, during the 1950s in particular, pursued a policy similar to the import-substituting industrialisation of developing countries (GATT, 1971). Nevertheless opportunities were provided for exports from other countries. A long-term arrangement on cotton textiles from 1962 was negotiated under the auspices of GATT, contrary to two of its own fundamental principles: non-discrimination and the freeing of trade from quantitative restrictions. Cotton textile industries in industrial countries were allowed to decline, but at a slower rate than market forces dictated. Total output of woven cotton piece goods in the UK halved between 1951 and 1965, and textile employment fell by 5 per cent per annum (Little *et al.*, 1970). Diversification in the textile centre of Lancashire proceeded rapidly: whereas 11 per cent were employed in textiles in 1959, by 1964 the proportion had fallen to 8 per cent. Perhaps three-quarters of the cotton mills closed between 1951 and 1964 were reoccupied for other purposes.

Similarly, in response to oil imports, the coal output of the EEC declined from 1956, and in the UK from 1958. European OECD hard coal production fell 12 per cent between 1958 and 1965. The increase in consumption of oil in competitive uses was one-half of total coal production in 1965. New imports of oil rose 15 per cent per annum, and coal employment in Belgium, France, the UK and West Germany fell by 6 per cent per annum. Even so, oil was not allowed to compete freely; oil usage was heavily taxed and financial assistance was given to displaced miners.

The remarkable dependence of the United States on oil imports by 1973, which dominated international relations in the 1970s, came about quite contrary to the professed intentions of trade policy. Imports of crude oil into the United States were limited by a quota system in 1959, whereby imported oil could not exceed 9 per cent of consumption (Yergin, 1991, pp. 538–40; GATT, 1972). The quotas were allegedly intended to provide an incentive for making the United States independent of crude oil imports in times of emergency. Higher oil prices induced by the quotas were expected to encourage a more intensive search for domestic oil. But such was the oil glut of the 1960s that the quotas were insufficient to raise oil prices. In January 1969 the price of oil relative to other wholesale products was 10 per cent lower than it had been eleven years earlier (McKie, 1976, pp. 73–4). Falling real domestic prices of oil and a high growth rate of the economy during the 1960s rapidly expanded the demand for oil in the USA and elsewhere. Then American imports during the early 1970s were driven up by three factors: the curtailment of natural gas supplies by regulations; the unexpectedly low rate of growth of nuclear power; and a variety of environmental restrictions on the strip mining and burning of coal, and on the development of domestic oil supplies. America's indigenous oil supplies nevertheless meant that American oil imports in 1972, at 4.74 million barrels a day, were similar to the Japanese. They were small by comparison with western Europe's 14.06 million barrels, even though western European oil consumption was less than that of the United States (Dornstadter and Landsberg, 1976, p. 21).

Trade policy of non-industrial countries

The oil embargo and quadrupling of oil prices in 1973/4 is perhaps the best-known example of the trade policy of the non-industrial countries. Posted prices rose from $2.90 a barrel in mid-1973 to $5.12 in October 1973 and $11.65 in December. Oil, however, differs from most other non-industrial country commodity exports by virtue of being imported, instead of exported, by most less developed economies. The wresting of control over the oil trade from western multinational companies by LDC governments also distinguishes oil from most other commodity exports. The oil companies pursued restrictive practices in selling to LDCs for longer than in other markets. Until the mid-1960s, the West African Supply Agreement for supplying the small markets of western Africa eliminated all competition and kept prices high (Odell, 1975). Only when some west African countries began to refine their own oil did the arrangement end. Similarly, the price rigging of oil supplies to the Indian market only ended when in 1960 the Indian government was offered cheaper oil from the USSR, and formed a national oil company able to handle imports independently. LDCs attempted to reduce dependence and oil import costs by building refineries even where the market was smaller than the 2 million tons per annum minimum efficient size. Generally speaking, the large countries were successful and the smaller were not.

Exploitation by the oil companies were negligible in comparison to the later exploitation by the Organisation of Petroleum Exporting Countries (OPEC). Oil companies took $1–2 a barrel at the most for their services, whereas from 1973 the oil-producing countries took royalties of $10 or more on 10–30 cent production costs. Foreign exchange reserves were just not available in many poorer economies to pay for this increase. OPEC's ability to raise all oil export prices so dramatically in 1973–4 stemmed from the breakdown of the 'working relationship' of the eight major oil companies with the entry of independent oil companies into the Middle East oil fields. Further assistance came from improvements in oil-exporting countries' collection and interpretation of information that affected their negotiating position (Vernon, 1976, pp. 7–8; Sampson, 1975; ch. 7). Libya learned how to use Occidental, an independent oil company, to put pressure on the 'majors'. A common desire of the main oil-producing countries to defeat Israel, and the ability of some, especially Saudi Arabia, to cut their oil production if necessary, were the trigger for the price increase in October 1973.

The success of non-industrial countries in raising oil prices encouraged the United Nations Conference on Trade and Development (UNCTAD) to press harder for schemes of commodity price management which they had always advocated. The developing countries in the United Nations formed UNCTAD in 1964 to advance their particular concerns about international economic relations. The only major international commodity price stabilisation scheme that operated through the post-war period did not justify UNCTAD's enthusiasm. The International Tin Agreement (ITA) of 1956 set up a buffer stock and a management that was to buy or sell tin on the world market to keep the price within an agreed range for most of the time. Unfortunately, the buffer stock was not large enough to maintain the price within the desired limits; producer countries did not want to bear the costs of financing an adequate stock (Desai, 1966; Smith and Schink, 1976).[9] The ability of ITA to hold the tin price above the agreement floor in most years required the buffer stock operation to be supplemented by tin export quotas. The agreement only slightly reduced the instability of tin prices and producer incomes.

Before the oil price rise of 1973, successive UNCTAD conferences were largely ineffective. The most constructive change brought about by UNCTAD was the Generalised System of Preferences (GSP), whereby manufactured exports from poor countries received non-reciprocal tariff preferences in developed country markets. The impact of these preferences on LDC exports was probably small because developed countries usually excluded the crucial textile sector (see above) and were also liable to impose import quotas (Baldwin and Murray, 1977).

It was much easier to pursue national policies to combat export instability than to try to secure international agreement on what should be done. Import-substituting industrialisation (ISI) was the most common national response to export instability, for this policy also promised to solve the other problems of non-industrial countries' exports: the tariff discrimination and other restrictions on their manufactured exports to developed countries, and the limited scope for economic growth based on the expansion of markets for (non-oil) primary products (which, not being competitive with developed countries, were therefore not discriminated against). As Table 13.4 shows, developed countries taxed manufactured imports from poor countries more heavily than others. This was because of the heavy concentration in textiles and clothing, industries that were

Table 13.4 Rates of protection of manufactured imports.

	Nominal rate of protection		Effective rate of protection	
	Developed country total imports	Imports from less developed countries	Developed country total imports	Imports from less developed countries
Pre-Kennedy Round	10.9	17.1	19.2	33.4
Post-Kennedy Round	6.5	11.8	11.2	22.6

Source: Little *et al.* (1970, table 8.1).

losing competitiveness in the West and which therefore clamoured for protection. The tariff reductions of the Kennedy Round, although benefiting LDCs, did not alter this discrimination.

Even higher than nominal rates of protection were effective rates. The effective rate is an attempt to measure the extent to which a domestic manufacturer can increase its processing costs (or value added) as compared with a foreign competitor without exceeding the price of the imported product. Employing certain restrictive assumptions, the effective rate takes account of the effects of tariffs on the prices of (importable) inputs, as well as output. The protective effect of a tariff on domestic manufacturing is therefore larger the lower is the duty on the (importable) raw materials used in the manufacturing process. The higher effective than nominal rates in Table 13.4 can then be generated by nominal rates on final goods generally exceeding those on intermediate goods.[10] Tariffs and quotas in foreign markets nevertheless did not prevent the small open economies of south-east Asia from achieving remarkable performances in increased manufactured exports. Hong Kong, South Korea and Taiwan, whose combined manufactured exports rose from under $500 million in 1959 to at least $20 billion by 1976, were especially successful (Batchelor et al., 1980, p. 25).

Another reason for supposing that import-substituting industrialisation policies were not essential is that fluctuations in exports were not necessarily as harmful as had been thought. The implications for export instability of the permanent income hypothesis of consumption and saving suggest that this instability might actually be beneficial for economic development (Yotopoulos and Nugent, 1976, pp. 328–40).[11] Transitory changes in income contribute more to saving and less to consumption according to this hypothesis, than 'permanent' or long-run and predictable changes. An economy with substantial unpredictable variations in export earnings will therefore save more than an economy with stable earnings, other things being equal. If the financial system is efficient enough to channel these additional savings into investment, the first economy might reasonably be expected to experience a higher rate of economic growth. An empirical test for thirty-eight countries between 1949 and 1967 supports this theory. Consistent with the conclusion for the nineteenth century (see Chapter 6), export instability is unlikely to have offered an insurmountable obstacle to economic development, because in the new liberal trade era, although a lack of diversification in exports was a cause of instability, it was less significant for LDCs than for developed countries (Batchelor et al., 1980, p. 222).

Instruments for promoting ISI policies included heavy taxation of primary produce exportables and the use of the revenues so obtained to subsidise domestically produced manufactures. Tariffs and quotas on manufactured imports were also applied to induce the same effect: a shift of domestic prices against the agricultural sector in favour of manufactures. Protection of domestic manufactures by tariffs and quotas on imports in pursuit of ISI was often carried to extreme lengths. The 118 per cent average rate of effective protection in Brazil ranged from 41 per cent on machinery products to 8,480 per cent on perfumes and soap (Little *et al.*, 1970). Domestic farmers were hurt by the higher prices that they had to pay for manufactures, and by the lower prices they received for their own produce. In Pakistan for much of the 1950s and 1960s the prices of manufactures in relation to farm prices were twice as high as the world market averages. Incentives to produce food, and even to stay in the countryside, were therefore greatly curtailed. The result was often food shortages and accelerated migration to urban areas. Indian agricultural production failed to increase in the fifteen years after 1950, despite the population explosion and the average growth of GDP per head at 1.5 per cent per annum. The ISI policy against the agricultural sector in LDCs meant that, before the quadrupling of oil prices in 1973–4, LDC imports of food were increasing at almost twice the rate of exports of food (Little *et al.*, 1970, pp. 102–3). Fifteen per cent of all LDC imports were foodstuffs in the early 1970s, and 30 per cent were primary products.

Diverting resources from agriculture to manufactures not only increased food imports, but also reduced exports, both primary products and manufactures. LDCs lost market shares in the 1950s and 1960s in three-quarters of their primary product exports, exclusive of fuel, in which substitute or competitive commodities were produced in developed countries (grains, fruits, vegetables, and oils and oil seeds) (Kravis, 1970). But about one-third of the non-oil-exporting LDCs for which statistics are readily available achieved export growth rates in excess of 6 per cent per annum during these decades. The first thirteen achieved higher export growth rates that the industrial countries as a whole. All were small economies. Successful exporters' incomes also tended to grow more rapidly. Export growth must have been caused either by world markets or by supply conditions, whether spontaneous or brought about by policies favourable to trade, the opposite of ISI. If supply conditions are the explanation, successful exporters must have increased their relative shares in traditional exports and diversified the commodity composition of their exports. GATT computed world market growth, diversification and competitiveness factors for LDC export performance between 1959–61 and 1964–5, classifying countries into three groups by export performance. The successful performers were distinguished from the less successful primarily by increases in their shares of world markets for their traditional exports, rather than by a rapidly expanding world demand for their particular exports. The successful exporters tended also to gain more exports by diversification, although the margins of superiority here were much smaller. As has been seen in previous chapters, India's manufactured exports increased their competitiveness during the early twentieth century. With the adoption of ISI policies on independence, from being a leading exporter of manufactures in 1953, India lost 18 per cent of total exports in 1965 through failing to maintain its share in world exports. The system of import and investment licensing gave virtually every Indian firm a monopolistic position, and served the critical link between profitability and economic performance (Krueger, 1975, p. 108).

Trade patterns and spheres of influence

The trade policies of the industrial and non-industrial countries, together with a three- or four-factor Hecksher–Ohlin model, can explain much of the pattern of trade during the two booming decades after the Korean War, but not all. Apart from the formation of regional customs unions, commercial policies under GATT were not supposed to discriminate between countries. It was true that British Commonwealth tariff preferences persisted until the entry of the United Kingdom into the EEC in 1973, but by 1948 the preference rate for British goods averaged only 7 per cent – although since then and the early 1960s, changes in the area and commodity pattern of Commonwealth trade had tended to raise the margin (MacDougall and Hutt, 1954; *Board of Trade Journal*, 1965a, 1965b). Sterling area controls during the early 1950s also tended to create discriminatory trading patterns. The US–Canadian Automotive Trade Agreement of 1965 worked in the same direction, as did a number of other administrative controls and conventions. The net effect of these forces, historical and cultural ties between traders, the tying of aid, the establishment of multinational subsidiaries, and preferential tariffs, as well as transport costs, tended to bias the manufactured goods trade flows from the industrial countries of the world in a way which would not be expected from Hecksher–Ohlin theory.

Industrial countries tended to market their weakest sectors of manufactures disproportionately in areas which could be thought of as 'spheres of influence', a pattern which cannot be explained solely by physical distance between traders (Roemer, 1976). For instance, in ships, furniture, domestic electrical appliances and iron and steel, the United States in 1971 held at least seven times the market share in Canada that it maintained internationally for those sectors. The US world market share in each of these sectors was less than half of its world market share in manufactures as a whole. These results cannot be explained by the Hecksher–Ohlin theory; in every area market, a given country was competing with the same adversaries, and their relative factor endowments, although varying for different trade sectors, remained unchanged for the same sector between market areas. Neither can multinational subsidiaries account for the phenomenon because, for countries other than the USA, intra-firm trade in manufactures was a small fraction of total trade in manufactures (for Japan about 5 per cent). For the United States, eliminating intra-firm trade did not alter the relationship. For most countries other than the UK and the British Commonwealth, sphere of influence and economic distance were correlated, but transport costs were a small proportion of output value and therefore could not account for the trading patterns. An exporter's strong sector commanded about the same market share in all areas. On one interpretation the importers were a captive market. On another the trade may have been less imperfect than trade between arbitrary pairs of countries because channels of communication and information were more highly developed. It was cheaper and less risky for importers to buy a known product than to pay the costs necessary to compare products of various exporters.

The captive market interpretation is usually most favoured in accounting for the dominance of the imperial power in colonial trade. Hence the granting of independence to European colonies in the 1950s and 1960s, by freeing the markets, must have lost the former imperial powers trade. In the period after independence, the imports of former colonies did grow much more slowly than those of other less developed areas (Kleiman, 1976). Thus it seems likely that, had colonial rule persisted, the imports of the colonies would have been greater and the colonising power's share higher than they actually were after independence. On this basis the United Kingdom's export loss in 1972 from decolonisation was 8.5 per cent of exports and the French loss was 13.1 per cent of French exports (Kleiman, 1978). The gains to the UK and France in 1960 from not decolonising earlier were 8.2 and 16.2 per cent respectively of their exports, although the magnitude of the export loss varies significantly with the choice of control group. In the above case, imports of former colonies are assumed to have grown at the same rate as world imports as a whole in the absence of independence. If instead import growth into other less developed areas is chosen as the control, then the losses to the UK and France are radically reduced. There is a problem in applying a common import substitution factor to all ex-colonies despite differing sizes of domestic market, stages of industrial development, and economic structures (Livingstone, 1978). Furthermore, many of the factors 'distorting' trade under colonialism were part of a process which continued under independence; comparative advantage and market forces are not easily abolished by a change of political regime. The Associated State status of some former colonial African countries, giving them privileged access to the EEC market, was another continuing distortion, although only the small, low-income Commonwealth nations were granted this status when the UK joined the Common Market in 1973.

The effects of colonial status on economic structure and international trade in the 1950s and 1960s are agreed, at least qualitatively. Colonial status appeared to raise the ratio of trade to output among the least developed countries, and to depress the share of manufactured goods in total exports, and of chemicals and capital goods in imports (Batchelor *et al.*, 1980). Achievement of independence more than reversed this effect on the composition of exports, but it did not seem to change the relationship between total trade and output, although a weakening of trading links with the former colonial power followed the weakening or severance of political links.

There is evidence of a difference in experience between colonial powers. Between 1960–2 and 1968–70, the decline in the British share in former dependencies could for the most part be explained by the UK's generally reduced role in the trade of the world's less developed areas (Kleiman, 1976). By contrast, the French share in the trade of former dependencies declined no less rapidly, despite the considerable increase in France's share in the trade of other less developed areas. The greater effect of decolonisation on the French group is consistent with the conclusion based on the situation in 1960–2: that the degree to which the share of the imperial power in the colonies' trade can be ascribed to colonial role was inversely related to the importance of the colonial power in world trade. This is also suggested by the account in Chapter 6 of trade and colonisation in the last two decades of the nineteenth century.

If colonisation raised the share of the colonial power in the colonies' trade, decolonisation cannot be attributed to a declining significance of this trade, because the importance of the dependencies in the trade of their imperial countries increased in the three decades before their independence began to be considered. In any case, relative to the domestic product of the imperial countries, any potential gain from colonial trade could have been of only limited significance for them. The colonial territories, on the other hand, with a combination of high imperial trade shares and high ratio of trade to GDP, were very vulnerable to the behaviour and development of their imperial powers, as had been demonstrated in the inter-war years (Chapter 10). Before and after decolonisation, the industrial countries influenced these territories through the terms of trade between primary and manufactured products, among other ways.

The terms of trade between primary and manufactured products

Investment in primary products triggered by the Korean War boom proved sufficient to expand supply so that the terms of trade of primary products with manufactures declined gently from 1952 until the end of the 1960s, despite massive growth in the production of, and trade in, manufactures (Spraos, 1980). Nevertheless, primary product prices in money terms had risen. At the end of 1971, market prices in sterling for staple commodities other than oil were between three and four times as high as in 1939 (Cairncross, 1974). By the spring of 1974 they had tripled again. Most of this rise was concentrated in the eighteen months from the autumn of 1972, over which period the rise was half as great as in the preceding century. The increase in oil prices, after a gradual recovery from the low levels of 1969–70, was more violent, raising import costs fourfold within a year. Other commodities – zinc, wool and sugar, for example – matched the rise in oil prices, but lacked the economic importance of oil and did not provide comparable gains to their producers. The price increases did little more than restore the price relations with manufactures of the early 1950s. Taking British import and export unit values as an imperfect measure of these relations, the shift in relative prices was about 22 per cent in favour of British exports between 1953–5 and autumn 1972, and 27 per cent against British exports over the next year and a half. Fuel imports costs in 1972 were no higher in sterling terms than fifteen years earlier, whereas manufactures were selling at much higher prices.

Apart from OPEC's actions, there were three main explanations for the price explosion: the 17 per cent expansion of world production between 1971 and 1973 as all industrial countries reflated their economies together; delays in the development of new primary product sources of supply; and a speculative boom, touched off by international monetary uncertainties, that made the holding of commodities preferable to currency. These price increases raised the value of imports in the first quarter of 1974 for Italy by 92 per cent, for Japan by 85 per cent (February to April) and for the UK by 60 per cent (February to April). They created a transfer problem comparable to the German reparations problem of the 1920s (Chapter 11). In the British case, even allowing for a 20 per cent rise in exports prices, the implied transfer of resources to the countries from which the imports came was equivalent to 8 per cent of GDP, or a 10 per cent drop in consumption. In order to make the transfer, exports would have had to increase by nearly 50 per cent in volume within a year. Where some of the oil exporters were concerned, the transfer was complicated by the inability of the recipients to spend all their new income on oil importers' exports, or on anything else, because their new earnings were so large.

No group of primary product producers could completely emulate OPEC as proponents of the new order hoped, but the International Bauxite Association came close to it. Jamaica began the bauxite offensive in March 1974, successfully tying its income to the prices of aluminium ingots, and raising its tax revenues from $25 million to $170 million. Over the following months, all four other Caribbean producers and Guinea instituted similar taxation methods. Of the main bauxite producers, only Australia failed to follow suit.

By mid-1974 commodity prices (excluding food) were falling, adding to the difficulties of the oil-importing developing countries. Speculative activity had declined and world industrial production growth had decelerated.

Communist-system trade policy

Largely independent of free markets, the communist bloc economies were less affected by these massive swings in prices. In the centrally planned economies, supplies and demands were balanced instead by physical controls. Prices remained fixed for long periods of time (in Russia, from 1955 to 1967) and did not reflect relative scarcities or consumers' wants. Inflation was therefore suppressed and was measured by shortages of goods, rather than price rises (Holzman, 1976).

The absence of an internal pricing system whereby prices reflected costs required intra-bloc trade through the Council for Mutual Economic Assistance (CMEA) to be based on historical world prices. Raw materials were thereby obviously underpriced, and a form of bilateralism tended to be introduced: the exchange of raw materials for other raw materials, and similarly for manufactures. The USSR, primarily a raw material exporter to the CMEA, lost from that trade. In 1980, at the prevailing rouble–dollar exchange rate, the Polish import price for Soviet crude oil was 52 per cent below the average import price from the West (Vanous and Marrese, 1982). Securing gains from trade with such irrational prices was problematic, and the communist countries consequently were more trade averse and had much lower trade/GNP ratios than the West: in 1967 the communist ratio was 11.3 per cent whereas the West's ratio was 69.3 per cent.

In some respects, trade within the communist bloc (CMEA) encountered difficulties similar to those of market economies unwilling to accept the specialisation imposed by market forces: Romania, for example, resented the demands of the more advanced CMEA countries that it should concentrate on low-technology production. But unlike the trade of western economies, intra-bloc trade was clearly motivated by political objectives. The Soviet Union would have gained more materially from exporting its energy and raw materials to the West and buying manufactures there, rather than conducting this trade with eastern Europe. The trade subsidies that the Soviet Union granted eastern Europe averaged $5.8 billion between 1976 and 1978, rising to $10.4 billion in 1979 and $21.7 billion in 1980. These payments can be regarded only as a means of obtaining important military, strategic and political benefits associated with the control of eastern Europe. The subsidies are unlikely to have compensated fully for the loss of economic sovereignty. But full compensation was unnecessary because of the high cost of a popular rebellion to eastern Europe, as was demonstrated in Hungary in 1956, in Czechoslovakia in 1968 and, with the introduction of martial law, in Poland in 1981. Moreover, a restructuring of internal and external economic relations, which the removal of Soviet domination would have allowed, would almost certainly have raised the low productivity of the eastern European economies by many times more than was necessary to offset the hypothetical loss of the trade subsidy.

The absence of prices reflecting opportunity costs, and the use of quantity planning, created a number of difficulties in communist trade with the West. Tariffs in centrally planned economies did not protect because they did not affect prices or the availability of goods. Hence tariffs were used solely as bargaining devices in negotiations with the West for most favoured nation treatment. Dumping, the selling of goods in foreign markets at prices below the domestic costs of production, could not be defined for a communist bloc economy, yet from the viewpoint of western producers in import-competing industries such sales could be extremely harmful. As Table 13.1 shows, East–West trade was a rapidly expanding, though small, proportion of world trade.

Economic interest sometimes transcended ideology: Poland was admitted to the Kennedy Round negotiations in 1965, and Romania and Hungary joined GATT in 1971 and 1973. Romania was the most dissident of the satellite states in expanding trade with western Europe. Between 1959 and 1968 such trade as a proportion of the Romanian total rose from 20 to 45 per cent. Politics was, however, always liable to take priority in East–West trade, as in intra-bloc trade. The history of the US–USSR Trade Agreement, approved by Congress in 1974, amply demonstrates this proposition. The following year the agreement was annulled by the USSR because the revaluation of gold and oil improved the Soviet foreign exchange position. The Soviet authorities therefore concluded that it was no longer necessary to concede rights for Jewish emigration with which the agreement was linked.

International labour mobility

As the above example shows, communist countries differed from the western economies in their attitudes to international migration. Whereas the centrally planned bloc attempted to stop people leaving, the market economies were concerned to prevent new arrivals settling. Despite the greater sense of national exclusiveness exemplified by immigration controls, the return of international full employment in the 1950s allowed migration largely to resume the nineteenth-century pattern: a movement from the populous lands of Europe to the regions of recent European settlement, and within Europe, from agricultural regions to the rapidly developing industrial regions. There were important differences though. Indian and Chinese migration ceased to follow the nineteenth-century pattern, but in greatly reduced volume some migration from Asia (and Africa) now went to Europe. Latin America, which had formerly been a net receiver of migrants, now lost population on balance as relative prosperity diminished, and the United States became a destination for Latin American, especially Mexican and Caribbean, migrants. Immigration no longer made a major contribution to the population growth of the United States, but it continued to do so in Australia, Canada and New Zealand. Fifty-nine per cent of Australia's population growth between 1947 and 1973 was due to immigration (R. L. Smith, 1979). Immigration to Canada accounted for up to 35 per cent of Canadian population increase in years of high immigration (Scott, 1971). The precise contribution of immigration to the US population is masked by the unknown number of illegal immigrants. By the 1970s between 2 and 12 per cent of the US labour force came into this category, and of these one-half to three-quarters were Mexicans, mainly working in south-western agriculture (Piore, 1979).

A slowdown of European migration overseas was accompanied by an intensification of international migration within Europe, and by the emergence of new migration into Europe from other areas, notably from northern Africa and Turkey, and from some previously colonial countries overseas. By contrast with previous history, Europe as a whole had an outward migratory balance of fewer than 3 million during 1950 and 1960, and fewer than 400,000 during 1960 to 1970 (United Nations, 1974). West Germany became a destination, as it had been before the First World War, and France took even more immigrants (see Table 13.5). By 1973 there were around 11.5 million legal and illegal foreign workers in western Europe (Paine, 1974, p. 1).

Encouraged by the integration policies of the EEC and EFTA, the countries of the Mediterranean, first Italy, then Greece, Spain and Portugal, and later Turkey and Yugoslavia, became a reservoir of labour for the industrial nations. Until 1950 the majority of Italian emigrants left Europe, but subsequently, with the European boom, most stayed on the continent (Rieben, 1971). Switzerland took almost half of Italian European migration. As the Italian domestic demand for labour expanded rapidly during the 1960s, Italian migration declined. A similar pattern, but later in timing, is shown by Greek emigration. Until 1960 Greek emigrants went mainly to America and Australia, then the majority were drawn

Table 13.5 Approximate net international migration, 1960–70 (million)

Destinations		Origins	
USA	3.9	Latin America	1.9
Australia	0.9	N. Africa	0.6[a]
Canada	0.7	Portugal	1.3
France	2.2	Italy	0.8
W. Germany	2.1	Yugoslavia	0.7
Switzerland	0.3	E. Germany	0.6
Sweden	0.2	Spain	0.5
Belgium	0.15	Greece	0.45
		Poland	0.3
		UK	0.15

[a] Excluding the million or so French settlers who returned to France after Algerian independence.
Source: United Nations (1974, p.23).

into West German industry, and after 1968, with the growth of Greek industry and the slowing of west European growth rates, Greek emigration in total subsided.

In addition to migration inspired by pecuniary motives, the post-war decades saw other major population movements. Ten million refugees left Bangladesh for India during the fighting of 1971 and subsequently returned. The following year the Asians of Uganda were expelled. Indo-China, the Middle East, Algeria, Hungary, Cuba and Czechoslovakia all supplied migrants for political reasons who did not return.

By redistributing labour from lower-productivity economies to higher-productivity economies, commercially motivated international migration should have raised total world production, and both sets of economies should usually have gained a share of this increased output. Concern about the economic costs of migration to both the host and the donor countries often originated in the operation of national fiscal systems. A donor country could be concerned at the emigration of skilled labour educated at the expense of the taxpayer, and a host country could be worried about unskilled immigration becoming a net drain upon the social services. The United Kingdom received a net inflow of largely unskilled New Commonwealth immigrants, which reached a peak between 1960 and 1966 before being restricted by law to avoid racial disturbances such as had occurred in Notting Hill in 1958. New Commonwealth immigrants were sometimes accused of making more demands on the social services than were offset by their work contribution. In fact, their demands on the social services were less than those of the indigenous population, and indigenous living standards probably did not suffer because of immigrants' capital needs (Jones and Smith, 1970, pp. 159–61). Immigrants bore the brunt of deflationary fluctuations, suffering greater than average unemployment during slumps, although when general unemployment was low, immigrant employment rates did not differ significantly from general unemployment.

Nevertheless antagonism to foreign labour led Switzerland to attempt to 'stabilise' its migrant labour force in the spring of 1971, and then to introduce measures to reduce it in July 1973. In 1972 and 1973 France, Holland and West Germany introduced restrictions on new entrants, culminating at the end of 1973 in a ban by the latter two countries on the entry of non-EEC workers. The oil crisis was probably merely a trigger for a decision that had already been reached.

The tightening of immigration controls tended to reconfigure the international division of labour. Controls discriminated against the manufacture of those goods whose production demanded a high proportion of immigrant-type labour. Immigration restrictions favoured importing labour-intensive goods and mechanisation of the production of non-traded goods, such as construction, where the proportion of foreign workers in Europe was highest. Countries originally supplying labour might 'inherit' activities abandoned by the countries of immigration, either through direct investment or through loans financed on the international capital markets. The use of immigrant workers to manufacture in West Germany either delayed the transfer of the activity to other countries, or more likely delayed increasing mechanisation of immigrant-intensive industries such as car production.

A favourable view of the impact of migration on a land-abundant economy underlay the very accommodating Australian immigration policy that lasted until the late 1960s. But more important were strategic considerations. Near-invasion by Japan during the Second World War seemed to prove the vulnerability of the vast and relatively empty Australian continent to the overcrowded countries of Asia. In order to ensure enough people to provide troops and to support an economy capable of providing advanced armaments, it was estimated that a population of 25 million was required, yet the 1945 population was only 7.3 million (R. L. Smith, 1979). Even in 1970 national security was used to justify official immigration policies. The practical consequence of this policy was that more than 60 per cent of new arrivals received government financial assistance. Between 1945 and 1976 about 3.3 million settlers went to Australia. A little under half of all immigrants were British. The 'White Australia' immigration policy became an increasing embarrassment as Australian relations with Asia developed. Between 1970/1 and 1975/6 the proportion of immigrants from Asia in the total rose from 5 to 14 per cent as a result of modifications to this policy.

The effects of migration on an indigenous population can be analysed with an elementary 'sources of growth' model. The contribution of factor inputs to the growth of an economy's output is assessed by weighting the actual growth rate of the inputs by the share of respective factor payments in national income.[12] The implications of a hypothetical elimination of immigration can be derived by cutting the growth rate of the labour force. Applying this approach to the study of skilled migration into Canada between 1950 and 1962, Canadian GNP would have grown at 3.3 per cent per annum instead of 3.8 per cent per annum, and GNP per head would have grown faster, at 2 instead of 1.8 per cent per annum (Scott, 1971). The calculation has to be modified to take into consideration the capital that immigrants brought with them, and the capital and technical progress induced by migration. Immigrants to Canada were more skilled on average than the indigenous population; they added disproportionately to the stock of human capital. In 1967 the average stock of education of the labour force was 3 per cent higher than it would have been without immigration. Immigrants' needs for physical capital may have offset the beneficial impact on the indigenous population of the human capital they brought, although this seems unlikely.

More often the donor country is thought to suffer from skilled migration, and the host country to benefit. Such a reflection contributed to the change in immigration regulations in a number of major recipient countries during the mid- and late 1960s. The United States, Australia, Canada and the UK reduced discrimination by country of origin and promoted immigration of skilled labour. Professional and technically trained labour emigrating to the United States from LDCs increased. The evidence for the UK and Canada is mixed, however, perhaps because of diversion of this labour to the United States (Bhagwati, 1976, pp. 6–7). The LDCs will have suffered from this 'brain drain', unless their skilled labour was underemployed, and emigration acted as a 'safety valve'.

Any evaluation of unskilled migration must take into consideration that aspirations and wants are often changed by migration. Particularly, this is true of the second generation, who may become dissatisfied with accepting their parents' lot at the bottom of the wage and status scale. Puerto Rican migration, concentrated in the 1940s and 1950s, exemplifies the process. New York by the 1970s possessed a second-generation Puerto Rican community, the youth of which generally refused low-wage menial jobs and were prone to rioting and unemployment (Piore, 1979, p. 165). The riots in the English conurbations during 1981 can be attributed in part to the same phenomenon at work in the West Indian communities. In such cases the host country acquired a problem that the liberal doctrines of free trade and factor mobility, with unchanging or exogenous preferences, did not predict.

Some gains to the donor economy from migrant workers were as obvious as they were for nineteenth-century Ireland (Chapter 2). For Turkey, repatriated earnings amounted to 57 per cent of the import bill and about 7 per cent of GNP in 1973, when just over 5 per cent of the active population was working abroad (Paine, 1974, ch. 3). On the other hand, savings and remittances out of earnings abroad were spent mainly on consumption, not on agricultural or industrial investment. Few returned migrants learned new skills abroad, and those who did so tended not to use them on their return. Because the poorest did not migrate from Turkey and because land purchase was the most common use for the savings of returned migrants, the distribution of income was more unequal.

Welfare and economic growth

If migration to new environments changed preferences and thereby made the evaluation of that migration problematic, economic growth created similar difficulties. Rapid economic development transformed environments in ways that were not always planned, and perhaps also altered tastes, causing doubts to be expressed towards the end of this unprecedented economic expansion as to its benefits. Even though national income per head increased, it was argued, people were not necessarily better off in view of the stress, overcrowding and pollution of the modern economy. A further reason for doubting the benefits of rising incomes was the confusion of means with ends in national income accounts, such as the inclusion of the costs of the journey to work as an item of final consumption, instead of as an intermediate good.[13] However, the use of social indicators of well-being, such as the suicide rate, protein consumption per head, doctors per head of population, and students as a proportion of the population, suggests that between 1951 and 1969 for twenty advanced industrial countries there was a positive association between economic growth and welfare indices computed from seventeen social indicators (although the strength of the association diminished over time) (King, 1974). The correlation of welfare with national income was not perfect when countries were compared. A social well-being index ranked the United States below the Scandinavian countries and Canada in 1969. On the basis of income per head, the USA should have been ranked above these countries.

The social indicator approach does not show that people judged themselves to be happier in the 1970s than in the 1950s, even if they were, by 'objective' standards, better-off. According to Scitovsky (1976), the psychology of consumption implies that it is novelty which gives gratification, and therefore a constant increase in consumption is required to maintain a given level of 'utility'; the television sets, motor cars and central heating that spread through the industrial world may not have permanently raised welfare. Hence any inability of the national economies to maintain a constant growth rate, and thereby continue to supply new consumption experiences, would be liable to cause industrial unrest and wage-push inflation, such as occurred during the 1970s.

There are a number of processes by which the international economy could have contributed to this economic growth. The two-gap theory, discussed in Chapter 14, focuses on the relaxation of the import constraint by foreign capital, but it can be extended to show how a larger share of exports in national product can also remove the constraint and promote economic growth. Where output is held back by inadequate domestic investment, the opposite relationship holds: a lower ratio of exports to national product allows a large investment component and therefore greater output. This latter relationship seemed to hold for western Europe, excluding the UK, in the 1960s (Batchelor *et al.*, 1980, pp. 202–4).

Figure 13.2 shows the direct contribution of exports and other components of total expenditure to overall growth in the 1960s. Exports contributed substantially

288 *A History of the World Economy*

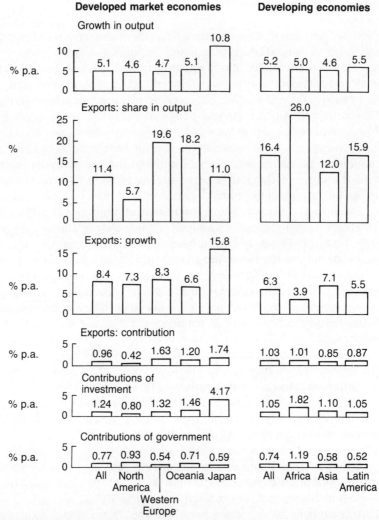

Note: 'Direct contribution' of aggregate demand component to growth is share of output multiplied by growth of component.

Figure 13.2 Direct contribution of exports and other factors to economic growth, 1960–70 (Batchelor *et al.*, 1980)

to output only in western Europe, Oceania and Africa. In no case did trade growth account for more than about a third of output growth, and the comparatively high trade contributions stemmed from differing circumstances. Exports from Africa during the 1960s increased slowly, but accounted for an exceptionally high proportion of total demand, whereas in western Europe and Oceania they rose rapidly. For the advanced countries generally, exports appeared to be equally significant with investment as an explanation of differences in growth performance. This was true also for LDCs where, probably because of foreign exchange constraints, capital imports were strongly correlated with export receipts. In the bigger semi-industrial countries, investment emerged as the more important factor, but in the smaller it was exports, in particular manufactured exports.

Figure 13.2 also shows divergences in growth rates, of which the most extreme is that of Japan. A comparison of developed market economies with developing economies shows very similar growth rates for the 1960s, despite their different rates of population growth. Because of the faster growth of population in less developed countries, this similarity meant a lower growth of income per head on average for developing countries. Within the LDC group, the divergent growth experiences of the oil countries and the newly industrialising countries of eastern Asia, Brazil and Mexico on the one hand, and the remaining LDCs on the other, were wider than the discrepancies in mean growth rates for the total developed and developing groups. Technology gap theories of growth imply that the growth of income per head in poor countries should have been higher instead of lower than in the rich countries, other things being equal. The discussion of economic policy showed that other things were not equal, and that different policies significantly determined economic performance. In addition, the experience of the OPEC economies showed that national resource endowments remained an important ingredient of economic growth, even if less so than in the nineteenth century.

Summary and conclusion

Judged by the increase in trade and incomes, the new liberal trade order was an enormous success, more so than the original liberal achievement of the nineteenth century, despite the failure to eliminate or even reduce international income inequality. The economic experience of the 1950s and 1960s cannot, however, be attributed solely to international economic relations. The United States, by virtue of its size and technological leadership, was largely independent of these relations, although most other economies were greatly influenced by the American economy. Even large countries with a low ratio of trade to national income probably gained substantially more than their 'openness' suggested from the international flow of new technologies. Transfers of ideas and the experience of skilled and professional workers were not necessarily reflected in the international exchange of goods and services. All types of economy were subject to this process – advanced market economies, less developed countries and the communist states. The means by which technology was transferred varied though: multinational companies, official development assistance, licensing, exports of capital equipment and the interchange of ideas all played different roles. How much advantage economies took of the new technologies depended primarily on their economic system, their investment rates and the size of their technological lag behind the United States.

Trade patterns reflected the new technological dynamism. Research and development and human capital became major determinants of trade in manufactures between industrial countries. Manufactures trade between these countries was boosted by widespread tariff cuts implemented in particular by the Kennedy Round. This liberal commercial policy led by the United States gave a further lease of life to the long post-war boom. America's acceptance of international economic leadership, already noted in the redirection of the world economy in the immediate post-war period, contrasted strongly with US policy in the inter-war years. Then there was no obvious external enemy against whom the market economies of the world had to be unified. Soviet expansion after 1945 and ideological competition between the two superpowers showed that there was in the later period. Nevertheless the United States was not uniquely responsible for the burgeoning of western economic relations. Liberalisation and the expenditure-stabilising role of western governments were probably more important in encouraging the investment and innovation that underlay the expansion of the world economy. Even where commercial relations were concerned, the new order of the 1940s developed its own momentum independently of the United States. Western European co-operation in iron and steel flowered into the Common Market as a means of reducing tariff barriers, once GATT appeared to have become ineffective.

The Common Market introduced the greatest western divergence from free trade in the form of the Common Agricultural Policy. The United States came a close second with its extreme agricultural protection, which contrasted with its liberal trade policy in manufactures. Although expensive to domestic consumers and potential foreign suppliers of temperate zone primary products, these policies were justified in Europe by their reduction of the number of workers obliged to make the transition from rural to urban life. Despite protection, technical progress in western agriculture was sufficiently fast to reduce the farm population while farm output increased.

Less developed countries (LDCs) typically pursued the opposite policy of taxing domestic agriculture to support urban manufacturing industries. They maintained that their economies' former pattern of exporting agricultural produce or raw materials offered no prospect for long-term economic development. The policy was usually not altogether successful. Generally, greater food imports were needed, while the subsidised manufacturing sector failed to increase international competitiveness. Independence from colonial rule allowed an increasing number of LDCs to adopt these policies and reduce their dependence on traditional manufactured imports from the colonial powers. LDCs did not win political independence because they had become economically redundant to the imperial powers. On the contrary, colonial shares in the trade of the imperial economies were increasing; political forces were fundamental to independence.

After the Korean War, throughout the 1950s and 1960s, the declining terms of trade of primary products relative to manufactures gave some substance to the claims of less developed countries that the export of primary products was not a viable road to economic development. The simultaneous expansion of the industrial economies in 1972, and the use of the oil weapon by OPEC the following year, ended this trend, bringing price relations back to Korean War levels.

The communist economies were subject to different, though related, problems both in intra-bloc trade and investment, and in relations with market economies. The central planning system experienced difficulties in encouraging innovation across all sectors, and therefore the communist economies became more dependent on western technology imports. Lacking a rational pricing system, communist countries attained much lower trade/GNP ratios than western economies, and engaged in the heavy subsidisation of trade. Some subsidisation, in particular that by the USSR of eastern Europe, was in effect a payment for political advantages.

Outside the communist economies, a resumption of international factor mobility accompanied the increase in world trade, even though the two processes are usually thought to be substitutes for each other. Tight controls limited labour migration once the boom of the 1960s came to an end (and even earlier in some cases), for fear of friction between immigrants and the indigenous population. The pattern of migration showed the strength of western European economic growth. By the 1960s, domestic expansion substantially eliminated net emigration by providing an alternative demand for labour to that of the regions of recent European settlement. Nevertheless, the latter areas continued to take considerable numbers of Europeans and others. These regions shifted their policy to encourage skilled labour regardless of origins, as befitted the new direction of the international economy.

Overall, the new international regime contributed to a period of remarkable economic progress. Even though there were offsets, often associated with urban conditions, surveys of the urban poor showed that they usually thought their conditions better than in the countryside from which they had come, and where they were so much less noticeable (United Nations, 1974, p. 35).

Notes

1. At the end of 1964, 1.935 billion persons lived in areas that were originally malarious. Of these, 41 per cent were living in areas from which malaria had been eradicated, and only 19 per cent lived where no specific anti-malarial measures had been taken.
2. This phenomenon was identified by Friedman (1968).
3. On surplus labour, see Kaldor (1966); Kindleberger (1967); and Cripps and Tarling (1973).
4. For a formalisation of intra-industry trade stemming from product differentiation economies of scale, see Dixit and Norman (1980, pp. 281–95). These authors conclude from their model that there are not necessarily gains from trade. Trade may increase biases in product selection and encourage the establishment of too many firms.

5. On technology gap trade theories and product cycle, see Posner (1961); Vernon (1966); and Wells (1972).
6. On the trade creation and trade diversion effect of a customs union, see Robson (1980, chs. 2–4).
7. The subsidy is calculated by valuing output at prices received by farmers and at import prices for the same year, and expressing the excess as a percentage of the value of output at national prices (GATT, 1971).
8. Johnson assumes that the lowest elasticities of supply and demand likely are respectively 0.15 and 0.2 in absolute value.
9. On commodity schemes in general, see Brown (1980).
10. The effective rate of protection can be represented by the expression

$$e = \frac{n - mi}{v}$$

where e = effective rate, n = nominal rate on the final good, m = nominal rate on the input good, i = coefficient of material input, v = proportion of final output accounted for by value added. Suppose the nominal duty on steel was 20 per cent, and steel was 40 per cent of the manufacturing cost of cars, on which there was 10 per cent duty. Then $n = 0.1$, $m = 0.2$, $i = 0.4$, $v = 0.6$ and $e = 3.33$ per cent on cars. The formula assumes perfect substitutability between imports and domestic products, that all goods are traded at fixed world prices and that there are no changes in the input–output coefficients.
11. The survey by Stein concludes that the relationship between export instability and growth could be in either direction depending on particular national circumstances, although the relationship is a weak one in either case.
12. If the economy were perfectly competitive and had constant returns to scale, then the share of wages in national product would measure the elasticity of output to the labour input. This type of measurement was popularised by Denison (1962).
13. One of the most published exponents of this view was E. J. Mishan (see Misham, 1977).

14 The Bretton Woods system and its transformation

International monetary liberalisation complemented the new liberal trading order. Both supported the great post-war boom. But abolition of controls did not proceed at anything like the pace which the Bretton Woods architects planned. The regime formally differed from the earlier gold exchange standard mainly by allowing the possibility of changing exchange parities. Actual international monetary practice, even when the IMF began operating as intended from the mid-1950s, deviated substantially from the rules of the system. Fixed exchange rates among the major trading powers lasted a remarkably short time by historical standards. Yet a distinctive characteristic of the system – consultation between the largest economies that recognised their interdependence – allowed change and flexibility in response to the crisis of the 1970s, preventing a repetition of the 1931 international collapse. This co-operation continued throughout the period, despite alterations in the relative strengths of the nations involved.

During the 1950s and 1960s, the greatest policy efforts in international economic relations were directed to enhancing world liquidity. This chapter therefore first explains the pressures to raise international reserves and liquidity, the necessity for these increases and the means by which they were implemented. While the 'adjustable peg' exchange rate system lasted, greater liquidity was not intended to remove the need for nations eventually to adjust their economies to balance of payments disequilibria. The pattern of such adjustment under the Bretton Woods system is discussed next. The end of the Bretton Woods exchange rate system came in the early 1970s, but the transition to floating rates described in the third section was soon overshadowed by the disruptions of the oil shocks of 1973. The recession that began that year was the severest that the international economy had experienced since 1929. A comparison between the two slumps, attempted next, illuminates why, nevertheless, the later recession was not as deep as the earlier. Increasing unification of world capital markets contributed to the environment in which the old monetary system was no longer viable. Hence the new institutions and forms of short- and long-term private investment, and their consequences, are considered in the fifth section. Private capital flows to the less developed countries were substantially supplemented by official flows, raising different questions, discussed in the following section. Finally, the problems created by the accumulation of foreign debt are described and analysed.

The growth of international reserves and liquidity

As a creditor nation with a balance of payments surplus and with little dependence on foreign trade, the United States at Bretton Woods had been little concerned with international reserves. Instead it focused on the necessity to ensure the rapid adjustment of deficit nations to keep them within their incomes. Keynes, representing the United Kingdom as a deficit nation, had argued strongly in favour of a more liberal reserves and adjustment policy, but had been overridden. As described in Chapter 12, the American stance was soon modified in the face of Europe's plight and the Cold War. With the change in the US balance of payments, the American view on international reserves during the 1960s began to converge with the British position. Reserve and credit facilities were enlarged and a new form of international reserves, special drawing rights (SDRs), was eventually created.

At the end of 1958 western European currencies became fully convertible into dollars on current account as required by the IMF. The willingness of the rest of the world to hold US dollars as a reserve currency, as an international store of value and medium of exchange, meant that the balance of payments deficit was not then a problem for the United States. In any case, US total dollar liquid liabilities in 1958 were only 80 per cent of US gold reserves (Argy, 1981, p. 34). From the viewpoint of the rest of the world, however, there was a potential problem – the shortage of reserves and credits. Sudden calls on the IMF in the wake of the Suez crisis towards the end of 1956 depleted the Fund's holdings of US dollars by nearly one-third (Strange, 1976, p. 196). An emergency solution was found by selling IMF gold to the US Treasury. The fear was not that a lack of liquidity would oblige the Fund to suspend operations, but that, without a continuous injection of dollars, a general economic recession might ensue. The problem was to decide what determined the size of the injection of dollar reserves needed to avoid a slump, or what determined the demand for reserves.

The demand for international reserves

The demand by governments for reserves must be distinguished from the need or desire of governments for reserves. The availability of reserves affected governments' policies by changing the perceived relative costs of different means of responding to balance of payments disturbances – primarily financing, as against adjustment of balance of payments disequilibrium (Crockett, 1978). If the objective was to avoid unemployment, as it was in the 1950s and 1960s, an increase in the growth of liquidity was required, so financing was preferred to deflating demand to reduce imports. If the objective was to restrain demand and inflation, as it was from the mid-1970s, a contraction of the rate of growth of liquidity was called for, but the result of this contraction might instead be increased resort to devaluations or import restrictions.

The demand for reserves was necessarily more unstable than a national demand for money because there were only 133 members of the IMF by 1978, and the two largest reserve holders alone accounted for about 25 per cent of world reserves. Consequently, there was substantial variability in the relationship between payment imbalances and the level of world trade. Nevertheless, a reasonably stable demand relationship for gross reserves (unadjusted for foreign liabilities) did seem to exist both for the fixed exchange rate years and for the floating exchange rate period from 1973, in which demand depended on the average propensity of an economy to import, the size of the country's imports and the variability of imports (Heller and Khan, 1978).[1] The more open the economy, the greater the demand for reserves because the more susceptible was the country to foreign disturbances. The costs of holding reserves generally did not affect demand significantly. Higher imports required more reserves. Greater variability in the balance of payments created more uncertainty and increased the demand for reserves. With precise measurements of the influence of these variables, it was possible to say what demand would be for different variable values.

Shifts in the composition of reserves

The composition of reserves changed with the differing elasticities of supply of reserves as the demand expanded. Table 14.1 clearly shows that the American insistence on maintaining a fixed dollar price of gold throughout the 1950s and 1960s reduced the role of gold in international monetary relations. The artificially low price reduced production and stimulated commercial use, while the United States' currency became the international medium of exchange. Had the price of gold been determined in the free markets, or had the IMF provided a suitable substitute reserve asset in the appropriate volume, the dollar would not have been the dominant reserve currency for so long. However, the United States obtained

Table 14.1 Percentage composition of international reserves, 1956–73.

Year	Gold	Foreign exchange	Reserves in IMF	Special drawing rights
1956	64.2	31.7	4.1	—
1965	59.3	33.0	6.7	—
1970	40.2	48.2	8.3	3.3
1973	23.7	66.4	4.1	5.8

Source: IMF, *International Financial Statistics*.

substantial benefits from the reserve currency role of the dollar, and therefore resisted any change that might supersede it. Nevertheless, like sterling in the years before 1914, the reserve currency role of the dollar contained the seeds of its own destruction. Foreign confidence in the exchange value of the currency declined.[2] The failure of confidence stemmed from the large and persistent US balance of payments deficit, which raised the ratio of US dollar liabilities to gold reserves as other countries held more dollar reserves. An attempt by foreigners to convert all their dollars into gold by 1967, when external US liabilities were three times gold reserves, and when official US liabilities were 1.5 times gold reserves, would have required a devaluation of the dollar against gold. A more suitable reserve asset in which there was more international confidence might have allowed the Bretton Woods system to survive longer. As it was, the value of American gold reserves was held down by the fixed gold price, while dollar liabilities rose.

Holding $35 an ounce proved increasingly difficult. Between 1954 and 1960 the gold price in the London market kept within the limits prescribed by the Fund, the dollar price fixed in 1934. Upward pressure on the price of gold was countered in 1961 by the formation of the London Gold Pool. The UK, the Common Market and Switzerland agreed to provide half the gold necessary to maintain a market price of $35 per ounce, and the United States agreed to provide the other half (Horsefield, 1969, vol. 1, pp. 484–5; Strange, 1976, pp. 77–9). This arrangement was ended in 1968 after the long-awaited devaluation of sterling shifted speculative pressure to the dollar as the next most likely currency to change its exchange rate. Speculation took the form of selling dollars for gold and driving the gold price of dollars upwards. The London gold market was forced to close temporarily in March 1968 to stop the drain on official monetary reserves (de Vries, 1976, pp. 403–4). Unable to maintain the Gold Pool, the central banks instead established a two-tier price system with gold exchanging between themselves at $35 per ounce, but trading elsewhere at the free-market rate. Henceforth, speculation could only be against the exchange rate of the dollar with other currencies, and foreign central banks could be expected to bear the cost of countering such speculation.

There were savings for the world as a whole from the increasing use of a fiat money, such as the dollar, which cost little to produce, rather than a commodity money, such as gold, which cost a great deal. The resource cost of a gold standard has been estimated at $2–5 billion for the 1960s (Williamson, 1973). The seignorage benefits to the United States arising from other nations' willingness to hold dollars solely because of their usefulness as international media of exchange have been variously calculated at $420 million in 1963, and $1.8 billion in 1961 by different authors and methods. The reserve currency use of sterling in the 1960s may have earned the United Kingdom £100–165 million. Whereas the international use of the dollar rarely constrained American economic policy, British policy was severely restricted by sterling balances (non-British holdings of sterling), and by speculation about the sterling exchange rate caused by these balances.

Although there were real resource savings from fiat reserve currency, they accrued to the reserve currency countries as seignorage. By contrast, a revaluation of gold would have conferred benefits on the gold producers, of which the largest were South Africa and the USSR, and on gold hoarders, especially France, which was determined to reduce the international role of the dollar. In 1967 General de Gaulle converted all French dollar reserves into gold with severe repercussions on the world gold market. An additional benefit to the United States, but not to the rest of the world, from the reserve currency role of the dollar was that the United States did not need to bother about its own balance of payments. Instead America could require the rest of the world to adjust their balances of payments.[3] From the mid-1960s until the end of the fixed-rate system in the early 1970s, this enabled America to pursue expansionary economic policies, and thereby to export inflation to some countries through higher import and export prices and low interest rates.

Measures to increase international reserves

During the earlier period of the Bretton Woods system, it was probable that the American deficit was mainly demand determined: the rest of the world's demand for a larger supply of reserves required them to hold dollars in view of the lack of a suitable alternative. Although the IMF quotas were increased in 1959, this enhancement of international liquidity quickly proved inadequate for an adjustable peg regime in a world of increasing capital mobility. This regime provided those who had to make large and frequent international payments, or who merely wanted to make, or avoid losing, money from foreign exchange transactions, with a one-way bet. Either a government would fulfil its obligation to maintain its exchange rate, or it would be unable to resist pressures for a change, which in any period would always be in one direction.[4] The Basle Agreement, the General Arrangements to Borrow, and the introduction of special drawing rights were all ways of trying to provide suitable supplementary reserves to dollars in order to resist speculative pressures to change exchange rates.

In the late 1950s and early 1960s it was obvious that the growing relative strength of the German economy was increasing the probability of a revaluation of the mark. Speculators therefore bought marks knowing that they could not lose and might well gain, and thereby created more pressure for a revaluation. The revaluation of the mark in March 1961 shifted speculative pressure to sterling. At Basle in 1961 the central banks of the richest industrial countries had agreed to provide automatic short-term support for any currency whose exchange rate was threatened by foreign exchange market pressures (Horsefield, 1969, vol. 1, pp. 483–7; Strange, 1976, pp. 83–9). They would do this either by accumulating sterling (if that was the threatened currency) and holding it for at least three months, or by recycling the hot money inflows, depositing in the central bank of the country of origin an equivalent sum in the refugee currency. Under this agreement $900 million was supplied to the United Kingdom between March and July, but the UK still needed to draw $1,500 million from the IMF in August. This drawing practically exhausted the Fund's resources and obliged the IMF to sell $500 million of its gold holdings.

The Group of Ten industrial nations (independently of the IMF) in December 1961 made public their agreement, the General Arrangements to Borrow, to create a fund of $6 billion for members of the Group from October 1961 (Horsefield, 1969, vol. 1, pp. 511–15). The General Arrangements to Borrow, first used in 1964 by the UK, were thought inadequate to the growing demands for liquidity, at least by the United States, and this led to the creation of special drawing rights, a new reserve asset. In 1969 a $3.5 billion issue was agreed for the first year and $3 billion for each of two subsequent years (Argy, 1981, p. 55). The IMF had concluded that reserve needs were $4–5 billion on the basis of reserves/imports and reserves/imbalance ratios, and that other reserve growth would amount to $1.5 billion.

In contrast to a reserve currency, the SDR seignorage accrued to the SDR participants, not to the issuer of the reserve currency. Interest initially at 1.5 per cent was paid by holders of SDRs. When a participant decided to use SDRs to finance a balance of payments deficit, the IMF designated another participant who would accept these SDRs and provide foreign exchange in return. A participant who had used some of its SDR allocation continued to pay charges at 1.5 per cent on its original allocation, but received interest on its depleted holdings. Conversely, the SDR recipient received interest on its now larger holdings of SDRs and paid charges only on its original allocation. The acquisition of SDRs had no initial impact on the money supply, appearing as a balance sheet adjustment of the national monetary authorities, although SDRs may have allowed the creation of domestic money.

In the event of crises in the gold and foreign exchange markets, national governments and central banks reacted promptly, whereas when faced with the need to adapt international institutions, such as the introduction of the SDR, change was much slower (de Vries, 1976, p. 188; Strange, 1976, p. 255). The creation of SDRs took place only because of the policy reversal of the United States in response to its balance of payments position, and because of the persistence of negotiating officials who thought that, in ways not then clear, SDRs would alleviate the frequent monetary crises. The slow pace of reform was also attributable to the changing political and economic environment of the period of negotiations, and to the technicalities involved.

As it turned out, by the time the SDR reforms were implemented they were unnecessary, for balance of payments disequilibria were endogenously generating increases in reserves; in particular, expansionary American monetary policy in the late 1960s and early 1970s, by worsening the US balance of payments, increased world reserves by more than was desirable. Apart from the special position of the United States, the growth of international capital markets allowed other countries also to increase reserves. A price increase for an important traded commodity,

Table 14.2 Growth in world reserves and world trade. 1950–76 (% p.a.).[a]

	Reserves	Exports
1950–4	3.1	6.2
1955–9	1.6	6.1
1960–4	3.7	8.6
1965–9	2.5	9.8
1970–3[b]	29.7	16.5
1973–6[c]	9.7(3.6)[d]	23.7

[a] In SDRs.
[b] Through the first quarter of 1973.
[c] Beginning with the second quarter of 1973.
[d] World reserves less those of OPEC.
Source: Crockett (1978).

altering current account flows, may have led the surplus country to place newly acquired reserves as short-term deposits in a Eurobank (which accepted deposits denominated in foreign currencies), rather than changing the exchange rate. These deposits counted as an increase in reserves. If the deficit country did not wish to see a decline in its reserves, it could borrow the funds deposited by the surplus countries. These borrowings were not counted as offsets to reserves. Gold and SDRs could not be increased, but national currencies were substitutes for them. Hence the IMF could not control the supply of international liquidity, as shown by the erratic relationship between reserves and export growth, especially the explosion of reserve growth in the primary commodity boom period of 1970–3 (Table 14.2).

Balance of payments adjustment under Bretton Woods

Under the fixed exchange rate Bretton Woods system, economies were forced to adjust eventually to balance of payments disequilibria whatever the availability of reserves. Imbalances occurred because of differing income elasticities of demand for imports and export growth, because of shifts in the terms of trade or recessions, or because of monetary expansion different from that of the rest of the world, unwarranted by differential productivity growth. The cost and availability of reserves and the cost of adjusting determined the optimum pace of adjustment back to a balance of payments equilibrium. Either output needed to increase or domestic expenditure had to be reduced so that more goods were made available to foreigners. Expenditure-reducing policies consisted primarily of monetary and fiscal policy. Expenditure-switching policies were designed to divert foreign expenditure to domestic goods, and domestic expenditure from foreign goods. They entailed altering the price of domestic relative to foreign output, by devaluation, tariff imposition or multiple exchange rates. Where there were unemployed resources, expenditure-switching measures might have been adequate, but not in conditions of full employment.

The monetary approach saw the adjustments as coming about through an elimination of the excess supply of money. A contraction of the supply of money (which includes foreign exchange reserves), or an increase in the demand for money, was required to eliminate a balance of payments deficit. A devaluation raised domestic prices, increased the demand for money, and thereby improved the balance of payments.[5] A study of eighteen independent devaluations by less developed countries, small open economies unable to influence world prices, between 1959 and 1970 showed that devaluations were quite successful in improving their payments position, although the effects were reduced by the usually simultaneous trade liberalisation (Connolly and Taylor, 1976). A moderate decline in the rate of growth of credit was sufficient to ensure the success of the devaluations.

Until the mid-1960s, E. M. Bernstein, a former Research Director of the IMF, thought that the adjustment process had been working well. When an effort had been made to restore the balance of payments either through a change in parity or through domestic policies, it had been successful (Bernstein, 1973). France in 1958/9, Italy in 1963/4 and Germany in 1965/6 all achieved prompt turnabouts from payments deficits to surpluses. The United States managed a remarkable increase in its trade balance between 1960 and 1964, although this was offset by the enormous rise in US foreign investment. Failure of the adjustment process was a development of the later 1960s, seen in the large and persistent deficits of the United States, in the payments difficulties of France, in the enormous surplus of Germany and Japan on the goods and services account, and in the recurrent exchange crises. This was also the period when the dollar began to dominate international reserves, these reserves began to grow rapidly, and international inflation started to rise.

The 1967 devaluation of sterling was one of the most important balance of payments adjustments under the Bretton Woods system, and the way the process worked is vital to an assessment of the fixed rate regime. In the year following the devaluation, British import prices in sterling increased almost proportionately to the change in the exchange rate, by 14.3 per cent against the dollar (Artus, 1975). The case for a devaluation rests to a large extent on the hypothesis that hourly labour earnings in money terms will be adjusted by significantly less than the full amount of the change in the exchange rate. This will improve the international competitiveness of domestic firms. In the British case, the rise in hourly labour earnings caused by the devaluation was slow to appear, and was significantly smaller than the increase in the price of foreign exchange. Primarily this was because food, rent and public service components of the cost of living index were not sharply increased by the devaluation. The incomes policy introduced in March 1968 also delayed adjustment of labour earnings to the cost of living.

The increases in hourly earnings in manufacturing, adjusted for overtime, were abnormally low in 1968/9 considering the rise in retail prices and the level of employment. In 1970 real earnings more than caught up with the rate of increase in output per man-hour, when the period of wage restraint ended and the devaluation effect took place. Feedbacks of the 1967 devaluation on export prices of British manufactures were large. By 1971 they accounted for an increase in export prices of British semi-finished and finished manufactures of 6.5 and 9 per cent, respectively. Prices of British finished manufactures competing with imports rose by only about 1 per cent because of devaluation, and by 5.9 per cent because of higher labour earnings by 1971.

High price elasticities of demand make it more likely that a devaluation will succeed, because the large initial switch in demand away from foreign traded goods and services and away from traded goods and services will start the adjustment process on the right path. The British long-run import price elasticity for finished manufactures was −1 and for semi-finished manufactures −3.4 according to Artus (1975). Even for food and beverages, the Cambridge Growth Project found an elasticity of −0.3. For British exports of semi-finished and finished manufactures respectively, Artus estimated long-run price demand elasticities of −2.5 and −1.4. Previous work that had found trade flows unresponsive to relative price changes was marred by biases in aggregating different commodities with different price changes and elasticities. The impact of the devaluation on the current balance of payments is shown in Table 14.3.

The improvement in the full employment current balance by 1971 was almost £1,300 million. Such calculations assume that a sufficient resource transfer into the balance of payments can take place without additional inflation because real domestic absorption is cut by the improvement in the current balance plus the terms of trade effects: about 3 per cent of GNP in 1971.

Just before devaluation the government expanded rather than contracted demand. When added to the 'J curve effect' (the initial tendency of the balance to deteriorate because of the immediate impact of higher import prices and the delayed impact of lower export prices), the trade balance worsened so much that there was another speculative attack on sterling. Thereafter domestic demand was reduced nearly to a standstill until mid-1971, when an actual surplus on current balance of £1,093 million was obtained, compared with a deficit of £316 million in 1967.

Successful adjustment then required domestic policies which raised unemployment. As it happened, much of this adjustment was vitiated by the international

Table 14.3 The effect of the 1967 devaluation on the British balance of payments (£m)

	1968		1969		1970	1971	Final effect at the scale of flows in 1971
	I[a]	II[a]	I	II			
Trade balance	−134	140	354	533	726	940	709
Current balance (includes services plus other variables)	77	360	592	795	1,036	1,271	996

Source: Artus (1975).

monetary changes of 1971, when the pound was repegged at a higher parity with the dollar. In June the next year, sterling was forced to float (Tew, 1978, p. 315).

The IMF and national adjustment policies

IMF surveillance policies were usually responsible for ensuring that appropriate domestic policies were pursued, that credit growth was restrained, that trade was liberalised, and often that devaluation took place. The Fund took over many of the functions of the nineteenth-century gold standard in enforcing international monetary discipline, and incurred much opprobrium thereby in a number of countries. Some accused the IMF of deliberately frustrating the very type of financial discipline and production adjustments that were most badly needed in less developed countries, by maintaining these countries' openness to international money. The Brazilian slide from democracy to military dictatorship in the early 1960s has been attributed to the insidious influence of the Fund (Payer, 1974, p. 210, ch. 7).

In 1958 a $300 million loan from the United States to Brazil was made contingent on an agreement with the IMF on stabilisation measures. The President of the Bank of Brazil refused to acquiesce in a credit squeeze that threatened to depress the private sector. Coffee growers protested when the coffee purchase programme was cut back, and radical nationalists accused the Brazilian President, Kubitschek, of selling the country to the United States and the IMF. Negotiations were broken off and finance was instead obtained from high-cost private foreign sources, the repayment problem being left to the next Brazilian President, Quadros. Quadros immediately came to terms with the IMF, reformed the exchange system, abolished exchange auctions and substituted instead a dual exchange rate which effectively devalued the currency by 50 per cent for the rate at which 'necessaries' were imported. The other exchange rate was left to the free market. Credits were obtained, but inflation continued and Quadros resigned after only eight months. Under his more left-wing successor, Goulart, economic growth levelled off in 1962. The last serious attempt at stabilisation under the democratic system was made in 1963, designed with the hope of securing IMF approval, so that the foreign debt burden would not take 45 per cent of Brazil's export earnings, as it threatened to do. An agreement for $398.5 million was signed with the United States, conditional upon Brazil pursuing an agreed stabilisation programme. The contemplation of a 70 per cent pay rise for civilian and military government employees signalled the departure from this programme, and aid was suspended. Inflation reached 100 per cent, and the laws about the remittance of profits by multinationals were made more restrictive.

In early 1964 negotiations with the IMF and European creditors were again begun, but Goulart also announced that in response to peasant agitation large tracts of private land would be expropriated and redistributed, and that all private oil refineries would be nationalised. The military deposed Goulart and formed their own government, which quickly obtained new IMF credits, the rescheduling of foreign debts and large sums of American aid – $1.6 billion between 1964 and 1968. In return, the military abolished import subsidies on wheat and petrol. Industrial production fell by 7 per cent in 1965, and living costs increased by 45–60 per cent. The right to strike was virtually abolished, trade union leaders were imprisoned and the income distribution became more unequal. Nevertheless, even if Brazil had been a closed economy quite independent of the IMF or the United States, the same political pressures and their manifestation in the form of high rates of inflation would almost certainly still have been present. International institutions and foreign powers merely provided a convenient scapegoat for domestic difficulties.

When the pegged exchange rate system was abandoned and OPEC seemed to enhance the power of poorer countries, the IMF conceded them some privileges. Although the share of less developed countries in world trade fell during the 1970s, their share in IMF quotas was raised. Under the original Bretton Woods formula, LDCs would have had a 17.6 per cent share, whereas they achieved roughly half as much again (Willett, 1979, pp. 38–9). In addition, the so-called Wittveen facilities of 1977 mainly benefited the LDCs (Erb, 1979, pp. 398–9). Without the facility, IMF resources were considered inadequate to cope with the very large imbalances associated with the oil crisis and the depression. A number of governments therefore agreed to establish a special $10 billion supplementary financing facility.

Not only less developed countries objected to IMF loan conditions. France's drawings of 1958 were conditional, whereas those of the UK in 1957 and 1958 were not, encouraging beliefs that the Fund was run primarily for the benefit of the UK and America (Strange, 1976, p. 54). Latin American countries were typically subject to a much more detailed list of conditions. In ensuring that these conditions were met, the Fund was increasingly concerned from 1959 not to give overt offence to debtor governments, while keeping its power as a creditor. The use of objective performance criteria, such as domestic credit expansion by members with standby arrangements, were one means of achieving this end. These standbys provided the right to draw on the Fund for an agreed amount over an agreed period, subject to the pursuit of agreed policies. The conciliatory stance of the IMF in the 1960s made it easier for the United Kingdom among others to seek such conditional assistance. The Chancellor of the Exchequer wrote Letters of Intent specifying the policies he intended to pursue in 1964 and 1965, when the UK drew $1 billion and $1.4 billion.

Biases in adjustment

In addition to IMF surveillance and the constraint imposed upon national economic policies, the Bretton Woods system was criticised for forcing deflation and devaluation onto reserve-losing countries, without providing a corresponding incentive to revalue and reflate for reserve-gaining countries. A delay in devaluing leads to a rapid loss of competitive strength, but a postponement of revaluation benefits export industries which often have the ear of governments, so it was argued (see, for example, Yeager, 1976, p. 104).

Most parity changes by Fund members were by LDCs which resisted the decision to devalue because of the political impact of higher import prices, opting instead for overvalued exchange rates and exchange controls. Hence a devaluation bias can best be sought among industrial countries. Between 1960 and mid-1971 there were among the industrial countries four decisions to devalue (Canada in 1962, the UK and Denmark in 1967, and France in 1969) and five to revalue (Germany and the Netherlands in 1961, Germany in 1969, and Austria and Switzerland in 1971). In addition, the Canadian authorities in 1970 and the German and Dutch governments in May 1971 allowed their currencies to float upwards. The devaluations ranged between 8 and 14 per cent and the revaluations between 5 and 9 per cent.

This experience can be summarised by weighting the parity change by the share of exports of the industrial countries supplied by the country whose parity was altered. Largely because of the heavy weight given to the two German revaluations by this method, there was a near-balance between devaluations and revaluations: the net weighted change to mid-1971, excluding the floats from 1970, was –0.65 per cent (Katz, 1972). This negative sum was reduced by the inclusion of the post-1970 floats and increased by the inclusion of the French devaluations of 1957–8.

The near-balance in exchange rate decisions between devaluation and appreciation contradicts the devaluation bias hypothesis. An explanation may be found in the increased access to official credits available to deficit countries during the 1960s. The continuing use of such finance nevertheless proved a costly strategy, by which government policies were tightly constrained as a condition of access to funds. The British decision to devalue in 1967, after borrowing for three years to resist the change, was taken partly for fear of the conditions that might be attached to further loans.

The Fund could exercise much less influence over surplus countries. Revaluation undoubtedly reflected a desire to avoid imported inflation, especially by Germany, whose currency had already been destroyed twice by inflation in the twentieth century. Japan reacted differently, however, being more concerned with export competitiveness.

For much of the Bretton Woods period, from 1950 to 1962, and again from 1970, Canada was distinguished from other developed countries by operating a floating instead of a fixed exchange rate regime, and thereby being spared some of the adjustment problems of other countries. Canada needed such a regime, the government stated, because of its close connections with the massive American economy and the need to maintain some independence of economic policy. Later Trudeau, the Canadian Prime Minister, compared Canada's problem to sharing a bed with an elephant: however good relations were, even a slight movement of the partner could cause a disaster (Drouin and Malmgren, 1981). Canada could adopt floating rates and yet remain a member of the IMF because of its special political relation with the United States (Strange, 1976, pp. 45–6). The regime was very successful. When subjected to temporary disturbances, short-term private capital flows played a role analogous to changes in foreign exchange reserves, but the Canadian government did not have to hold these funds or make judgements about the equilibrium exchange rate (Helliwell, 1975). The floating rate in fact operated much like a fixed rate, with a depreciation of the exchange rate causing a net capital inflow, thereby tending to stimulate an offsetting movement of the rate.

The end of the fixed exchange rate regime

Despite, or perhaps because of, the variety of measures undertaken to enhance international liquidity, the international monetary system partly broke down in 1971. The proportion of reserves held as foreign exchange and the level of these reserves both increased greatly from the mid-1960s. The United States balance of payments deficits showed the symptoms expected from financing the Vietnam War and the Great Society social programmes substantially by monetary growth. The second half of the 1960s was also the period when the balance of payments adjustment process ceased to work well, and inflation began to rise. An obvious inference was that the supply of reserves was excessive, but only Germany at the time adopted this position. The creation of SDRs was probably in a form calculated to raise inflation rather than to enhance liquidity, because the interest rate payable on SDR holdings was less than on dollars, offering an incentive to sell these reserves rather than to hold them (Johnson, 1973). Johnson asserts that the low interest rates on SDRs were fixed specifically to prevent these assets becoming a substitute for dollars and thereby reducing American seignorage. The IMF view was that the SDRs were unlikely to increase significantly the imbalances of countries in deficit, and that stabilisation measures in these countries might otherwise be frustrated by the defensive measures of other countries if no SDRs were created (de Vries, 1976, p. 221).

The failure of the United States to take measures to adjust to its deficit in the spring of 1971 caused a flow of dollars to Germany, and the Germans allowed the mark to float upwards (Strange, 1976, pp. 334–44; de Vries, 1976, pp. 520, 531–3). The Japanese, however, resisted the speculative pressure to appreciate the yen against the dollar to avoid reducing the competitiveness of Japanese exports. In a bid to force the rest of the world to revalue, and thereby reduce the American deficit, President Nixon announced in August the closing of the official gold window and the levying of a 10 per cent import surcharge. Immediately an atmosphere of crisis pervaded monetary relations. Ultimately the United States needed to adjust to its deficit and, as an indication of a willingness to do this, Nixon also imposed price and wage controls.

The Smithsonian Agreement of December 1971 settled new exchange rates and a 10 per cent devaluation of the gold price of the dollar, an official gold price of $38 per ounce. These changes did not solve the basic problems of the fixed rate regime, partly because the basic US deficit remained.

Early in 1973 higher US inflation with the removal of the price and wage controls caused a new run on the dollar and then a 10 per cent devaluation against gold (International Monetary Fund, 1973, pp. 2–6). The three major currencies that were already floating – the pound sterling, the Canadian dollar and the Swiss franc – continued to do so and were joined by the yen and the lira. From 2 to 19 March the major European central banks and the Bank of Japan closed their official foreign exchange markets in the face of the dollar glut. The EEC decided to float jointly against the dollar, limiting the fluctuations between their currencies to 2.25 per cent. Currencies in the European narrow margins arrangement, or closely associated with those currencies, appreciated against the US dollar by 9–18 per cent from early May to mid-July 1973 (International Monetary Fund, 1974, p. 16). The December oil price rises led to a rapid appreciation of the dollar under the influence of expectations that most of the surplus oil money would be invested in the United States. Removal of capital controls and re-evaluation of the oil funds position pulled down the dollar between January and May of 1974. The effective exchange rate for the dollar in June 1974 was about the same as in April 1973 before the oil shock. Effective rates for sterling, the franc, the lira and the yen showed depreciation, and for the mark, appreciation.

Large countries whose economies were more diverse and whose dependence on foreign trade was less were most inclined to floating. Conversely, smaller countries usually opted to peg their currencies. Eleven currencies, accounting for 46.6 per cent of the trade of Fund members in 1975, were floating independently, out of 122 (International Monetary Fund, 1975, p. 23). Seven currencies, the European groups, accounting for 23.2 per cent of trade, floated jointly. All floating currencies were subject to some official intervention between 1973 and 1975, and for this reason among others the IMF guidelines on floating of 1974 were necessary to define and avoid competitive exchange alteration. The legal problem was that, under the Bretton Woods Agreement, countries were obliged to maintain par exchange rates. The 1976 Jamaica Agreement, by removing this obligation, acknowledged officially the demise of the adjustable peg system.

Exchange rates were very volatile from 1973 because of the new uncertainties of the economic environment. In June and July 1973 some European currencies appreciated against the US dollar by 4 per cent a day, and by 10 per cent in little more than a week. Changes in the current balance were normally small in relation to potential changes in capital flows, which consequently exercised the most influence on exchange rate movements in the short run. Because of delays in the adjustment of trade flows, alterations in the current balance did not provide much resistance to exchange rate changes resulting from capital movements. Relative yields on assets held in different countries, and expectations about future exchange rate movements, largely determined capital flows.

Movements in the effective rates of the US dollar and the mark were generally in the same direction as alterations in their market rates against each other, but the amplitude of their swings was only about half as great (International Monetary Fund, 1975, pp. 29–30). This reflected each country trading substantially with other countries whose currencies were linked to their own or that followed an independent course. A greater stability of effective than market rates was also characteristic of other currencies, whose market rates generally changed less than the dollar–mark rate.

The most important systematic factor in the longer-run trends in exchange rates was the rate of price inflation in different economies. Between 1973 and 1975, Italy, the UK and Japan experienced price increases faster than the average for industrial countries. Germany managed a markedly lower rate, and the United States, France and Canada were clustered more closely about the average (Figure 14.1). These differential price movements were associated with changes in exchange rates. The two countries with the fastest inflation, Japan and Italy, experienced a substantial depreciation of their exchange rates which seemed more than sufficient to compensate for price trends.

Despite its earlier opposition to floating rates, the IMF concluded optimistically from the 1973–5 experience that exchange rate flexibility appeared to have enabled the world economy to surmount a succession of disturbing events, and to accommodate divergent trends in national costs and prices with less disruption of trade and payments than a system of par values would have been able to do. On the other hand, rate fluctuations had been in some cases much greater than could be justified on the basis of changes in underlying economic conditions. For example, the Swiss–United States exchange rate between 1973 and 1979 overshot the longer-term equilibrium rate in response to a monetary change by a factor of about two, taking two to three years to settle at the purchasing power parity rate (Driskill, 1981).

Developing countries experienced new problems arising from pegging to one currency, but these uncertainties were probably no greater than those that would have been involved under a par value system. Traders between developed countries at least could protect themselves against the exchange risk of any particular transaction by using the forward exchange markets. But forward rates were subject to much the same volatility as spot rates, and therefore such use did not reduce the variability of a firm's receipts. Hence there must have been some discouragement of trade. However, since uncertainty was part of the economic environment, fixed rates arguably would have led to more international trade than was ideal. There is no reason why foreign trade should have been subsidised by governments' bearing the costs of holding exchange reserves when domestic trade was not similarly subsidised.

The crash of 1974

Floating exchange rates eased the great adjustment that economies needed to make to the depression of 1974–5 and the preceding commodity price rises. The

308 *A History of the World Economy*

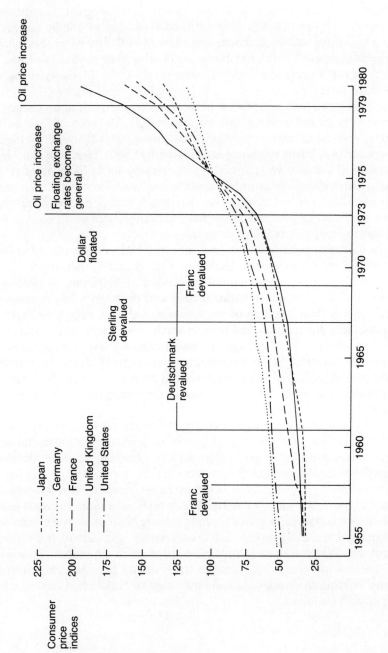

Figure 14.1 Price levels under the Bretton Woods system and after

only other occasion in the post-war period when both western Europe and the United States were simultaneously depressed was in 1958, and for much of western Europe this recession meant a decline in growth rates rather than a fall in their real national incomes. Fixed exchange rates – hedged with controls – were therefore not much of a handicap then. In explaining this pattern of fluctuations it is important to distinguish the role of general economic conditions from the effects of Keynesian demand management policies before 1974. The greater instability of the United States in the 1950s is suggestive; military expenditures especially were responsible for cyclical reversals (Lundberg, 1968, pp. 26, 128, 133–6).[6] By contrast, in other countries, especially the UK, Japan, the Netherlands, Denmark and Italy, policy changes can to a large exent be explained by balance of payments problems. In the UK, where restrictive actions were directed mainly at government expenditure, fixed investment and durable good purchases, this policy may have restrained supply growth, as well as demand growth. Denmark experienced problems similar to the UK's until 1957, but thereafter, with a rapid rise of industrial exports, Danish economic performance improved. Similarly, the better performance of the Dutch economy, compared with the British economy, may be attributable to rapidly and continuously expanding Dutch exports. In France, until 1958 persistent foreign exchange problems were dealt with not by restricting domestic demand as in the UK, Denmark and the Netherlands, but by import restrictions and devaluations. They therefore did not constrain demand and supply in the same way as those in the United Kingdom.

The balance of payments difficulties themselves should be regarded not as exogenous influences on policy, but rather as a consequence of the pursuit of policies which generated an inflation that was kept in check only because of the need periodically to balance foreign payments. When floating exchange rates removed these constraints, inflation could and did accelerate initially. Thereafter the depression of 1974–5 owed much to the simultaneous and severe contractionary policies pursued by the governments of the industrial world to bring inflation down again. Even more was this true of the depression following the 1979 oil price rise.

Although the depression of 1974 was the worst since the one that began in 1929, it was shorter and much less severe. A major difference between the two depressions in the key country, the United States, was the course of prices, which rose between 1973 and 1975 but fell between 1929 and 1931 (Table 14.4). The real money stock declined in the later depression, but rose in the earlier one. Stock prices fell greatly in both depressions, as did residential construction (Temin, 1976b). The instability of foreign exchange markets, the difficulties associated with large capital movements, and the problems of countries experiencing sharp changes in the terms of trade, were all important in the 1930s and present in the 1970s. But the floating exchange rate regime seemed less vulnerable than the fixed rate gold exchange standard, and the strength of autonomous consumers' expenditure steadied the economy in the 1970s so that GNP then fell by much less than in the 1930s. The collapse of consumer expenditure in the first period may

Table 14.4 Percentage changes for the United States economy in some macroeconomic variables, 1929–31 and 1973–5.

	1929–30	1930–1	1973–4	1974–5
GNP deflator	−3	−9	+10	+9
Real M1	−1	+3	−4	−4
Real M2	+1	+3	−1	−1
Real stock prices	−16	−26	−36	−7
Real residential construction, 1958 prices	−39	−19	−27	−22
Real consumption expenditures, 1958	−7	−4	−2	+1
Real GNP, 1958 prices	−10	−8	−2	−3

Source: Temin (1976b).

have been attributable to the stock market crash, which reduced household wealth and raised the ratio of their debts to their assets. In 1974 inflation rapidly reduced the real value of debt. The bank failures of the 1930s were not repeated because a lesson of the 1930s had been learned. The Federal Deposit Insurance Corporation was able to contain the failures of the United States National Bank of San Diego in 1973 and the Franklin National Bank of New York in 1974.

Balance of payments adjustment to the oil price increases, to the economic recession and to the partial recovery worked rather slowly because of the enormous volume of financing available through the private markets. From 1974 to 1976 foreign borrowing undertaken by OECD deficit countries and non-oil developing countries amounted to over $175 billion (OECD, 1977). Private banks recycled almost $100 billion of this sum to deficit countries. In any case the OECD area's oil import bill rose in 1974 by $65–70 billion, but the deterioration in the current balance was little more than half this amount because of increased OPEC imports and the widening deficit of the non-oil rest of the world. The recession of 1975 reduced OECD imports and, combined with big increases in exports to OPEC, the deficit nearly disappeared, only to reappear again with the beginning of recovery in 1976.

The most advanced industrial countries had little to fear from the balance of payments effects of the oil price increases, because their highly developed financial institutions ensured that they would receive most of the unspent oil money back in the form of savings. Of the 1974 OPEC financial surplus, 20 per cent was invested directly in the United States, another 13 per cent in the United Kingdom and over 40 per cent in Eurocurrency markets. Direct OPEC investments in the UK exceeded additional oil payments, and British foreign exchange reserves rose marginally in 1974 (Argy, 1981, pp. 195–207). Most major countries borrowed on international capital markets to ease their balance of payments. So also did non-oil LDCs, which supplemented these funds with additional aid, commercial credit and some official borrowing. The external debt of these countries rose from 13 per cent of GNP in 1973/4 to 18 per cent in 1977/8. Another boost to foreign indebtedness occurred when the Iranian crisis helped increase oil prices by a further 50 per cent in 1979. Between 1974 and 1978 the OPEC current account surplus fell from $60 billion to about $10 billion as a result of a 20 per cent fall in their terms of trade, the growth of OPEC imports, and the decline in oil imports from OPEC by the industrial countries that came with the development of non-OPEC oil supplies in the North Sea, Mexico and Alaska.

From 1974, the slump was worldwide. The principal factor in the collapse of economic activity was the sharp decline in the rate of growth of real money balances (monetary growth less the inflation rate). Anti-inflation policy slowed monetary growth particularly severely in Australia, Japan and the UK. The lagged effect on inflation of earlier monetary growth, together with oil prices pushing up the price level, reduced the growth of real money balances relative to nominal monetary growth. The oil price increase also directly cut back demand by a significant amount in all countries, except Australia and Canada. Total expenditure was switched from home demand to imports to pay for the oil, and increased demand for domestic goods for export to OPEC countries did not compensate. Fiscal policy, unlike monetary policy, contributed little to the downturn, except in the United Kingdom. In Australia, Canada, France, Germany and Japan the downturn was accompanied by more expansionary fiscal policies as Keynes might have prescribed (Bell, 1973, pp. 8–12).

Because monetary policy did not accommodate the rise in wages to cover the higher prices, unemployment grew as the real value of wages was pushed up faster than productivity. Among the OECD countries, only the United States and Sweden were free of this wage–productivity gap in 1975. Although unemployment after 1976 was higher than at any other time in the post-1945 period, there was no attempt to reduce it with fiscal policy, and monetary policy was switched towards the announcement of monetary targets which gave little room for any significant reduction in unemployment. The main fear had become inflation.

International private capital mobility

The international mobility of private capital proved a two-edged sword for the Bretton Woods system. It helped destroy the adjustable peg exchange rate regime, but also assisted the adjustment of the modified system to the events of 1973 to 1975. During the 1950s and 1960s banks extended their international operations in parallel with the largest companies of the industrial world. In so doing they contributed to the emergence of the Eurodollar market (a market in dollar deposits held in banks outside the United States), which was among the most efficient vehicles of greater capital mobility. The origins of the market can be traced to the attempt of the British government of 1957 to restrict the use of sterling to finance foreign third-party trade. The British banks therefore began to use dollars instead (Bell, 1973, p. 54). There were advantages from not repatriating these dollars stemming from an interest rate differential between the United States and western Europe. Federal Reserve Regulation Q, designed to prevent stock market speculation like that of 1929, limited interest payable on deposits in member banks of the Federal Reserve system below what could usually be earned in Europe.

The Eurocurrency system expanded in response to American capital controls of the early 1960s, especially the Interest Equalisation Tax of 1963 (a tariff on new securities), which in turn were an attempt to stop the United States balance of payments deficit enhancing world liquidity. The rapid growth of this market continued into the 1970s helped by the desire of the Soviet and OPEC governments to hold dollars where they would not be confiscated or measured. American and foreign borrowers demanded more Eurodollars as a result of restrictions on the transfer of funds by Americans to Euromarkets. In 1968 and 1969, when United States financial conditions became tight, American banks borrowed heavily from their foreign branches operating in the Eurodollar market, and thereby partly neutralised the contractionary monetary policy (Bell, 1973, pp. 8–12).

The capacity of the Eurodollar market for creating liquidity was very limited, unlike a domestic banking system, because the proceeds of loans by Eurobanks were rarely redeposited in another Eurobank (Argy, 1981, ch. 7). The existence of the Eurodollar market facilitated the financing of deficits and so, effectively, increased the availability of reserves. This probably more than offset the effects on the balance of payments of any increase in capital mobility allowed by the market, but it reduced the effectiveness of monetary policy by limiting control over domestic interest rates as a policy instrument; sterilisation became less feasible.

It was the mobility of long-term American capital that stimulated French concern about the United States balance of payments after 1960, as much as the belief that the burden of adjustment was forced on the rest of the world. The balance of American private long-term capital was massively negative, a little over $3 billion in 1960, $4.2 billion ten years later, and after the depression of 1974–5, $8.5 billion in 1977 (OECD, 1979). Most of the stock of American foreign direct investment (73.7 per cent in 1976) was in developed countries, and in particular in Europe (40.7 per cent in 1976). The French were worried that American multinationals were coming to dominate all the advanced technology sectors of European industry (see especially Servan-Schreiber, 1968). Because of their enormous home market, these firms achieved economies of scale, and in addition received the benefit of the large research and development outlays of the United States government on armaments and aerospace. The size of the firm was thought crucial to the ability to innovate and therefore to grow. This was the 'American Challenge'. With the benefit of hindsight, the growth of large firms bore no relation to firm size during the 1960s (although the variability of growth did) (Rowthorne, 1971). Hence, on this account, there was no reason to expect American capital to take over Europe, and American multinational investment in western Europe indeed turned out to have been no greater in the 1960s than European investment in the United States.

For the OECD countries as a whole, the capital balance with the rest of the world in 1960 was small (–$398 million) by contrast with the capital export and import of the OECD countries. Most foreign investment was directed to developed countries, as well as originating in them. For small, open developed economies, foreign investment could comprise a considerable proportion of total capital formation. Between 1949 and 1964 the flow of overseas capital to Australia, 90 per cent of which was direct investment, amounted to about 10 per cent of total savings (Dunning, 1970, p. 35). Foreign-owned enterprises in Norway in the 1960s accounted for one-third of all corporations, and in Canada, capital imports financed 15 per cent of total imports between 1955 and 1960.

The political impact

However, it was in less developed countries that hosting foreign direct investment was most controversial. As in the nineteenth century, foreign investment combined with a belief by the investing country that important national security issues were at stake could prove a dangerous mixture for host countries. Those who felt that foreign direct investment jeopardised national objectives were vindicated by the role of the United Fruit Company in Guatemala in the early 1950s. The United Fruit Company was the largest landowner in the country and objected to President Arbenz's expropriation of the company's uncultivated land as part of a land reform programme. By playing on fears about Soviet threats to free enterprise in Guatemala, the company persuaded the United States to intervene and overthrow Arbenz in 1954 (Schlesinger and Kinzer, 1982). The 1951 nationalisation of the Anglo-Iranian Oil Company's (later renamed BP) oil interests in Iran also ultimately precipitated a coup assisted by the American Central Intelligence Agency (CIA) in 1953. Matters were complicated by civil unrest, assassination and economic warfare. The coup returned the Shah to power and removed a chance that Soviet power was about to be extended, as it had been just before in eastern Europe (Yergin, 1991, ch. 23; Sampson, 1975, ch. 6).

The political activities of ITT in Chile, although apparently less successful than those of United Fruit in influencing the American government, appeared to be equally effective. ITT by the 1960s employed 6,000 people in its Chilean telephone company with assets valued (by ITT) at $150 million (Sampson, 1973). In 1970, ITT plotted to stop the election of the Marxist Allende to the Chilean presidency on the grounds that the election would jeopardise the safety of its assets; the company had offered to contribute a dollar sum of up to seven figures to the White House to stop Allende.[7] Allende nevertheless became President in 1970. Thereafter ITT regularly dealt with the CIA to try to create economic chaos in Chile and to encourage a military coup. In September 1973 this was achieved, and Allende was killed. Nothing directly connected the coup with the CIA or ITT, but the circumstantial evidence was compelling.

Long-term economic effects

In addition to political interference, potential LDC recipients of foreign capital were fearful of some economic effects, such as finding insufficient foreign exchange to pay the interest and dividends on the capital. Foreign investment in exploiting natural resources for export was sometimes thought to be less beneficial than foreign investment in manufacturing because the resources were not renewable. In other investment projects the foreign entrepreneurial input was greatest during the early stages. Subsequently, the payment of profits abroad tended to rise, while the necessity for foreign entrepreneurial talents diminished. For this reason many LDCs preferred to aim for a minimum foreign equity stake in a project consistent with obtaining the benefits of a multinational's technology.

Donor countries also were hardly favourably inclined to foreign investment. After 1960, United States balance of payments difficulties forced a reappraisal of the former policy of active encouragement of overseas investment. Similarly, the United Kingdom controlled overseas investment for most of the post-war period, although the extent of control varied with the balance of payments position. There was also a longer-term concern that, even if the private marginal return on overseas investment was higher than at home, the loss of tax revenues, the possible deterioration in the terms of trade, and the retardation of growth at home because of the diversion of capital overseas might outweigh the benefits of higher profits (Casson, 1979). The Reddaway Report of 1967 concluded that, for the UK, there was no evidence of any great resource misallocation from investment overseas, and that the expected gains from extending the international division of labour were therefore being obtained (Reddaway et al., 1967).

The distinctive post-war pattern of foreign direct investment (FDI) provides clues as to why it occurred. It also offers suggestions as to the long-term economic impact. Most FDI took place between developed capital-abundant countries. Such investment therefore could not be interpreted as a process of equalising factor returns in the same way as labour migration, which flowed to relatively labour-scarce countries. Furthermore, FDI could be undertaken without any international transfer of capital, if the multinational took over an indigenous firm with capital raised in the host country's capital market, or otherwise borrowed in the host country using the subsidiary's assets as collateral. The possibility of nationalisation of the multinational's assets gave an added incentive for firms to hedge in this fashion. If a change of regime brought expropriation, the head office of the company could tell the host country creditors to collect their repayments from their own nationalising government.

In Chapter 13 it was suggested that FDI transferred technology, and the concentration of FDI in manufacturing industry where new technology would be most important confirms this view: in Brazil foreign firms controlled 22 per cent of manufacturing output in 1972, but only 7 per cent of primary output and 2 per cent of the service sector (Casson, 1979, pp. 9–11). Even so, the question remains as to why multinational firms did not more often license foreign firms to use their technology for an appropriate payment, and avoid the problems of learning to compete in a different economic, legal and cultural environment. One answer is the difficulty of agreeing a price for the technology. If the buyer knows the market value of the technology, and therefore knows the fair purchase price, the chances are that it will also know what the technology is, and therefore will not have to buy it. The seller, on the other hand, knows the value of the technology and will be unwilling to lower the sale price to the level acceptable to the (ignorant) buyer. Another answer is that the existence of national boundaries with different tax and tariff regimes confers an advantage on a firm which can straddle the frontiers. Such a company can transfer intermediate goods between subsidiaries at national 'transfer prices' which minimise the firm's tax burden. Enterprises exporting to other firms at market prices across the same frontier pay higher taxes or tariffs, and are therefore less competitive than the multinational. These two answers have different implications for the dispensability of FDI. The first suggests that FDI was probably the most efficient means of transferring technology, whereas the second implies that FDI was primarily a device for tax avoidance.

International official capital flows

As noted in Chapter 12, the early 1950s effectively saw the rise of a new component of international capital flows, official aid. Although a small proportion of total flows in the ensuing two decades, for the less developed countries, receipts of foreign official development assistance in 1976 amounted to one-third of total net investment receipts. Because official aid was ultimately supplied by political, rather than business, organisations, the motives and impacts were likely to differ from commercial flows.

An early rationale for aid was that the structure of LDCs made foreign aid particularly valuable in raising their growth rates. The two-gap theory attempts to give an account of growth constrained by the availability of foreign capital, and in particular official aid (Chenery and Strout, 1966).[8] According to this theory, there is a minimum necessary additional amount of imports required to support a given increase in national output. Exports are not related to internal factors, but are determined by external demand, which increases at some given rate. For some projected future level of national output, imports may exceed exports, and this gap must be financed by foreign aid if the projected level of output is to be attained. At this level of output, the amount that people would be prepared to save may exceed the amount required to finance the investment needed to sustain the projected rate of growth, after subtracting the contribution of foreign aid to this financing. Hence aid is required not as a supplement to domestic savings, but as a supplement to foreign exchange earnings. The 'trade gap' (the excess of imports over exports) dominates the savings gap (the excess of savings over investment). Increased savings will not increase investment because they will not raise exports and permit more imports to be bought, a prerequisite of higher output. A cut in aid, therefore, reduces output by a multiple of its value.

The theory implicitly assumes inflexible domestic policies: that there should or can be no substitution of domestic for imported goods, or of exports for domestic supplies, to eliminate the current account deficit (Griffin, 1970).[9] It can therefore be a rationalisation and support for the post-war policies pursued by a majority of LDCs, which in many cases have had unfortunate consequences. For this reason, and others, some have doubted that aid flows made any positive contribution to LDC development in the 1950s and 1960s. These other reasons include the possibility that aid flows acted merely as a supplement to consumption. Even though often tied to investment projects, foreign aid may have released funds that would otherwise have been allocated to investment. These liberated funds may then have permitted higher government 'consumption' in prestige activities such as state airlines. Even if the consumption benefits did filter through to the mass of the population, there would have been a highly undesirable consequence if it had encouraged permanent international dependency. If aid did supplement investment, the results may still not have been worth while: donors typically required large and visible projects, like the Aswan Dam in Egypt, to minimise administrative costs and maximise prestige. Such motivation is likely to have reduced the productivity of investment by its emphasis on providing infrastructure. If domestic savings were also reduced, because the savings were now being made by foreigners, then together with reduced productivity of capital, the overall beneficial impact of the aid could have been extremely small.

The effect of foreign capital inflows in less developed countries cannot be settled by a priori arguments of this nature. It is possible to identify particular cases where aid probably did reduce domestic savings: in Korea in the mid-1950s, and in India and Pakistan. The supply of United States surplus agricultural stocks probably encouraged the neglect of domestic agriculture (Papanek, 1973). But overall, statistical evidence suggests that aid was an important beneficial influence. During the 1950s and 1960s, countries with higher investment, private foreign, official aid and domestic savings, experienced significantly higher growth rates. A 1 per cent higher ratio of aid to national income between countries was on average associated with a 0.39 per cent higher growth rate. There was no evidence of reverse causation, of higher growth rates encouraging donors, in these results. The strongest impact of aid flows was found in the Asian and Mediterranean countries.

The pattern of official capital flows

Aid was distributed inequitably. Moderately prosperous developing countries received twenty to a hundred times as much aid per head in 1970 as a number of other extremely poor countries. Small countries tended to be given more aid per head than large ones (OECD, 1969). Hence the desire to redistribute world income more fairly does not apparently explain aid. More consistent with the evidence is that donors expected to get something in return, such as political support or markets and sources of supply, which depended in part upon the size of the original flow of foreign aid to the recipient (Dudley and Montmarquette, 1976). Auxiliary variables in a plausible model include political ties, defined as

Table 14.5 The main recipients of British aid, 1974.

Country	Gross aid disbursements in 1974 by the UK (£000)	Population (m)	National income per head (£)	UK exports to aid recipients (£m)
Bangladesh	6,437	71.61	30	11.6
India	75,445	574.22	50	126.8
Malaysia	5,452	11.28	210	113.3
Singapore	4,470	2.69	690	153.5
Indonesia	8,866	124.60	40	46.8
Nigeria	5,966	59.61	70	222.2
Botswana	4,047	0.65	100	1.3
Kenya	16,435	12.48	80	79.3
Malawi	8,071	4.79	50	10.5
Zambia	8,397	4.64	170	63.7
Jamaica	4,788	1.98	370	50.1

Sources: *British Aid Statistics 1970–74* (HMSO); *Development Cooperation 1974 Review* (OECD); *Annual Statement of Overseas Trade of the United Kingdom for the year 1974*, vol. 4 (HMSO).

former colonial status, exports from donor to recipient, and a bandwagon effect whereby the donor residents evaluate the impact of this aid more highly the greater the aid given by the rest of the world to the recipient in question.

Estimation of this model suggested that the degree of distortion due to the division of the world's poor into countries of different size population was small (with the exception of French aid). The negative correlations of aid per head with country population was the consequence of the negative correlation between population on the one hand, and exports, aid from the rest of the world and political links on the other. Political and economic links were very important determinants, but bandwagon effects were less significant.

Inspection of the largest recipients of British aid in 1974 confirms the role of political and economic links in a more impressionistic fashion (Table 14.5). Ten of the eleven recipients were Commonwealth countries, which, as expected from the sphere of influence analysis, had the strongest non-oil trade and investment links with Britain of all LDCs. The inclusion of Indonesia in the group might be explained as an attempt to secure reliable oil supplies.

Official bilateral development assistance as a proportion of GNP of all industrial countries declined from 0.52 in 1960 to 0.34 in 1970. In an attempt to reverse this trend the Pearson Commission, set up by World Bank President Robert McNamara, in 1969 called for a commitment to a 0.7 per cent aid/GNP ratio from the developed countries, but to no avail (Prout, 1976). Similarly, the Brandt Commission (also first proposed by McNamara, in 1977) repeated the need to reach this target by 1985, and to reach 1 per cent by the end of the century, but it received an identical reaction. By 1980 the ratio achieved was only 0.38 per cent, falling to 0.35 per cent the following year (Independent Commission on International Development Issues, 1980).

Multilateral aid through the World Bank did little to correct the 'distortion' of bilateral official capital flows – and in any event contributed little more than 10 per cent of official flows. In none of the top ten recipients of World Bank loans per head of population was the total population as large as 5 million (Mason and Asher, 1973, p. 198). So for whatever reason, the poor of large countries received less of both types of aid than the poor of small countries.

Accepting World Bank loans could have political consequences similar to foreign private direct investment, for the Bank's perception of the development process influenced the conditions of its loans and those of the aid consortia organised by the Bank. The Bank managed to persuade the Indian government to change some balance of payments policies and reduce direct economic controls in the mid-1960s in order to get foreign aid (Bhambi, 1980). The Bank's recognition of the importance of agriculture, shown by the increase in the proportion of World Bank lending for agriculture from 9 per cent in 1960 to 18 per cent in 1971, also began to act as a partial antidote in some cases to indigenous policies of import-substituting industrialisation.

The European Community offered more indirect support to selected poorer economies, the forty-six associated states in Africa, the Caribbean and the Pacific. The STABEX mechanism of the 1975 Lomé Convention was a striking institutional innovation. The mechanism was intended to stabilise the product earnings of twelve major primary product exports of the associated states by providing interest-free loans (Hasenpflug, 1977, p. 6).

International debt

Institutions obliged to raise money in commercial markets, the commercial banks and the World Bank, needed to monitor closely the ability of their clients to repay. The rising burden of debt caused increasing problems in this respect. Between 1955 and 1962 the external debt of LDCs rose at 15 per cent per annum. In the mid-1950s debt service averaged 6 per cent of export earnings, rising to 12 per cent ten years later. Unable to service the foreign debt, Argentina rescheduled four times in eight years, and Indonesia rescheduled three times in three years.

The world recession of 1974–5 markedly increased the foreign borrowing of LDC governments. Less important was the quadrupling of oil prices and the demands of development finance. The private banking system provided the greater part of the debt build-up of the 1970s. As the risk of default increased with debt accumulation in 1975, the banks raised the rates they charged on loans to LDCs (Brittain, 1977).

Three-quarters of the world's deficits between 1974 and 1976 were financed by the banks (Sampson, 1982, p. 299). Unlike the IMF, they could not impose conditions on borrowing country governments to ensure they were repaid. These governments' policies and prospects were taken as given in assessing the risk of lending in the first place. Country risk assessment was difficult, and hence a herd instinct could predominate. Once a country borrowed substantial sums from international banks or institutions, the banks could be forced to lend more in order to avoid a default.

So anxious were banks and officials to avoid a default as the volume of debt accumulated that rescheduling became increasingly common as a means of debt relief (see Figure 14.2). Between 1975 and the end of 1980 there were sixteen official debt negotiations for $9 billion, whereas in the previous eighteen years there had been thirty, for debts of $7 billion (*The Economist*, 1982). The most surprising rescheduling, as well as the largest, was of Poland's $4.9 billion debt in 1981. Western capital, instead of driving a wedge between Poland and its Soviet master, pulling Poland towards the market economies, gave the trade union Solidarity a taste for freedom that almost brought down the tottering economy. Maintaining debt payments joined the interests of western banks with Poland's martial law administrators.

Nevertheless the performance of the private banks through the market seemed to be consistent with efficient lending behaviour. As the cost of borrowing increased, the amount of lending declined. Brazil, Taiwan, Indonesia and others undertook internal adjustment programmes because they found it increasingly costly and difficult to finance their external deficits (Erb, 1979, p. 393). Furthermore, the variety of alternative sources of finance gave countries more freedom to manoeuvre than if they were reliant solely on the IMF.

Summary and conclusion

The international monetary system after 1945 clearly differed from all preceding systems in the much greater contribution of governments to forming, modifying and operating it. The overriding goals of the great majority of these governments were to minimise unemployment and boost living standards. Greater governmental participation also drove up the international price level as well as encouraging economic growth. Analogously to the mainly sterling-based gold exchange standard of the years before 1914, international exchange became based on the dollar, which ceased to be convertible with gold at a fixed price from 1968. The fundamental difference between these two systems was that the nineteenth-century liberal doctrines of the balanced budget had been abandoned. The world therefore experienced a dollar glut by the late 1960s. Ironically, negotiations to enhance world reserves and international liquidity took place throughout the 1960s, despite rising prices. The United States ruled out an increase in the price of gold as a means of increasing reserves (a decrease in the gold price of the dollar). In part this may be explained by the benefits that America obtained from the increasing international role of the dollar.

As the dollar glut continued, the international trading community lost some of its former confidence in the purchasing power of the dollar, and in the viability of

320 A History of the World Economy

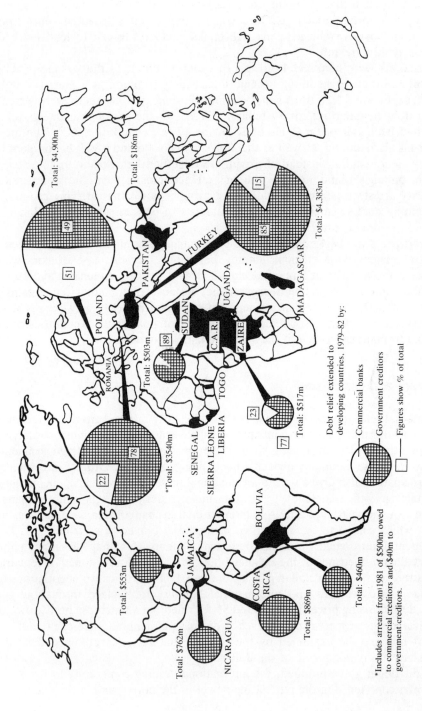

Figure 14.2 International debts rescheduled, 1979–82 (*The Economist*, 29 March 1982)

many fixed exchange rates. Speculative crises dominated the last few years of the fixed rate system, which may be said to have ended at some time between 1971 and 1973. The floating rate regime for the major industrial countries that followed was remarkably successful in dealing with the oil price rise of 1973 and the consequent redistribution of international income and saving. Recognising their interdependence, economies eschewed competitive exchange rate depreciation to export their unemployment. The depression of 1973–4 was in any case less severe than the collapse of 1929–32 because, despite a stock exchange crash of a similar order of magnitude, the American economy remained relatively buoyant.

Under the fixed rate regime, emerging balance of payments deficits required adjustments of the economy (except for the United States) to remove them. Adjustment might be avoided if disequilibria were temporary and sufficient international reserves were available to finance them. Inadequate international reserves could force countries to deflate unnecessarily, spreading their unemployment worldwide through a reduction in their demand for foreign goods. A devaluation of the exchange rate could also have this effect by reducing the demand for imports and increasing the competitiveness of exports. For this reason IMF surveillance of exchange rate policies was introduced in the Bretton Woods system. As a lender of international reserves, the IMF could exercise considerable power over government policies when a nation was in deficit and needed to borrow. Surplus or reserve currency economies were exempted from this discipline. The IMF often incurred opprobrium for imposing restrictive policies, but these policies were effective in correcting deficits, albeit at a cost. The relatively low rates of inflation, so long as the fixed exchange rate regime persisted among the industrialised countries, suggest that the discipline worked. On the other hand, the causal chain could operate in the opposite direction, as noted in earlier chapters: fixed rates may have worked only as long as countries preferred relatively low rates of inflation.

Capital mobility played an important role in the breakdown of the fixed rate regime. The adjustable peg system of the 1960s was less capable of generating stabilising speculation than the fixed rates of the 1900s, once capital restrictions were removed. Capital mobility also reduced the control that national monetary authorities could exercise over their own economies.

The greatest capital movements arose from the massive transfer of income, from oil consumers to oil producers from 1973. Unable to spend all their newly acquired income some of the richest oil producers lent their 'petrodollars' to the financial centres of the industrial world. For most of the West, this short-term investment largely offset the detrimental effects of higher oil prices on the balance of payments. For poor oil-importing countries, there was no such compensation and their balance of payments deteriorated under the combined influence of the recession and the cost of oil imports. Their foreign debt increased as they borrowed to avoid deflating. An increasing proportion of their debt was held by private banks which did not and could not exercise the same surveillance over these debtors to ensure repayment as the IMF or the World Bank. Rescheduling of debt payments to avoid default therefore increased in the second half of the 1970s.

Longer-term capital movements in many instances were as controversial as they had been in the nineteenth century. Bilateral official aid flows tended to reinforce national spheres of influence. Multinational investment sometimes carried political consequences and, where it did not, its benefits were frequently questioned by both donor and recipient countries. Some statistical analysis does, however, suggest that both official and private flows benefited recipient LDC growth rates in the 1950s and 1960s.

Most economies failed to continue to combine low inflation and low unemployment from the 1970s. The severity and duration of the world depression after 1974 were largely induced by restrictive policies which placed a greater value on lower inflation than on higher employment. Since activity in one economy was influenced by demand in its trading partners, interdependence was now holding back the world economy, instead of driving it forward as it had during the great post-war boom.

Notes

1. Floating is generally assumed to reduce the demand for reserves, although if under fixed exchange rates a peg acts as a focus for stabilising speculation, this will not be true. Currencies pegged to a single floating currency (as many Latin American exchange rates remained fixed to the US dollar) might actually have needed more reserves because of the added variability in payments balances caused by exchange rate movements between third countries and the intervention currency.
2. Triffin (1960) was the first to draw attention to this problem.
3. Giscard d'Estaing, at the time French Finance Minister and later Prime Minister, compared the United States and Italian deficits of 1963 and 1964. The United States deficit began earlier, but it was Italy that had to reduce employment, trade and growth to return to equilibrium, he asserted, in Mundell and Swoboda (1969).
4. See Williamson (1977, pp. 44–6) on the incompatibility of the discrete changes of the adjustable peg system with any probable process of formation of speculators' expectations, in a world of high capital mobility.
5. This was true of most consistent models in the long run. The early monetarist models (e.g. in Johnson and Frenkel, 1976) were distinguished by their prediction of very rapid adjustment. Later monetarist models postulated only that the balance on capital account depended on the excess demand for money. See the Stockholm Conference papers in *Scandinavian Journal of Economics* (1976).
6. See also Maddison (1977, table 5), which shows the remarkable absence of international recessions after 1945 compared with the period from 1870 to that date.
7. Kissinger, then Secretary of State, reports that he politely declined ITT's offer because he considered that sort of activity inappropriate for private enterprise (Kissinger, 1982, p. 389).
8. This model exercised a great influence over US aid policies in the 1960s.
9. Bauer (1971) offers a number of other more wide-ranging criticisms of aid.

15 The search for a new regime: the world economy of the 1980s

A regime that adequately replaced the pegged exchange rates of Bretton Woods needed institutions that could support a variety of conditions. They would encourage productivity improvement, price stability and full employment, and avoid the destruction of the environment. These were national problems as well, but in market economies 'open' to the world, inflation can be imported and exported and so can jobs. And even if a nation were not economically open, it would still be vulnerable to certain international environmental 'spillover' effects (sometimes literally). Recognition that the international impact of discretionary national policies deepened the recessions after the first two oil shocks added impetus to attempts to find an alternative to the Bretton Woods exchange rate system.

Predicting the major shocks that will hit the world economy may be well nigh impossible, but the certainty that they are part of the human condition places a premium on the robustness of the international economic regime. Another desirable attribute of a regime is its capacity to promote convergence among trading economies: transfers of technology, and factor price equalisation by goods and factor movements, tend to equalise incomes. Of course, the international regime is only a part of the story, a necessary but not sufficient condition. Domestic policies and institutions must also be appropriate. The first section of this chapter discusses international factor and product movements and how they have been shaped by policy during the 1980s. To assess the convergence achievement, the next section looks at productivity growth patterns among the technological leaders of the world economy, the USA, Japan and the EC. The following two sections shift to lagging economies, in Latin America and Africa, and those that caught up, in Asia. In each case some attempt is made to distinguish between domestic and foreign sources of performance.

A good many of the challenges faced by the world economy in the 1970s originated with OPEC policies. As the fifth section shows, the state of the oil market continued to determine the prosperity of OPEC members in the 1980s. With the third oil shock of 1986 OPEC's fortunes deteriorated. Centrally planned economies also diverged from the best-practice market economies during the

1980s, culminating in the collapse of the Soviet Empire in 1989. The sixth section shows how the Bretton Woods institutions, the World Bank and the IMF, found further scope for their activities in supporting the successor states through the painful transition to market economies. The next section looks at western Europe's regionalist approach, the move to monetary unification and the international tensions caused by the agricultural policy. Exchange rates, in particular the dollar, swung about wildly in the 1980s. How far international policy co-operation and co-ordination was a way of avoiding and correcting such misalignments naturally forms the next topic. International debt management was a particularly vital component of that co-operation. Finally, an appraisal is offered of co-operative policy on international common property resources in the 1980s.

Trade and trade policy

The 1970s ended the liberalisation of the great post-war boom, with two oil shocks, bursts of double-digit inflation and rising unemployment. Despite the poor precedent, the 1980s proved less uncomfortable for the world economy as a whole, although the position of certain regions was more painful. The decade was marked by falling oil prices, more stringent national fiscal policies and the most ambitious, albeit uncompleted, GATT round. Growth in the 1970s was often unsustainable, based on excessive monetary expansion and over-borrowing. This was not true of the 1980s. Asian trade as well as output was clearly the most dynamic, in particular because of the opening of China, but trade/GDP ratios rose the most in North America. In Africa, hit hard by falling oil exports and weak primary commodity prices, the ratio fell sharply.

Table 15.1 summarises regional international economic experience during the 1980s. Regions with the most rapid population growth, the Middle East and Africa, suffered the largest falls in exports and were importing no more at the end of the decade than at the beginning. Asia clearly led in trade expansion, but North America and Europe were not too far behind. Latin America passed through a painful adjustment, with imports per head falling but exports increasing.

Thanks to the oil shocks and misaligned exchange rates, 'managed' trade (managed by voluntary restraint agreements, VRAs) accounted for not much under half the world total by the end of the 1970s. The volume of merchandise world trade increased by one-half in the following decade, rather less than in the 1970s. However, the trade–output ratio rose, suggesting that some liberalisation was effective, if only regionally. Australia and New Zealand negotiated a Closer Economic Trade Agreement, Canada and the United States signed a Free Trade Agreement, and the European Community was enlarged to include Spain and Portugal. Within Europe the UK was the 'pivot' for the northern 'periphery' (virtually coterminous with EFTA), as Germany was for the EC 'core'. After thirty years of integration, the core of western Europe had become more

Table 15.1 Population and merchandise trade by region in 1990

	Population		Exports (US$)		Imports (US$)	
	1990 (m)	Ann. av. % change 1980-90	Value 1990 ($bn)	Ann. av. % change 1980-90	Value 1990 ($bn)	Ann. av. % change 1980-90
North America	275	1	525	6	641	7
Latin America	451	2	148	3	133	0.5
Western Europe	438	0.5	1,613	7	1,685	6
Central/eastern Europe/USSR	407	1	182	–	187	–
Africa	645	3	94	−2.5	93	0
Middle East	126	3	132	−5	103	0
Asia	2,903	1.5	791	9.5	765	8
World	5,245	1.5	3,485	5.5	3,608	5.5

Source: GATT (1990).

cohesive; the removal of some trade barriers had shaped the pattern of trade. The share of the original six EC members' exports that went to others doubled between 1958 and 1987. For West Germany the proportion reached 37 per cent. For Belgium–Luxembourg, with a higher ratio of exports to income, the proportion was 62 per cent (Wijkman, 1990).

Factor endowments, technology and incomes all drove different components of trade. The product composition slightly shifted from merchandise to income-elastic services. Trade in all commercial services by the end of the decade was twice as large as in all mining products. Surprisingly in view of the restrictions imposed by the Multi-Fibre Arrangement (MFA), clothing was one of the fastest-growing categories, behind office and telecom equipment. Clothing is a Hecksher–Ohlin good, trade in which depends on relative factor endowments. By contrast, the second category (office and telecom equipment) includes new high-technology products, such as personal computers, faxes and semi-conductors. Here trade is based on, and disseminates, new technology (GATT, 1990).

Clothing trade expansion was in part due to the growth of southern European exports at the expense of northern Europe, but China, the countries of the Association of South East Asian Nations (ASEAN) and southern Asia accounted for more. While sectors intensive in unskilled labour were under pressure, growing intra-industry trade demonstrated enhanced prospects for capital-intensive segments. Intra-firm trade also seems to have burgeoned in certain sectors (Casson, 1986). The Middle East suffered most from the rearrangement of world trade in crude oil, with its supply to the world market falling by about one-half. That western Europe increased its share of cereal exports from 24 to 38 per cent in the first eight years of the 1980s is a measure of the profligacy of the subsidies in the Common Agricultural Policy. Japan's share of world merchandise

Figure 15.1 Yen–dollar exchange rate, 1970–92

exports rose from 6.6 per cent in 1970 to 9.8 per cent in 1987, while the US proportion fell from 14.9 per cent to 10.6 per cent. Japan's rising competitiveness and persistent trade surplus became a continuing source of concern for Europe and the United States.

Slower economic growth and recession created unfavourable political environments for trade expansion at the beginning of the 1980s. Moreover, an overvalued US dollar was a recipe for stronger US protectionist sentiments. Although the yen rose strongly against the dollar during the later 1980s (see Figure 15.1), in the earlier years the yen was much lower than competitiveness warranted. Trade bloc tensions between Japan, the EC and the United States regularly surfaced over allegedly unfair trading practices. US legislation became more aggressive, and increasing regionalism indicated a declining faith in a multilateral trading system. All in all, in view of the pressures on the regime, what GATT continued to achieve was remarkable.

The world economy in the 1980s experienced the benefits of liberalisation achieved in the Tokyo Round. This GATT negotiation, lasting from 1973 until 1979, agreed tariff cuts of a similar order of magnitude to those of the Kennedy Round. They were phased over the first eight years of the 1980s. Only industrial products were covered by the agreement; agriculture continued to be highly protected, no concessions were made on sensitive industrial products such as textiles, and non-tariff barriers became increasingly important.

In 1980 and 1981 American car manufacturers and workers successfully lobbied for the 'voluntary limitation' of Japanese car exports to the United States. More Japanese car imports meant less consumption spending allocated to domestically produced goods, and in particular to buying American cars. Fewer American car sales meant fewer jobs and lower profits. Japan's voluntary limits on car exports to the USA pushed car prices up and improved Detroit's cash flow, but did little for sales or employment.

Over a longer period, non-tariff barriers were agreed for the highly cyclical steel trade in three phases. The first consisted of bilateral negotiations between Japan and the USA, and Japan and the EC, which restricted Japanese steel exports during the 1970s. The European Coal and Steel Community opened negotiations with the Japanese Ministry for International Trade and Industry as early as 1965. Later the EC Commission created Eurofer, a producers' cartel, and negotiated acrimoniously with Japan, culminating in a 1978 agreement on steel prices and quantities.

The second phase in the late 1970s to mid-1980s was a conflict between the USA and the EC. New tariffs and quotas on special steels were established in July 1983 with the overall goal of holding imports to 18.5 per cent of the US market. The EC subsidised steel hugely, and the competition was judged unfair. Unwilling to acquiesce, the EC undertook deliberate and direct retaliation in February 1984 against US import restrictions.

In the third phase, LDCs were persuaded to limit steel imports into the USA and the EC. Fifteen VRAs were agreed by 1985. The Japanese held steel exports to the United States to 5–6 per cent of the US market, satisfied by the higher profit margins they could earn (Conybeare, 1987, ch. 9). They also bought equity stakes in American steel producers.

Steel and cars were only two commodities that aroused the ire of competitors in Japan's export markets. Industrial machinery and video cassette recorders were also subjects of formal complaints. In exchange for deeper Japanese export penetration, the wide-ranging Structural Impediments Initiative talks between Japan and the USA issued a series of proposals for liberalising Japan's domestic economy, many of which were adopted. Japan proposed a number of American changes, including to US fiscal policy, which were also broadly accepted. The Japanese Action Programme reduced or eliminated tariffs on over 1,800 items from January 1986 (Yamazawa, 1990; Agraa, 1988). Appreciation of the yen between 1984 and 1988 was likely to have been more effective in reducing Japan's trade surplus. These policies of 'leaning against the wind' in the EC and USA could sometimes be justified in the short run by the difficulties of pricing in industries with high fixed costs and cyclical demand. Competitive pricing at marginal costs would fail to cover total costs, so surviving competitors in a depression would be those with the greatest financial resources. In practice, the policies were liable to deter industrial modernisation and restructuring, instead promoting 'industrial sclerosis'.

Japan's industrial policy was traditionally concerned with encouraging domestic production at the expense of imports, but that does not imply that such policies enhanced Japanese well-being. The closure of the Japanese semi-conductor market allowed that industry to develop in Japan and to gain a foothold in export markets. But Japan would have been better off without the industry (Baldwin and Krugman, 1988b). Analogously, Europe lost by subsidising the wide-bodied medium-range commercial aircraft, the Airbus A300, in competition with the Boeing 767. Airbus competition lowered the price in that market and so conferred gains on consumers, but they were insufficient to offset the subsidy paid by taxpayers (Baldwin and Krugman, 1988a).

The most ambitious attempt at multilateral trade liberalisation, the eighth GATT round, began in Uruguay in 1986. Scheduled to end in 1990, the round included services, agriculture and the MFA in its remit. However, more players than ever before, and complex linkages, reduced the chances of success. A well-past-the-last-minute agreement (in 1993) on agriculture at Blair House was quickly repudiated by France but finally in 1994 a compromise was reached. In the middle of this round in 1988, the United States passed an Act designed to 'level the playing field'. One clause, 'super 301', allowed the USA to name trading partners engaged in 'unfair trading'. Once named, these countries were called on to negotiate bilaterally over changes to their behaviour. If no agreement was reached within eighteen months, retaliation might be taken. Japan, Brazil and India were soon named and eight countries were placed 'on remand'.

The positive side of the new bilateralism was the US–Canada Free Trade Agreement from 1989. Over ten years all tariffs were to be eliminated, as were many non-tariff barriers. Three-quarters of the trade was already duty free, so the expected gains were not large. Calculations of these, and all other, gains from trade liberalisation are larger when descriptively accurate, imperfectly competitive markets are assumed. More foreign competition lowers price–cost margins and creates larger, lower-cost, firms than under perfect competition.

The North American Free Trade Area (NAFTA) was planned to extend the agreement to include Mexico. Fears of manufacturing jobs moving south to low-wage Mexico threatened US ratification. Mexico had already liberalised substantially, joining GATT in 1986, and by 1992 reducing tariffs from 22 per cent to 9 per cent, and scrapping import prohibitions and quotas. Mexican motor manufacturers expected to be able to source components from all over NAFTA, and hoped thereby to increase competitiveness. Foreign investors were certainly optimistic. Foreign direct investment in Mexico doubled from $2.6 billion in 1990 to $5.4 billion in 1992. A strong currency sucked in $40.6 billion of US imports. Mexican consumers were targeted by US supermarket chains, and Mexican industry was restructured in the face of import competition: cement, cars, glass, steel and some petrochemical businesses were modernised. Other sectors, textiles, paper and electrical appliances, suffered from foreign competition without restructuring. Mexican unit labour costs were about 35 per cent lower than in the USA; low wages were not the enormous source of advantage they appeared.

One estimate of the total impact of NAFTA was that Mexico would grow 0.5

per cent faster in the first three years and thereafter faster by 1 per cent (*The Economist*, 1993). NAFTA may be expected to increase the agricultural exports of both the United States and Mexico, by some 15 per cent on 1988 levels. Because initial protection was higher in Mexico (25 per cent as against 5 per cent), US exports would increase by more. The two countries would specialise further, with the USA exporting more grain and importing more horticultural products. Intra-industry trade in livestock and meat products would expand (Grennes and Krissof, 1993).

The United States' apparent loss of faith in multilateralism and opting for the regionalism of NAFTA need explaining. Canada was regarded as an exporter of raw materials and a natural candidate for integration. The Canadian treaty was not seen as a move away from multilateralism. As neighbouring countries, Mexico and Canada had considerable foreign policy importance to the USA, perhaps more so than their joint bilateral trade warranted (less than one-third of the US total). American trade unions objected to the Mexican link, unlike the Canadian connection, yet the administration overrode them. This contrasted with the general increase in American protection through administrative measures directed at the rest of the world, as exemplified by 'super 301'. Since American broader economic interests were almost certainly better served by multilateralism, the shift of policy may have been a sign of policy capture by special interests, constrained only by US foreign policy objectives (Krueger, 1992). Alternatively bilateralism may have reflected the US administration's frustration with lack of progress in multilateral negotiations.

Multilateralism had been abandoned earlier for LDCs. Revisions to GATT in 1979 exempted these economies from the requirement that only full customs unions or free-trade areas were legal under GATT. By far the largest grouping of LDC traders was ASEAN (Brunei, Indonesia, Malaya, the Philippines, Singapore and Thailand). More visionary was the South–South general trade preferences agreement concluded in 1986 by the Group of 77 (developing countries). They hoped that the agreement would compensate for the slowing down of North trade, which was unhelpful for their exports. But if it was eventually to do so, the gains took a considerable time to be realised.

Migration

An alternative to the movement of goods, as a means of promoting convergence, is the international mobility of people. Hence restricting trade encourages migration. As we have seen in previous chapters, there are many other causes of migration in the source and host countries, and immigration controls inhibit migration. The movement of guest workers from the South ceased around 1973 when tighter controls were introduced in the face of rising unemployment. The next phase was family reunification. During the recession in the earlier 1980s, European migration again stagnated. Thereafter gross migration flows rose. By the end of the 1980s, migrants were mainly young and temporary.

Southern European countries, formerly sources, became destinations for immigration. Their economic growth and the lack of jobs at home for the burgeoning population of northern Africa were contributory factors. Foreign residents were most abundant in Italy, Spain and Portugal. For European migrant labour, West Germany in the 1980s was a sort of core, as it was for trade. In 1983–6 West Germany overtook the USA as the principal recipient of European foreign skills. In the end year, West Germany employed 7.5 per cent of total European foreign residents.

Immigration into the USA from poorer countries increased with the passing of the Immigration Act of 1965. During the 1980s the five leading source countries were Mexico, the Philippines, China, Vietnam and Korea. Even though entries were restricted by visas, incentives affected the number applying for visas and the number of illegal immigrants. Less skilled Puerto Ricans were more prone to emigrate to the USA than more skilled. Mexican income per head was about one-tenth of the United States', yet only the ending of the guest worker programme for Mexican farmworkers seemed to trigger massive immigration. Five per cent of all Mexican-born people lived in the United States by the 1990s.

Fertility adjusted very quickly to the host country pattern, but wages remained lower where the source country was poor. Immigration increased the supply of less skilled workers, so raising US earnings inequality. Immigrants to the USA had higher labour force participation rates than natives. Their greater use of social services was only marginal (Greenwood and McDowell, 1986; Freeman, 1993).

By expanding labour-intensive sectors, trade restrictions increase the demand for foreign workers. At the same time they reduce job opportunities abroad for labour absorption in labour-intensive export sectors (Faini and Venturini, 1993). Across sectors in the USA, the share of immigrant labour was negatively associated with the export/output ratio and positively correlated with the import penetration ratio. In sectors such as clothing and textiles which were well protected, the share of foreign workers reached 10.4 per cent. In France a similar result was found, although the figure was higher, at 13 per cent. Comparable, though less significant, findings emerge for Spain. These percentages, and immigration policies, indicate that convergence by means of trade and migration were resisted by the richer countries.

Convergence and productivity slowdown

Convergence with the USA was becoming easier in both the 1970s and the 1980s because US productivity growth was slowing down. After 1973 US productivity growth averaged half the trend established over the preceding eighty years (Maddison, 1991, p. 71). No less striking was the leap of Japanese productivity ahead of those of the USA and Europe. The Japanese developed a new organisation for manufacturing, more effective than the American Fordist model.

Advanced nations of the world came to share similar technologies. But the exent to which they could take advantage of them depended on the skills and experience of the labour force, appropriate organisation and investment.[1] US savings were increasingly diverted to finance the federal budget deficit. Investment fell as a proportion of GNP. As trade barriers came down, economies smaller and less well endowed with natural resources could take advantage of mass production in the same way as the USA had done before 1914. That was the opportunity. Once Japan and Europe invested sufficiently in R & D and science and engineering education, they also acquired the social capability to take advantage of the opportunity in advanced technology sectors (Abramovitz, 1989).

The US lead in high-technology industry exports held up better than the productivity advantage. The same is true of patenting. Hi-tech industries are defined as those exceeding a certain percentage of R & D. Particularly after 1983, it was US imports of hi-tech that rose rapidly rather than US exports that fell, causing a deterioration in the US hi-tech trade balance. Motor vehicles and machine-tools sectors, in which Europe performed quite well and the USA did not, are excluded from the hi-tech sector. In 1987 the USA still maintained its lead in scientists and engineers engaged in R & D per 1,000 workers, although Japan had almost caught up (Nelson and Wright, 1992).

At the level of the industry, as for whole economies, catching-up was apparent, but so too was overtaking. In 1970 actual unit car costs adjusted for quality were lowest in Germany – 34 per cent below those in the USA and 3 per cent below Japan's. Higher US labour costs explained more than half the difference (Fuss and Waverman, 1992). Germany was 27 per cent more efficient than Japan and 17 per cent more efficient than US car producers. During the 1970s average total factor productivity growth was 1.7 per cent per year higher in the Japanese industry than among German, US and Canadian motor manufacturers. This was due primarily to technical change, but also to capacity utilisation. Even with an appreciating yen, the Japanese were able to maintain their cost advantage over the USA. The German motor industry lagged the Japanese by 8 per cent in 1980. Relative to the USA, a 10 per cent Japanese technical disadvantage in 1970 became a 17 per cent advantage by 1984. Japanese plant and technology management, organisational methods (*kanban*), quality circles, engineering innovation, broad job specifications and limited hierarchy proved more effective. In many manufacturing sectors these characteristics were diffused by Japanese foreign direct investment. Over 1 million Japanese cars were assembled in North America during 1989. Within three years that total had more than doubled. In Indiana, which received far less Japanese FDI than California, or many other states, there were more than eighty Japanese-owned plants by the beginning of the 1990s.

Convergence or catching-up, predicted by the Hecksher–Ohlin view of the world economy, so long as it remained open, was striking in some regions and quite absent in others. Taking into consideration the heterogeneity of the economic structures and specialisations of converging and stagnating groups, the evidence indicates that outward-looking domestic policies, rather than the world economy, were the key difference.

The non-convergers: Latin America and Africa

For Latin America and Africa generally, the 1970s were difficult thanks to the oil shocks, but international borrowing at low rates eased their position. With high interest rates and slower growth in the industrial economies that were their markets, the 1980s were very much worse. Domestic economic policies misallocated resources between industries and depressed exports. Social and geographical inequality increased. Between 1973 and 1981 Latin America's debt/export ratio rose from 1.4 to 2.5, with interest payments representing 23 per cent of the value of exports by the end of that period. Export growth decelerated with stagnation in the West. Debt service required that government expenditure be cut, and trade surpluses generated through export promotion and import restraint. National currencies were devalued particularly strongly in Argentina, Bolivia, Chile, Ecuador, Mexico and Uruguay. The effects on exports were often disappointing because so many other competitors devalued at the same time, and because of the higher prices of imported farm inputs such as tractors, harvesters and fertiliser. Brazil, Chile, Costa Rica and Peru raised import duties, while Mexico and Venezuela among others introduced quantitative controls and prohibitions. Central government deficits fell from 2.1 per cent of GDP in 1982 to 0.2 per cent in 1985. Real interest rates of more than 5 per cent per *month* were maintained in Argentina, Bolivia and Brazil for extended periods (Food and Agricultural Organisation, 1991). Consumption was intended to fall much as it did, but so too did savings and investment. Latin American investment as a proportion of GDP fell from 23 per cent in 1981 to 15 per cent in 1984. Imports collapsed, but so to a lesser extent did exports. Current account deficits fluctuated around 9 per cent of GDP after 1983.

A temporary improvement in 1984–6 disappeared in the acceleration of inflation between 1987 and 1990. The debt service export ratio remained unsustainably high at 26 per cent in 1990.

Agricultural output rose faster than other sectors at 2.4 per cent during the 1980s compared with 0.5 per cent for industry and 1.2 per cent for the Latin American economies in total. Export products flourished while supplies to the internal market did not match population growth. Unlike the rest of the economy, agriculture did achieve small labour productivity growth. During the first half of the 1980s, agriculture reaped the harvest of heavy investment and production incentives of the 1970s. The second half was a period of market liberalisation together with the exercise of some environmental concerns. The 1980s saw GDP per head fall by a cumulative 10 per cent in Latin America and the Caribbean, and the terms of trade fall by 20 per cent.

Sub-Saharan Africa was hardest hit by the economic shocks of the early 1980s.

Even in the 1960s, the most favourable period of the world economy, the sub-Saharan economies only managed to average 1 per cent GDP per head growth. In the following decade, income per head hardly grew at all. Food production per head actually declined, and the economies of the region increasingly switched from being net exporters to net importers. When drought came in the first half of the 1980s, famine struck twenty-five countries, exacerbated by the undeveloped state of irrigation. By the mid-1980s commercial food imports cost almost 20 per cent of the region's export earnings. Food aid doubled in the second half of the 1970s, by the mid-1980s reaching four times the level of ten years earlier. Yet availability of nutrients per head was no higher in 1985 than in 1970 (2,160 calories and 53.5 g protein daily). Over the 1980s sub-Saharan export earnings from agriculture lost about one-fifth of their capacity to finance imports of manufactures and oil, through a combination of declining terms of trade and stagnant export volumes. Debt service in 1988 was 4.3 per cent of GNP compared with 4.7 per cent for Latin America and the Caribbean. But since subsistence production was more prevalent in Africa, the debt service ratio was a proportionately greater burden on the money economy than the above percentages imply. Eight of the forty African countries were involved in major military conflicts over the decade, and many others suffered civil unrest and instability. Population growth, already the highest of any region in the 1970s at 3 per cent per annum, accelerated to 3.1 per cent during the 1980s, while for all other regions it fell. Urbanisation absorbed this natural increase despite marked declines in urban–rural income differentials. Policies usually affected agriculture adversely. So far as generalisations are possible, mechanisms to regulate internally traded food products were largely ineffective. They did little more than provide public employment and add to transactions costs. Domestic price interventions tended to be dropped in the 1980s to the benefit of government budgets, but without obvious increases in domestic supplies. Subsidies on food imports were eliminated, most dramatically in Nigeria, when imports of cereals and vegetable oils were banned in 1985 and 1986.

Catching-up in Asia

No more extreme contrast with African economic performance in the 1980s could be found than in Asia. In the later 1970s policy shifted towards the market in the state-planned economies of China, Laos and Vietnam, as well as in India, Thailand and elsewhere. With real GDP growth of nearly 7 per cent per year, this was easily the fastest-growing region during the 1980s. The 'four little dragons' – Taiwan, South Korea, Singapore and Hong Kong – by the end of the decade exported manufactured products equal to about one-half of Japan's or the United States', and averaged annual growth rates of 8 per cent (Vogel, 1991). Each of these economies could look to the Japanese 'model' of development in which exports were necessary to obtain food and raw materials. All had been occupied by the Japanese in the Second World War. Post-war US aid, the destruction of the old order, a sense of political urgency and literate workforces swelled by immigrants all contributed. Just as the 'dragons' started to industrialise, consumer electronics created labour-intensive opportunities that required relatively modest capital. The train of events beginning with the 1842 Treaty of Nanking (Chapter 1), which effectively created Hong Kong as a commercial settlement, yielded extraordinary consequences. Political stability, nineteenth-century liberalism and Chinese culture brought tiny Hong Kong's exports to the same level as massive China's by the beginning of the 1980s. Then the growth of the Chinese economy increased Hong Kong's re-exports sixfold (between 1979 and 1989), while Hong Kong's domestic merchandise exports 'merely' doubled. Textiles and clothing dominated the Hong Kong-China trade. By the end of the 1980s, Hong Kong accounted for perhaps two-thirds of foreign direct investment in China. In the adjacent Guangdong province, Hong Kong businesses were estimated to employ 2 million Chinese workers compared with a total of 2.8 million in Hong Kong itself.

Agriculture accounted for 30 per cent of the continent's GDP, and the region grew nearly 90 per cent of the world's rice. In China and India 60 per cent of the world's agricultural labour force produced food for 2 billion people on only 20 per cent of the world's arable land. By the mid-1980s India was self-sufficient in food and had built up sufficient reserves to withstand the drought of 1985-7. As much as 80 per cent of the increase in India's agricultural output was attributed by the World Bank to the combined use of irrigation and high-yield inputs, and 60 per cent alone to irrigation. On the other hand, the availability of pulse per head declined. Investment in agriculture fell throughout the 1980s. High levels of industrial protection drew resources away. The downside of agricultural output achievements was soil erosion, waterlogging, salinity, deforestation and other environmental problems throughout Asia.

Thailand's trade policies in the 1980s demonstrated the potential of allowing the market free play. Rice export duty was cut from 30 per cent in 1980 to 5 per cent in 1981, and was abolished in 1986. The baht was depreciated against the dollar by 30 per cent. Lower import tariffs brought cheaper fertilisers and pesticides, and encouraged their greater use. Competition among exporters increased with the elimination of export licensing. All these reforms raised the prices received by rice growers and persuaded them to increase supply, so that Thailand provided more than one-third of the world's rice exports. Even though Thailand's manufacturing sector accounted for about one-quarter of its GDP and two-thirds of its exports, agriculture continued to employ nearly 60 per cent of the labour force.

South-east Asia by the end of the 1980s was a major exporter of textiles, clothing, footwear, semi-conductor devices, computer parts, furniture and plastic products. Unlike the newly industrialised economies and Japan, south-east Asian economies also engaged in agro-export diversification into frozen fowl, canned fruit and furniture.

China began liberalisation with agricultural reform in 1978, in the early 1980s shifting to the household responsibility system. Grain production rose by one-third in the six years after 1979. With the 'open door' policy, China approved direct foreign investment at the end of the 1970s, primarily through joint ventures. The Special Economic Zones in Guandong and Fujian were also approved, with the official aim of absorbing foreign technology and management. Some enterprises were given greater freedom to deal directly with foreign firms. Prices still did not reflect resource costs in the mid-1980s, giving the wrong signals for investment, but in the countryside, household production on leased land was replacing state farms. Imports were controlled when they grew too fast. Special Economic Zones were extended. The partial nature of the reforms led to a burgeoning of 'economic crimes' – 'accounting errors', black-market trading and the use of official positions for private gain.

OPEC and oil

Oil exporters' performance in the 1980s was the reverse of the Asian economies'. In the preceding ten years, oil had been what gold was to the sixteenth and seventeenth centuries: politically, the most desirable and influential commodity in international trade. By limiting the rate at which oil was sold, OPEC controlled the price. An economic analysis of the optimal depletion rate of known oil reserves requires that oil should be pumped out at a rate that equates the percentage price rise with the rate of interest. Then holding the resource in the ground is as attractive an investment as any other. The oil price *level* should be fixed to balance supply and demand in any period, given the price increase objective. However, new discoveries by increasing proven reserves reduce the utility of this formulation. Actual spot (as against long-term contract) prices proved extremely sensitive to inventory changes, in turn a reflection of highly volatile expectations of future oil availability.

During the 1970s, control of oil supplies and prices conferred the power to redistribute world income, international debt and current account surpluses. When in 1973/4 and 1978/80 oil prices jumped, economic growth in industrialised countries slowed, and exports of poorer economies and new industrialisers stagnated, cutting earnings available for imports and foreign debt service. Until the 1974 crisis there was no intergovernmental framework for containing emergencies in this vital market. Multinational oil companies negotiated with producer and consumer states individually. Oil supplies fell by 7 per cent from October to December 1973, and by March 1974 were still 5 per cent below October levels. France and the United Kingdom tried to pressurise their oil companies to give them preferential treatment. Italy, Spain and Belgium restricted oil exports. American, Japanese and German companies bid up oil in the spot market. The oil companies saved the OECD companies from the full consequences of this myopic self-interest. They tried to reduce shipments to all customers in roughly equal proportion. The USA called an energy conference in

Washington in 1974, as a consequence of which the International Energy Agency (IEA) was established later that year. By 1978 the Agency was capable of facilitating emergency oil sharing. The secretariat was empowered to identify oil reductions to one or more members and initiate reductions for all members so as to cushion the impact on those originally affected. The IEA also monitored oil markets and undertook long-term planning.

The revolt against the Shah eliminated all oil exports from Iran by Christmas 1978. The price shot up from $13 to $34 a barrel (Yergin, 1991, p. 687; Keohane, 1984). Inventory build-up of an extra 3 million barrels a day added to the price impact of 2 million barrels of lost supplies, amounting to about 10 per cent of consumption. Saudi Arabia pumped more oil so that in 1979, despite the Iranian shutdown, OPEC supplied 3 million barrels a day, 10 per cent more than in 1978. Production actually exceeded consumption in 1979, yet panic buying and hoarding doubled prices. By June 1980 the price per barrel was almost three times that of a year and a half earlier. The emergency oil-sharing system was not activated. The IEA actually negotiated down a Swedish request to do so. It feared that declaration of an emergency could worsen the shortage. At the Tokyo summit of 1979, the USA successfully proposed country import targets for 1980. The summit also agreed that private stockpiling was disruptive. These measures proved effective during the oil crisis of the following year.

In September 1980 Iraq attacked Iran. The war removed 15 per cent of OPEC output, amounting to 8 per cent of Western demand. Spot prices jumped to $42 a barrel. The IEA persuaded MITI to restrain Japanese companies which were stockpiling. The Saudis increased supply again. Non-OPEC oil from the North Sea and Mexico was flowing more abundantly. World oil supplies fell by 2.4 million barrels in the last quarter of 1980, rather more than in the first quarter of 1979. Yet spot prices rose by much less, and had fallen back by July to only 5 per cent above pre-war levels. Stocks in this crisis were drawn down at double their normal rate. Actual US imports in 1980 were 23 per cent below target. The IEA created its own self-fulfilling prophecy.

As Figure 15.2 shows, this was the peak of real oil prices. Thereafter world prices continued to rise faster than the nominal price per barrel of oil. OPEC agreed a price cut and quotas for all except the 'swing producer', Saudi Arabia, in 1983. Saudi Arabia was then still willing to accept variable revenue and market share in order to stabilise and support the market. In the same year, oil market functioning was improved when the New York Mercantile Exchange introduced futures trading in crude oil. Oil was completely deregulated in the USA in 1981, and the industry began restructuring according to returns on underlying assets. However, oil trouble was not yet over. The aftermath was the financial difficulties of banks that had lent on the assumption that high real oil prices were permanent. Continental Illinois, the largest bank in the mid-West and the seventh largest in the United States, was the most notable casualty. Falling prices ruined it in 1984. The US federal government ran the largest financial rescue in history, with $5.5

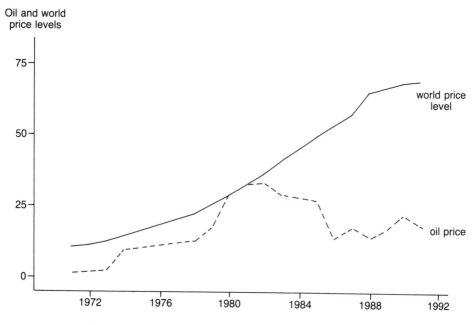

Figure 15.2 Oil prices and the world price level, 1970–92

billion of new capital and $8 billion in emergency loans. The bank was temporarily nationalised.

The third oil shock struck when, in 1986, Saudi Arabia and Kuwait wanted more revenue and therefore increased their oil market shares. Oil prices fell sharply. Spot prices for crude approached $10 per barrel in April, 60 per cent below the December 1985 price, but trading was primarily in the range $15–17. Stock reductions pushed prices down further. West Texas Intermediate oil fell from its peak at the end of November 1985 to one-third in 1986. The USA did not want prices as low as $10 a barrel for fear of the impact on US domestic producers, and the Saudis obliged by fixing an $18 price.

The transfer to industrial countries in 1986 amounted to about $63 billion a year, or 0.7 per cent of their GDP (Table 15.2). However, this fall did not reverse the gains and losses of the earlier rises. The North Sea fields were by the 1980s a major source of supply, and the industrial countries were in any case less dependent on oil. Among the industrialised countries, Norway stood to lose the most government revenue, 10 per cent, the Netherlands lost 7 per cent and the UK 3–4 per cent. OPEC economies, geared to massive and expanding oil revenues, found themselves in dire straits. Crude oil prices recovered to around $20 a barrel at the beginning of 1990, but OPEC exceeded quotas, and a slowing of world activity allowed prices to sag again. Then in the summer of 1990 Iraq invaded Kuwait for extra oil revenue to support the huge Iraqi military

Table 15.2 Income transfer from an $11 per barrel fall in oil prices in 1986, petroleum only ($bn, p.a.).

	Producers	Consumers	Net
OECD	−69	132	63
OPEC	−70	11	−59
Non-oil developing countries	−32	34	2
Centrally planned economies			−6

Source: OECD (1986, p. 2).

expenditure. The price doubled in September, but the Saudis increased output by 3 million barrels a day and demand was anyway weakening. The world oil market was by now so transformed that the political crisis did not once more become an energy crisis as well.

Centrally planned economies and the transition to the market

Like the OPEC economies in the second half of the 1980s, the centrally planned economies found themselves diverging from best practice in the market economies. Their armed forces recognised the necessity for economic liberalisation if they were to have access to advanced weapons technology. Central direction of economies failed to provide appropriate incentives for the workforce at every level. The system was poorly co-ordinated and could operate only because of an illegal army of 'fixers', supplying missing or surplus products through black markets. Lacking adequate computing power, which might have mitigated some of the problems inherent in central planning, planners attempted to balance supplies and demands by a repeated trial-and-error process, the method of material balances. Computers were on the strategic embargo list of the Co-ordinating Committee for Multilateral Export Controls (Cocom),[2] and Soviet planners had not given their development and production the same priority as weaponry and aerospace. By the 1960s west European countries applied the embargo only to military goods, but the United States, the principal computer supplier, continued to be much stricter. This enforcement gave rise to a severe disagreement between the blocs over a Soviet natural gas pipeline to Europe in 1985.

In compensation for the lack of co-ordination and incentives in the Soviet planning system, production unit targets were specified so as to render fulfilment difficult. This was 'taut planning' – building excessive demand into the system. Ruling élites resisted substantive economic reform during the 1970s and much of the 1980s, rightly fearing that their prerogatives would be threatened. Hungary was the most successful at combining the market and planning, because of a more favourable political environment. Comecon remained unable to provide mutual economic assistance on any significant scale. The Soviet Union merely required, in exchange for its raw materials, the machinery and other items which were likely to be permanently in short supply in the Soviet Union (Csaba, 1990, pp. 186–7). During the second half of the 1970s, growth slowed in the USSR, Czechoslovakia, Hungary and Poland, and it did so even further in the early 1980s. Labour and raw materials became scarcer (Adam, 1989, ch. 7).

Eastern Europe became increasingly reliant upon the West for technology and finance (Brus, 1986a). Advertisements for western consumer products were beamed by television to a populace unable to afford any such luxuries, and uncompensated even by political freedoms. From 1985 Gorbachev loosened the grip of the Soviet communist apparatus, and four years later eastern Europe was effectively allowed to throw off communism and Soviet control. Even the Russian Empire disintegrated into the 'Commonwealth of Independent States' (CIS) in 1990.

Eastern Europe, with a more recent tradition of markets, fared better than the USSR after the break-up. The West perceived considerable investment opportunities in the economy that had already experimented with 'market socialism' during the 1980s. By mid-1991 Hungary was estimated to have received $2.15 billion of foreign investment, four times the volume of foreign capital committed to direct investment in Czechoslovakia or Poland. Intra-eastern trade collapsed, but eastern European trade with the West, unlike CIS trade, continued to grow strongly. Bulgaria, Czechoslovakia, Hungary and Poland all showed that they could meet western quality standards for consumer goods and engineering products (United Nations, 1993).

Developments in eastern Europe outside East Germany raised western exports by only a small proportion. Even a doubling of east European imports boosted world trade by only 5 per cent, after multiplier effects were taken into account. Reduced military spending might have conferred greater gains.

Removal of repressive central authority unleashed disruptive nationalism. More than 200,000 were killed by the end of 1992 in the fighting after the break-up of the Soviet Union and Yugoslavia (85 per cent in Yugoslavia), and around 3 million were displaced. Output continued falling through 1992, except perhaps in Poland. In Armenia, subject to an economic blockade by Azerbaijan, the collapse may have been 40 per cent. The small open Baltic economies of Estonia, Latvia and Lithuania were almost as hard hit, by severing trade ties with the former Soviet Union and renegotiating higher oil and gas import prices. Unemployment was rife, with minimal 'social safety nets'.

In 1991 some $2 billion of western aid was given to the CIS. Three-quarters of humanitarian aid and 65 per cent of all financial assistance came from the European Community (Brusnikin, 1992). IMF loans for 1992 were not disbursed because the CIS failed to satisfy the stringent conditions attached. Economic progress was slow because of inadequately defined and alienable property rights, in turn blocked by the *nomenklatura*, party bureaucrats still maintaining positions of power. In this respect, the problem was comparable with eastern Europe's nineteenth-century abolition of serfdom, but now it was property that needed liberating. Another Marshall Plan could therefore not be expected to work in the same way in the CIS and eastern Europe as the original had in western Europe forty-five years earlier.

European economic integration

Western Europe had already closed a good deal of the gap with the United States in preceding decades. More political effort in the 1980s was directed to integrating economies of member states ('deepening') than to integrating the EC with the wider world. From the early 1970s, European productivity growth slowed, and unemployment and inflation rose ('Euro-sclerosis'). Unlike joblessness in the rest of the OECD, EC unemployment rose secularly, reaching 11 per cent by 1986, four percentage points higher than in the United States. Much of the reason was to be found in national labour market policies (Lawrence and Schultze, 1987).

Trade was no constraint on expansion. Sweden, France and Germany could keep trade balanced while matching 3 per cent GNP growth rates in the rest of the OECD. The UK could manage only half that rate without an increase in oil exports. Each 1 per cent rise in British GDP increased imports by 2 per cent. With unitary import and export price elasticities, the change in the terms of trade necessary to balance trade with growth of 1 per cent above the rate warranted by exports would cut real income by 0.5 per cent. Germany retained a surplus in hi-tech trade in 1983, and in France and Sweden hi-tech exports grew more rapidly than imports between 1970 and 1983.

The completion of the internal market in 1992 was to be one solution to 'Euro-sclerosis' under the Single European Act of 17 February 1986. In 1990 capital movements were liberalised. The single market was to increase competition, encourage industrial rationalisation and lower costs through greater utilisation of scale economies (Figure 15.3) (Cecchini, 1988). Over the medium term, European GDP was predicted to rise 3–6 per cent, and between 1.3 and 2.3 million extra jobs were expected. Other possible gains stemmed from external economies or from R & D co-operation across national borders. Imports from outside the EC were effective in holding down price–cost margins, whereas intra-EC imports were not. Freer world trade as a whole would therefore benefit the EC. EFTA gains from trade access to an integrated market were probably modest, due to scale advantages and competition in manufacturing, but there were likely to have been larger gains in services (Winters and Venables, 1991).

ERM, ECU, EMU

Another proposed solution to 'Euro-sclerosis' was monetary unification. Economic integration would be supported by exchange rate stability, but the weakness of the dollar in the later 1970s introduced uncertainty to all major exchange rates.

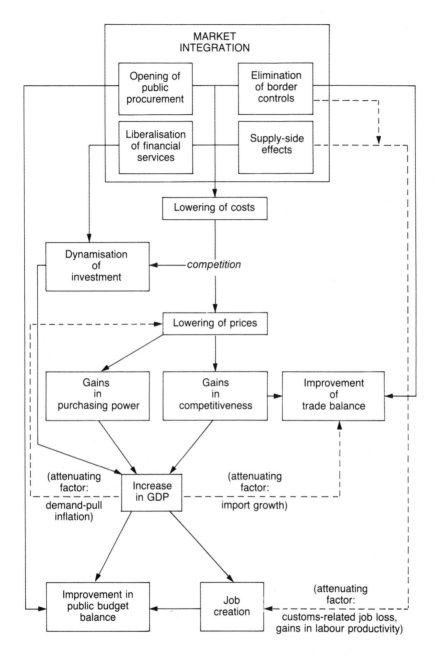

Figure 15.3 Impact of the completion of the European internal market (Cecchini, 1988)

France and Germany therefore agreed to create a zone of monetary stability (Feri, 1990). France wanted to safeguard the Common Agricultural Policy, and Germany wanted to avoid competitive devaluations. From 1979 the European Exchange Rate Mechanism (ERM), at first, excluding the United Kingdom, pegged member currencies within bands that allowed 2.25 per cent movement from the central rate. A wide margin of 6 per cent was allowed to Italy until January 1990. The ERM resembled the inter-war gold standard, Bretton Woods and the Latin Monetary Union in being a managed system (de Cecco, 1990). It was a very short-term credit mechanism for financing central bank balances resulting from exchange market intervention. More frequently, the system was supported by intervention on the part of the weak currency by the Bundesbank in the dollar market. Unlike its predecessor the 'snake', the ERM aimed to co-ordinate the macroeconomic policies of members. The Basle and Nybourg accords of August and September 1987 extended very short-term credits to include intra-marginal interventions. There was no obligation for the central banks of strong currencies to intervene within the margins. Co-ordination among ERM banks was pragmatic but effective. The expected inflation rate was higher outside the ERM, pushing the real interest rate in the UK above that inside the ERM (Thiel, 1990). The ERM accounting unit, the ECU (European currency unit), defined as a basket of currencies, became an important currency for lending. It was helpful for hedging by businesses with a range of activities in different European states.

At first the French franc and the Italian lira were periodically devalued against the German mark. Credibility that the exchange rates would be held was achieved for France in 1983 and for Italy in 1984. Favourable world factors – falling oil prices and the strong dollar – brought down inflation. The US dollar fell sharply against the ECU between 1985 and 1988 as a delayed consequence of persistent budgetary and trade deficits. Most European governments supported their pegged rates by pursuing deflationary policies, aiming to control inflation at the expense of unemployment.

Spain joined the ERM with wide margins in 1989, followed by the UK the next year. Italy moved to narrow margins in 1990. The ERM was credited with reducing the variability of members' bilateral exchange rates. Both greater policy co-ordination and continued capital controls may have played a part.

A new phase began with the Delors Report, or the final document of the *Committee for the Study of Economic and Monetary Union* (European Commission, 1989), and the subsequent report on *Economic and Monetary Union* (European Commission, 1990). These proposed the most radical change in macroeconomic policy in Europe since at least 1945. Monetary union required the irrevocable locking of exchange rates between currencies of member states, co-ordination of fiscal and monetary policies, and a European central banking system. However, conditions were not as propitious as they were assumed to be. Inflation convergence, with only one exchange rate realignment between 1987 and the partial collapse of the system in 1992/3, produced an undervaluation of the mark and an overvaluation of most other ERM currencies. German surpluses were massively offset by other European deficits. German unification in October 1990 introduced a new dimension. From that date Germany needed to invest vast resources, with a long-delayed pay-off, in the reconstruction of the east. This policy was made more difficult by the one-for-one exchange rate agreed for the two former German currencies. The initial exchange rate overvalued the ost mark by perhaps 200 per cent, condemning a large proportion of the eastern workforce and equipment to unemployment, in the absence of massive state support.

What Germany experienced with monetary unification, Europe as a whole could expect under the Delors plan. A 'Solidarity Pact' allowed for the continuation of very high support payments to the east German economy. This required the full integration of the east German *Länder* into the west German federal government revenue-sharing system by 1995. It was intended to raise revenue and cut spending so as to reduce the public-sector deficit, and convince the trade unions that their jobs would be preserved if they accepted wage restraint. The likelihood is that convergence of east and west German productivity will take thirty to forty years. Subsidies will continue for that length of time, placing very considerable demands on the German economy, and requiring high interest rates that will hold back the rest of core Europe (Hughes Hallett and Ma, 1993; Grauwe, 1991). The overvaluation of sterling in the 1920s pales into insignificance by comparison.

Unification at parity between the two mark currencies in July 1990 was a triumph of political will over economic common sense. The low productivity of eastern industry was well known, and the difficulties of east German market reorientation from the collapsing east to the west would have been severe enough without the additional handicap of a massive overvaluation. Between January 1989 and January 1992 about 5 per cent of the east German population migrated to west Germany, admittedly more before monetary unification. Effective earnings in the east were about half those in the west during 1992. Some east German regions suffered effective unemployment of 50 per cent or higher (Burda, 1993). Future migrants were likely to be the young, so the east would be left with an ageing population, requiring more social support payments.

Despite the emerging evidence from Germany, the European Council Agreement at Maastricht in December 1991 agreed financial and inflation 'convergence criteria' which needed to be satisfied before monetary unification. They included low inflation, a low budget deficit, a stable exchange rate and public debt no greater than 60 per cent of GDP. The Maastricht convergence criteria left no room for fiscal compensation for the European recession since few countries will meet the targets anyway by 1999. Higher interest rates, and the absence of labour market targets in the convergence criteria, ensured continuing high unemployment while the European Monetary Union (EMU) remained on the political agenda. Italy and the UK were forced out of the ERM in September 1992 by capital flight, and the Bundesbank slightly eased interest rates. The December 1992 Edinburgh summit adopted a small package to boost investment, but

Spanish and French depreciations in 1993 indicated that monetary unification was unlikely to keep to the Maastricht timetable.

The European Common Agricultural Policy and trade

What a common currency was to Brussels policy-makers in the late 1980s, the Common Agricultural Policy (CAP) was to the original Six. The CAP was the central pillar of the original European Economic Community. EC transfers between member states were very much influenced by their position in relation to this policy. The United Kingdom's small and productive agricultural sector placed it at a particular disadvantage in the web of taxes and subsidies of the CAP. During the early years of the EC, internal agricultural prices were held up by excluding non-EC produce. By the 1980s productivity had increased so much that the EC was subsidising exports to maintain domestic prices. The USA pursued an agricultural policy that was almost as protectionist, and regularly took issue with the EC's spoiling of US overseas markets.

Why agriculture should have been singled out for such special treatment can be explained by a simple political economy model. The first building block is the source of predominant political power. In an agricultural economy, landowners and/or agricultural workers are obviously influential. Even in an industrial economy, the agricultural voting block may well be larger, or geographically more effectively concentrated, than voters of other industries. Agriculture especially can disguise support policies as stabilisation policies. Cuts in food prices conferred only small individual benefits to the multitude of consumers and taxpayers, but holding up prices was of very considerable importance to the maintenance of a much smaller number of farmers' incomes. Hence farmers invest a great deal in defending the absurdities of the CAP, while consumers and taxpayers do not. As the number employed by agriculture in the West gradually declines, so too will their political influence, but this process will take several generations.

As exchange rate movements during the 1970s became more erratic, transfers between member states under the CAP were increasingly difficult to calculate. It was exchange rate fluctuations that prompted the introduction of notional 'green' exchange rates. Green rates deferred the agricultural price increases that would have followed from devaluation. But green rates broke the unity of the CAP, allowing agricultural prices in member states to differ markedly. The rise of the mark aggravated the difficulties of calculating transfers and common prices within the CAP, and the gyrations of the dollar disrupted world agricultural markets. When the dollar collapsed, the extent of the excess supply in the Community and the heavy costs of subsidising world exports became apparent. Imports fell and exports of cereals, beef, sugar and some dairy products increased. Demands upon the CAP budget increased.

The first enlargement, the admission of the United Kingdom, Ireland and Denmark in 1973, in effect gave notice of future pressure to reform the CAP.

Primarily because of the small size of the British agricultural sector, UK officials estimated that, on the basis of the Community's 1970 financial regulations, the United Kingdom would end up paying 31 per cent of the EC budget and receiving back 6 per cent. No formal safeguard was given beyond 1979, however, and the British were obviously going to press for changes. But so too were other countries. Greece entered in 1981 and could delay approval of entry for Spain and Portugal until the Greeks achieved a satisfactory deal. That entailed shifting the CAP budgetary constraint. The deal was that, when Spain and Portugal joined the EC in 1986, a new higher budgetary ceiling of 1.4 per cent of the VAT base came into operation. This was only a temporary solution, for expenditure on agriculture continued to rise, absorbing the additional resources more quickly than expected.

Lower internal EC prices would have reduced the demands on the CAP budget, but instead price increases in the 1970s more than compensated for inflation. This was no longer the case in the 1980s. EC price proposals of 1982/3 included guarantee thresholds for certain commodities to try to limit expenditure. A 1987 agreement effectively meant that farm prices in 1988 would fall by 18 per cent unless EC prices were raised by that amount. These measures were still inadequate, and from 1988, under the new 'set aside' policy, farmers were paid not to produce.

Political developments of the 1990s can only be expected to erode the viability of the CAP further. From October 1990 the former East Germany (GDR) was absorbed by the EC, and as productivity is brought up to West German levels, excess agricultural supply will be exacerbated. Yet the cost of generating this excess supply is already enormous. The total value of assistance to EC farmers in 1990 amounted to ECU64.3 billion or US$81.6 billion (over half the EC's budget), and the implicit tax on consumers was ECU50 billion or US$63.8 billion (OECD, 1991, pp. 157–9).

The 1992 reform proposals advanced by the Council of Ministers showed the remarkable durability and inflexibility of the CAP principles. Draft Council Regulation Document 7022/92 would impose a uniquely intrusive level of bureaucracy that would restrict farmers' responsiveness to market signals. Seventy-five per cent of the set aside area would be in three countries – France, Spain and the UK – while many producers in Italy, Greece and Germany would be untouched. In the UK an area roughly the size of Norfolk (600,000 hectares) was likely to be set aside.

The agreed package was extremely complex and would be expensive to administer. It ignored the proposal by the (UK) House of Lords Select Committee on the European Community in its report *Fraud against the Community*, which recommended that the Court of Auditors be consulted before any major policy change. Basing calculations on unverifiable numbers of hectares and animals creates great opportunities for overclaims and double payments, as the Court of Auditors noted. The non-application of milk quotas in Italy particularly concerned the British National Farmers' Union.

This expensive overproduction fiasco was entirely predictable. Open-ended subsidies to maintain high prices, productivity growth and fairly static demand were bound to lead to surpluses. Then politicians would try to avoid facing the bankruptcy of the basic structure of the CAP by tinkering. Yet good justifications are hard to find for taxpayers being required to increase the self-sufficiency in British temperate zone foodstuffs from 60 per cent at accession to over 80 per cent by 1988. Food is certainly plentiful in the Community, but it could have been supplied at lower prices without the CAP. The increasing complexity of the policy left behind the smaller and less well-educated farmers, those most in need of assistance (Tracy, 1989).

Multilateral trade negotiations made little progress in removing the trade distortions imposed by agricultural policies. In the Tokyo Round (1973–9) tariffs were cut on tropical products, but no EC-regulated products were directly affected. Although there were some bilateral concessions, they did not neutralise repeated subsequent conflicts between the EC and the USA over agricultural products. Japan also resisted any reduction in agricultural support (OECD, 1991). The GATT complaints procedure proved ineffective, particularly over export subsidies. The Uruguay Round begun in 1986 was specifically intended to liberalise agricultural trade. Meanwhile, Thailand was subject to a 'voluntary export restraint' agreement on manioc, which also covered sheep meat.

US subsidies were less than those of the EC but still considerable. The main policy instruments were deficiency payments on field crops, operated in conjunction with Acreage Reduction Programmes. Protection for the beef industry came from the Meat Import Act and associated voluntary restraint agreements on exports to the United States. Exports from the USA were boosted by an Export Enhancement Programme, which spent $312 million in 1990.

The economic consequences of agricultural policy were that the EC butter stocks by the mid-1980s were around a million tonnes, and beef and veal stocks had reached 600,000 tons. The USA held 38 million tons of wheat in 1984. Despite this effort to buoy up prices, relative incomes in OECD agriculture at best remained stable, or more often declined, between 1960 and 1984. Since subsidies were paid according to the amount produced, in the USA and EC over 75 per cent of the assistance was received by fewer than 25 per cent of the (larger) farms (OECD, 1987).

Monetary disorder

Exchange rate swings

Exchange rate swings which disrupted agricultural policy imposed penalties on a much wider range of activities as well. Sharply contractionary US monetary policy from 1979 raised interest rates and pulled up the dollar, even though inflation was slow to respond. The dollar remained high because expansionary US fiscal policy was coupled with contractionary German, British and Japanese stances in the early 1980s. Higher US interest rates drew in foreign capital, buoying up the (nominal) exchange rate. Excess demand for goods in the United States bid up domestic prices and appreciated the real exchange rate (increasing the volume of foreign products that could be bought for an equivalent expenditure on dollar goods). Consumers switched to buying cheaper imports, and exports were cut back, as they were priced out of foreign markets. Dollar appreciation reduced inflation in the USA, by about 1–1.5 per cent a year between 1980 and 1983, but also deepened the recession, losing perhaps 1.1 million jobs and 2.3 per cent of GNP over the same period. In Latin America the spillovers were disastrous (see below). In other industrial countries, export-led recovery was encouraged by the fall in US competitiveness, but so too was imported inflation. The United States and Europe needed opposite fiscal adjustments: a reduced deficit was required in America and a more expansionary stance was warranted in Europe.

Flexible exchange rates were and are ill-suited to asynchronised monetary and fiscal policy mixes. But the basic problem was the policy mix, not the exchange rate system (Dornbusch, 1984). US fiscal policy at the beginning of the 1980s was driven by the erroneous belief that tax cuts would increase revenue sufficiently to reduce the budget deficit. A happy consequence of the error was that the US swung out of the depression much faster than Europe or Japan. But the dollar appreciated in real terms by 40 per cent between 1980 and the peak in February 1985. The high dollar provoked US industry demands for protection from now cheaper imports. By 1985 there were 300 protectionist bills before Congress.

A remedy was sought at the New York Plaza Hotel in September 1985, where the five largest industrial nations agreed that the dollar was too high and that they would encourage its decline with appropriate monetary policies. By 1987 it had fallen back to the level of the early 1980s. The Group of 7 at the Louvre in Paris announced that the dollar had moved far enough. They agreed that fiscal consolidation in the USA and faster growth in Europe were needed. Japanese growth was to be backed up by Japanese fiscal expansion.

After the stock market crash of October 1987, the Federal Reserve loosened monetary policy and the dollar began to depreciate again. At the beginning of 1988 co-ordinated action by the Group of 7 stopped that. The US current account deficit peaked in 1987, two years after the dollar peaked. As US debt built up, due to the continuing current account deficit, so did the interest cost of servicing the debt. At a 10 per cent interest rate, the induced increase in the deficit was about $10 billion each year. For how long the USA would be able to finance this deficit remained uncertain. Undoubtedly needing to cut the deficit, the end of the Cold War and the break-up of the USSR in 1990 at first seemed to offer the USA a way out. The 'peace dividend' in practice proved hard to realise.

In view of the enormous capital flows that could pass through the exchanges if government aspirations were out of line with those of the markets, the dollar could only float against other major currencies. The best that could be hoped for was that consistent and co-ordinated policies would generally ensure that exchange rates were stable. Small countries lacked the transactions volume or financial institutions to warrant floating rates. They were instead obliged to peg their currencies to those of major trading partners. This brought other disadvantages. The thirteen sub-Saharan economies with their exchange rates linked to the French franc saw their rates appreciate with that currency in the late 1970s and into the 1980s. Until 1985, Liberia, with its currency linked to the US dollar, suffered the same experience. Botswana, Lesotho and Swaziland tied their currencies to the South African rand. To counterbalance these appreciations (and domestic inflation), nine African countries devalued by up to 400 per cent between 1983 and 1990, three others by up to 1,900 per cent, and another five by more than 2,000 per cent. Uganda's nominal devaluation was 20,000 per cent. Internal price rises offset only a portion of these nominal changes. All countries for which calculations can be made showed effective devaluations.

Capital movements and debt

The same macroeconomic policy that caused the overvaluation of the dollar also transformed the United States from the world's largest creditor into the world's largest debtor during the 1980s. Despite massive current account and government budget deficits, confidence in the ability of the USA to sustain these positions encouraged, and was supported by, massive inflows of foreign, especially Japanese, capital. When the domestic Japanese economy demanded funds to privatise the telecommunications network, NTT, Japanese capital returned home and the dollar tumbled. Japan meanwhile became the largest creditor, and Taiwan's foreign exchange reserves attained a level second only to Japan's.

High US interest rates, which held up the dollar and penalised Latin America, ultimately precipitated official (US) financial intervention on an unprecedented scale (Kucynski, 1988). In turn, bail-outs were triggered by the exposure of the major US banks. Latin American defaults could have paralysed the US banking system and thrown the whole economy into depression, as happened in the 1930s. Since creditor commercial banks were effectively being rescued as well as the debtor countries, the IMF made complementary additional bank financing a condition of assistance from the Fund (Snowden, 1985).

Until 1980–2 the value of debtor economies' exports was increasing as fast as debt. Bankers were willing to lend more even to allow debt service. When the world economy turned down at the end of the 1970s, deflated by restrictive policies in the major industrial economies, debtor export earnings fell. At the same time, interest costs rose and dollar appreciation made repayments denominated in dollars heavier. Mexico was forced to declare a moratorium on debt service in August 1982. Private bankers then refused to lend to other Latin American countries. As next-door neighbour to the USA, Mexico was of special diplomatic and strategic significance (Lipson, 1989). It had received special treatment in 1976. Once again it was bailed out. In August 1982 $1 billion came from the Treasury's exchange stabilisation fund, $1 billion from agriculture for food purchase and $1 billion for strategic oil purchases. The terms of the deals struck were, of course, favourable to the USA, but then no one else was able or willing to provide that sort of money. One of the interests of the USA was in preventing the additional waves of illegal migration from Mexico to the USA that a collapse of the Mexican economy would bring.

Maintaining the viability of the debtor economies was dependent on avoiding too great a drain of resources through their balances of payments in repayment of their debt. Capital flight, through loss of confidence, could exacerbate the position, requiring either impossible increases in exports or cutbacks in imports, if the exchange markets were to keep functioning. While new lending continued, smaller net foreign transfers were needed. US Secretary of the Treasury James Baker stated his intention to persuade commercial banks to renew lending in 1985. The so-called Baker Plan proved less than satisfactory by 1987. A variety of new and old ways were proposed for commercial banks to make new money available for the Argentine loan of that year.

A possible means of reducing the debt burden emerged with a secondary market for sovereign debt, in which loans were traded at discounted prices. 'Debt–equity swaps' then allowed an investor to swap the debt for domestic currency assets. Chile secured a major reduction in foreign debt in this way, by selling assets from its privatisation programme. Alternatively, debtors bought back their discounted debt on the secondary market and so obtained some debt relief. Bolivia in 1988 repurchased almost half its commercial bank debt at close to a 90 per cent discount.

US Secretary of the Treasury Nicholas Brady in 1989 proposed that the IMF and the World Bank make loans that would finance voluntary debt reductions, after extensive policy reforms. The first agreement, with Mexico, gave about 30 per cent debt relief but only about half of that in reduced resource transfer from Mexico to Mexican creditors.

The regions most heavily involved in debt rescheduling were those that experienced the real growth crisis of the 1980s, Latin American and in Africa. Rescheduling appeared to be addictive. Of forty-nine rescheduled between 1979 and 1982, only four did not repeat the process between 1983 and 1986. Turkey was among these four, growing rapidly and decreasingly dependent upon official capital, but this was a benefit of earlier concessionary loans, not of tight 'rollovers' of the 1980s. Major debtors, the middle-income countries were obliged to repay at market interest rates in full. Poorer, usually smaller, debtors were subsidised (Lindert, 1989). There were no large net resource flows. Nearly half of net repayments to private creditors in the first half of the 1980s came from Ecuador, Mexico and Venezuela – all oil exporters. Brazil paid little until 1985–6 and suspended payments in February 1987. Loans from governments partly replaced private debt in the 1980s, so that net transfers were less than transfers to private creditors. For ninety-four debtor states excluding the three oil exporters above, repayments were small or negative. The three repayers were close to the USA and therefore may have felt that their trade was vulnerable.

Policy co-ordination, collaboration and consultation

Exchange rate instability and possible capital flight placed a premium on policy co-ordination, as the Group of 7 recognised. They instituted economic summits in 1975 in reaction to the collapse of Bretton Woods and the oil crisis (Putnam and Bayne, 1987). Historically, co-ordination has been rare, unlike consultation and collaboration. Consultation is the exchange of information without the commitment to use it. International institutions, such as the IMF or the OECD, provide intermediate consultation. Periodic meetings of the Group of 7 (Canada, France, Germany, Italy, Japan, the UK and the USA) engage in 'multilateral surveillance' of their national policies under the auspices of the IMF.

Collaboration involves governments taking measures to achieve agreed objectives, but without mutually binding commitments on their national policies. Under the Bretton Wood system, certainly during the 1960s, there was co-operation based on close consultation with occasional collaboration. Working Party 3 of the OECD encouraged the search for consistent policies. The second amendment to the articles of agreement of the IMF, which ratified the shift to floating exchange rates, required collaboration to assure orderly exchange rate arrangements. Collaboration is often intended to preserve an international regime. Instances include Marshall Aid, the stabilisation loans of the 1920s, the joint operation of the Gold Pool from 1961 to 1968, and loans to Mexico and Brazil in 1982 and 1983. The Plaza communiqué of September 1985 was also regime preserving, intending to repress protectionist sentiment in the United States by a depreciation of the dollar. In the European Monetary System, commitments are explicit – much more than under the Bretton Woods system, where most access to IMF funds required negotiation.

Co-ordination entails mutual commitments to change policies to pursue a common objective or to help other governments pursue their own objectives. With co-ordination, mutual commitments may cover policy targets, as with the real GNP growth rates specified at the 1977 London summit; or they may pertain to policy instruments, as with the fiscal policy changes of Japan and Germany, and the United States' decontrol of domestic oil prices at the 1978 Bonn summit.

The need to co-ordinate national policies arises from the size and character of the 'spillover' effects of particular exchange rate arrangements. Large countries cannot ignore the impact they have on others. (In a world of n independent countries there are only $n-1$ independent exchange rates.) When there is a shortage of policy co-ordination, two remedies can be considered. The supply of co-ordination may be raised by altering government behaviour, or the demand may be reduced by changing exchange rate arrangements (Cooper, 1985; Kenen, 1989). With floating rates the demand for co-ordination is greater. If major nations will not co-ordinate policies then a fixed rate regime could compensate. Unfortunately, maintaining a fixed rate regime with unco-ordinated policies is likely to prove impossible in a world of high capital mobility.

How large the gains from policy co-ordination are estimated to be depends on the model in which they are measured. Moving from sub-optimal domestic policies to optimal non-co-operative policies produces the greatest and most substantial benefits. There are incentives for governments to try to do even better by abandoning co-ordination after a short while. When a government has pursued a particular policy for some time, it may be advantageous to change policies even though the government's commitments would be violated ('time inconsistency'). Typically, this is when the other players are slow to revise or modify their behaviour. In international relations there may be limited scope for reneging on agreements before other governments adjust their expectations. Then the problem reverts to the traditional 'prisoners' dilemma' (Keohane, 1984). Repudiating obligations may confer an advantage in one instance, but typically governments or economies must co-exist permanently, and negotiate over many issues. In a repeated prisoners' dilemma, non-co-operative behaviour will be punished and co-operation rewarded by rational players. Hence in a stable international environment co-ordination and co-operation, may be expected to emerge and persist. As discussed in Chapter 11, the inter-war years did not provide such stable conditions, even by comparison with the 1980s. US recognition of a direct interest in resolving the debt crisis of the 1980s avoided a single play of the prisoners' dilemma on that occasion.

International resource policy

International co-ordination and collaboration were nowhere more necessary than in formulating and enforcing a policy for protecting international common resources. The use of the environment created a problem for relations between national economies analogous to overfishing. The ocean's fish belong to no one, and therefore no one has a financial interest in preserving them. Or rather, although any individual fishing vessel may value the option of returning next year to the same fishing grounds for a profitable catch, there is no point in restricting operations this year because other fleets will not. Therefore, in the absence of binding international conservation agreements, the world's fish stock will be inexorably depleted.

A simple but often bloody case of collective resource depletion arises where two national economies share a river. The upstream economy can drain the river for irrigation or industrial use so as to impoverish the downstream economy. The Jordan, the Tigris and the Euphrates were all potential sources of conflict in the 1980s. Industrial emissions from one country can destroy the ozone layer of other economies. Tropical rain forests preserve a gene pool and help preserve the atmosphere for countries far distant from the nation within whose borders they lie. Global warming through the accumulation of industrial carbon dioxide in the atmosphere from burning carbon-based fuels constitutes another international economic 'spillover'. One economy's climate is influenced by another's fuel policy. Some may judge a warmer climate an improvement and some the reverse. Those countries burning most fuels are likely to conclude that the costs of cutting back exceed the climatic benefits, even though other states may wish them to reduce CO_2 emissions.

Discovery of an 'ozone hole' larger than continental United States over the Antarctic in 1985 gave added urgency to these abstract considerations. The ozone layer protects the earth from ultraviolet rays which can cause cancer. It was being destroyed by chlorofluorocarbons (CFCs), used for powering spray aerosols, refrigeration and cleaning computer chips. In 1989, twenty-nine countries ratified the Montreal Protocol on Substances that Deplete the Ozone Layer, agreeing a freeze on CFCs at 1986 levels. Very soon afterwards evidence showed that only the complete abolition of CFCs might allow the ozone layer to recover. Divergences of interest between the industrialised and developing countries over the priority to attach to this goal were resolved by a fund to finance the transfer to the poorer economies of alternative technology to CFCs (Mathews, 1990).

Undersea mining outside territorial waters proved ultimately to be a non-issue because of the investment required and sagging mineral prices in the 1970s and 1980s. Only in the 1960s was technology advanced sufficiently to suggest that the mining of minerals, in a form known as nodules, lying on the ocean floors, might be economically viable (Logue and Sweeney, 1977, Bulkley, 1979; Pontecorvo, 1986). At the consumption levels of the end of the 1970s, there may have been enough economically accessible nickel to supply total world demand for 2,000 years, enough copper for 140 years, manganese for 1,600 years and cobalt for 500 years. Since most nodules lie in the deep ocean outside the limits of national jurisdiction, the 'freedom of the high seas' principle embodied in international law gave ownership to whoever could gather the nodules. However, those currently producing minerals found in nodules could expect to become worse off as the new sources of supply depressed prices. Typically, therefore, LDCs would suffer.

In 1967 the UN set up an *ad hoc* Sea Bed Committee to investigate the issues, and in 1970 the UN General Assembly unanimously adopted the 'common heritage principle' for these minerals. But there was no agreement on the appropriate regime for exploiting the resources. Industrial countries could expect to gain from lower metal prices, and world income would be redistributed towards them and away from LDCs. Moreover, deep sea mining required very substantial technical knowledge and capital only available in the wealthier economies, to whom mining profits would therefore accrue. In 1982, a Law of the Sea treaty was signed by 117 nations, excluding the US, the UK and West Germany, among others. The treaty specified that the undersea mining was to be governed by a world authority, to whom private mining companies would be obliged to sell their expertise. Non-signatories objected that the regime would stifle development in order to gain a fairer distribution of a much diminished revenue. Continuing uncertainty over seabed ownership and rights may have obstructed exploitation of undersea minerals in the 1980s, but mineral prices were not sufficiently buoyant to warrant pressure for the clarification of international law.

Summary and conclusion

Despite the disruption of the first oil shock, OECD growth rates were still higher between 1973 and 1979 than at any time before 1945 (Maddison, 1982). It was the second oil shock and the deflationary policy reactions of the West that shaped the 1980s. Stagnating demand in industrial economies and tight money lay behind the debt crises that began with Mexico's default in 1982. Divergent national or even continental experiences thereafter demonstrated the key role of domestic policies and economic structure rather than the world economic environment. Latin America stagnated over the decade as it adjusted to the accumulation of debt in the 1970s. Sub-Saharan African economies actually deteriorated. Asia became increasingly prosperous. American productivity growth in particular slowed relative to Japan, fuelling US fears that their period of leadership of the world economy and polity was over. Western Europe grappled with the great experiment of widening and deepening the European Community.

With the oil price collapse of 1986, OPEC economies began to shrink, along with their oil revenue. In the West, lower oil prices boosted economic growth in industrialised countries. Centrally planned economies disintegrated under the weight of their economic failures. Their environmental pollution was considerably greater than in market economies, but major international economic issues arose in the 1980s over destruction of common natural resources such as the ozone layer.

Although the United States no longer dominated the world economy as it did at the beginning of the great post-war boom, it was still the largest single economy and continued to take the lead in addressing international economic problems. Where international debt was concerned, domestic banking stability was a focus for US interests, so that the USA operated effectively as an international central bank. American macroeconomic policy and a continuing massive budget deficit, destabilised the dollar and ultimately threatened US commitment to a multilateral trading system. An overvalued dollar boosted US protectionist sentiment and discredited floating exchange rates.

Unrestricted capital mobility, in a now highly integrated world economy, meant that a fixed rate regime was almost impossible for large, money-centre economies. Regional groupings around money centres, such as the ERM and NAFTA, remained a possibility. Such groupings begged the question of what determined an optimum currency area. Whether western Europe was the best size of region to share a common currency was not a question posed in Brussels, for politics set the agenda. It had done so for the terms of German monetary unification, with ominous implications for an extension of the experiment to the whole of western Europe.

Sustainable fiscal and monetary policies in the United States would have helped avoid fluctuations in exchange rates unwarranted by shifts in relative prices and competitiveness. Between the principal economic blocs, policy co-ordination to realign exchange rates and raise economic activity seemed to offer some hope for stabilising economic relationships. International policy co-ordination agreed at the Louvre and Plaza meetings of 1985 and 1987 attempted to hold the dollar and other major currencies at agreed levels. In 1987 the upswing ended with stock exchange collapses around the world. Computer selling at pre-set price limits was blamed for the speed of the decline. Folk memories of the 1929 US Stock Exchange panic prompted copious lending by western central banks. No collapse of aggregate demand ensued but excessive credit growth planted the seeds of another recession to squeeze out inflation.

Continuing themes were friction between the EC and the USA over trade policy, particularly over agricultural trade. Whether pressure groups will in the future be able to escalate the conflicts depends on, among other things, how many unfair trading practices there are to disagree about. Smaller, usually poorer, economies are typically not in a position to do other than acquiesce in the voluntary restraint agreements imposed upon them. In the economies that chose these restrictions, productivity growth may be expected to slow as they spread. Resources were being held in less productive sectors and industries with better growth prospects were being deprived of them.

Oil refused to disappear from the centre of the international stage. Iraq's invasion of Kuwait in 1990 and the US-led international response reflected the enormous economic and political power that Middle Eastern oil deposits continued to confer even in a period when oil prices had slumped.

What did not happen matters as much as what did for judgements as to how well national and international economic institutions worked. Despite major bank and stock exchange collapses and debt defaults, widespread economic and political disasters on the scale of the 1930s were avoided by determined action. Oil shocks were contained and, at the time of writing, so too has been the major economic error of German monetary unification: the parity of the East and West German marks. Even so deflationary policies, related in various ways to the international economy, contributed to the plight of the 30 million people unemployed in western Europe and North America, at the end of 1992. The figure was an increase of more than one-third in two and a half years. In Spain one-third of workers under twenty-five were jobless. As far as the world economy has contributed, these levels of unemployment do not warrant great satisfaction with its management and operation.

Notes

1. The recent literature analysing the statistical evidence for convergence includes Baumol (1986); DeLong (1989); Barro (1991); and Barro and Sala-i-Martin (1992).
2. Established in 1949, Cocom consisted of the USA, Canada, Japan, Belgium, Denmark, France, Greece, Italy, Luxemburg, the Netherlands, Norway, Portugal, Turkey, UK and West Germany.

Postcript

In ten years' time the outlines of the international economy should be recognisable from what we see today. The 1980s suggested that central planning was finished, at least for the next decade. Faith in the market as a means of organising economic activity continues to spread. China seems set to follow the industrialisers of south-east Asia, in the absence of major political changes. This will eventually shift the centre of gravity of the world economy further towards the Pacific. Sino-Japanese economic relations will then come increasingly to the fore.

The USA will continue to occupy the strongest position for shaping any new international economic regime, by virtue of its economic size and, in contrast to the European Community, its federal government. US commitment to multilateralism is likely to be weakened now that once more the world is safe for capitalism. The end of the Cold War has removed an incentive to support a free-market international economic regime that stands as an example to less committed nations. If coherent US fiscal and monetary policies are followed, the dollar will track competitiveness more closely and pressures for 'managed trade', particularly prominent in relations with Japan, may be reduced. It is hard to see any quick resolution of persistent intra-bloc conflict over subsidised agricultural trade. This will await either a further decline in European farm populations, or an eclipse of French power within the European Community. The only question is whether disagreements will be allowed to spill over into other areas of international economic relations.

Stable exchange rates are not identical with fixed, or pegged, rates. In the absence of capital controls, perhaps in the form of taxation, to slow down or reduce the enormous volume of funds that can shift between major currencies, it is not worth the economic blocs explicitly committing themselves to pegged exchange rates. The financial costs of failed support are extremely high, as a number of ERM governments were reminded in 1992 and 1993. Discretionary international co-operation in various forms has been shown to stabilise exchange rates and will occasionally be applied again.

Since 1850 the number of recognisable states between which economic relations are transacted has greatly increased. So has the volume of business and the number of people affected. The key position of nations continues, except perhaps in western Europe, despite what nineteenth-century liberals identified as their non-rationality. A national economy inherits from the past a set of institutions, customs and culture, often including a common language, that makes for greater homogeneity within the national borders than across them. What is conducive to transactions between 'insiders' is an obstruction for 'outsiders'. In eastern Europe, and within the territory of the former Soviet Union, nationalist pressures have multiplied the length of new state borders and made the frontiers of inter-war Europe seem relatively simple. The likelihood is that, in the absence of a creative supranational institutional response, the economic disruption will be at least comparable.

With the Maastricht Treaty of 1991 the European Community embarked upon an experiment which, if successful, would replace European national governments in their present form with a trans-European authority. Monetary unification, a common central bank, common external tariffs and a common agricultural policy, among other common policies, inevitably imply common taxation and spending, except within the narrow limits presently allowed to local government. There will still remain problems of regional, urban and local fiscal policy, but some instruments for their resolution will have been abolished. Economic specialisation by location, and spatial variations in unemployment, in income and in demand will persist, as will questions of what services to provide and how to pay for them. A nation, unlike a region, possesses the possibility of pursuing different exchange rate policies. It can tax, limit or subsidise foreign trade, restrict or encourage immigration or emigration, expand or contract the money supply, borrow or pay back government debt, spend money in support of desirable objectives, and so on. All this apparent discretion may be exercised well or badly. When the stewardship is poor there will be an incentive to find other authorities to whom these powers may be delegated. If economic development renders these policy instruments ineffective then the justification is gone for a government whose principal purpose is to use them. Outside Europe the consensus is that these instruments may still be useful.

Free-trade areas, such as NAFTA, are far less contentious than monetary, fiscal and political unification. They require much less agreement between contracting parties and are therefore easier to establish and less likely to break down. They do not require contentious free immigration, or government mergers. Yet, when coupled with suitable domestic policies, a similar tendency to equalise wages is set in motion. National distinctiveness, a contributor to many individuals' sense of identity and security, can be maintained within some bounds in a free-trade area. It is likely to be threatened in a more integrated economy, quite possibly with severe political repercussions.

The number of people living on the earth has multiplied almost five times since 1850. Most of them are in China and India now, as they were then. The filling-out of America has marked some change in the world population distribution, but one dwarfed by the income and trade shift towards North America. The extraordinary trade predominance of the United Kingdom has long since passed away. Can the earth continue to support these numbers or will resources be in increasingly short supply? Oil has ceased to pose a problem of scarcity, although the location of the deposits and their political vulnerability remain. It was oil that highlighted for the whole world a drawback of international economic dependence and interdependence. Lower energy costs were bought at the expense of political and economic vulnerability to disruption of supply. Whether oil will be allowed to exercise the same influence as it did in the last two decades is doubtful, but it will continue to be a vital and sensitive market. The longer-term challenge is to establish what are the vital common property resources and to construct an internationally acceptable framework that will prevent their excessive depletion. The 1989 ratification of the Montreal Protocol on substances that deplete the ozone layer shows what can be done.

Glossary

Adjustable peg exchange rate: Under the Bretton Woods agreement of 1944, countries were allowed to adjust the value of their exchange rates in terms of gold or dollars in the event of their balance of payments being in 'fundamental disequilibrium'. Some critics have suggested that the possibility of revaluation or devaluation (in contrast to the fixed rate of the gold standard) gave rise to destabilising speculation and contributed to the downfall of the Bretton Woods system.

Balance of payments: A measure of transactions between one country and the rest of the world. It includes the balance of trade (exports − imports) and balances of interest payments, profits, gifts and transfers, in total comprising the current account. The capital account records net capital movements. The sum of the two accounts must equal the change in government (or central bank) foreign exchange reserves, if any. If one country were drawing substantial profits or transfers from another, this would be recorded in a true balance of payments account.

Beggar-my-neighbour policies: Increasing national output and employment at the expense of another nation. Sometimes described as 'exporting unemployment', these policies were widely adopted in response to the Great Depression after 1930. Tariffs, quotas and currency depreciation priced foreign goods out of domestic markets, enhancing the demand for home-produced goods and therefore boosting the demand for the home country labour that made them. Abroad, unemployment in export industries rose. In the 1930s, countries retaliated in tariff and quota wars which further disrupted world trade and employment.

Bills of exchange: Means of financing internal and external trade. During the nineteenth century and earlier, transport and communications technology was such that a considerable period of time elapsed after production but before goods were sold to consumers, during which the goods were in transit. Bills of exchange allowed the producer to be paid by the selling merchant before the merchant had received payment. Bills therefore lubricated the wheels of commerce. During the twentieth century, as communications improved, bills were superseded by bank finance and the internal transactions of multinational companies.

Comparative advantage: According to David Ricardo (1772–1823), the basis of the mutual gains from trade. Under a market system, countries export goods with a lower domestic opportunity cost and import those with a higher. Comparative advantage should be distinguished from the principle of absolute advantage, which asserts that countries export goods if they can produce them absolutely more cheaply than the importers. According to comparative advantage, an exporter may be able to produce everything more cheaply than its trading partners, yet still, if domestic opportunity costs differ between the countries, all of them could gain from specialisation. Because their prices do not reflect opportunity costs, centrally planned economies have been loath to engage in trade even with other centrally planned economies because they cannot be sure that they will not lose from international transactions which yield financial benefits.

Customs union: A group of nations which abolish trade barriers among themselves and establish a common external tariff (CET) against the outside world. The Zollverein is the best-known nineteenth-century example. A customs union should be distinguished from a free-trade area which lacks the CET (hence the UK preference for EFTA, which did not penalise the Commonwealth) and from a common market which aims to harmonise all regulations governing economic activity as well as establishing a customs union. However, the Zollverein was intended by its architects to be a precursor of economic and political unification, and that may also be true of the European Common Market.

Direct investment: Investment abroad over which the investor maintains control, as distinct from indirect or portfolio investment where the recipient country controls the projects. It was increasingly important in the twentieth century, when multinationals flourished, encouraged by trade restrictions, economic nationalism and the political disruption of the capital markets that throve in the nineteenth century.

Dual economy: An economy divided into a high-productivity modern sector and a backward or traditional low-productivity sector. W. A. Lewis's model of economic development with unlimited supplies of labour represents the backward agricultural sector as providing migrants to the industrial/urban sector without a contraction of agricultural output, but with rapid growth in industrial output. In his *Growth and Fluctuations*, Lewis (1978b) applies similar concepts to the world economy 1870–1914. Expansion of trade could not confer much benefit upon the tropics, he maintains, because economy wide, wages were held down by the low productivity of the majority of the workforce in traditional agriculture. Similarly, non-European immigrants tended to force wages down to those supported by this low productivity wherever the migrants were allowed to work. This was the motivation for discrimination against the Chinese in the USA and elsewhere.

Equalisation of net advantage: The notion that workers, being interested in the pace and conditions of work, in religious and political freedoms and in the environment in which they live, will change jobs at home and abroad until the differences in wages between jobs reflect only the average value placed upon the non-pecuniary aspects of the work or location relative to others. This is a possible explanation for migration and international wage differentials.

Eurodollars: Originally dollar claims traded in London during the late 1950s in response to Soviet fears of political interference with their dollar assets and subsequently due to US banking restrictions, the term has come to cover any currency traded outside its own country: for example, yen traded in Hong Kong. In the wake of the first oil crisis, the Eurodollar market became a major instrument for recycling petrodollars from OPEC states back to oil buyers. The market is sometimes blamed for overlending by western banks to NICS, and thus for contributing to the debt crisis from 1982.

Exchange rate overshooting: The tendency of freely floating exchange rates to be excessively volatile in the face of shocks, such as unexpected political events or oil price changes. For most sets of economic and political forces, a long-run equilibrium exchange rate exists but a freely floating exchange rate will not adjust directly to this value, instead overshooting and then returning to the equilibrium. Although the theory was only worked out formally in the 1970s, judgements on the floating regimes of the 1920s and the 1930s by the architects of Bretton Woods reached similar conclusions. The experience of floating rates after the Bretton Woods adjustable peg system was abandoned certainly indicated excessive exchange rate volatility. At the beginning of the 1980s the sterling–dollar exchange rate halved almost to £1=$1 before returning to a longer-run rate more than 50 per cent higher. Such erratic behaviour imposes considerable costs on international traders.

Exploitation: Gaining at the expense of another. Neoclassical exploitation is the exercise of monopoly or monopsony bargaining power in a market to shift prices or wages from their competitive levels. A large country that imposes a tariff upon imports can drive down their price by buying less, and can thereby exploit the sellers if there are no alternative markets. Marxist exploitation is based upon an unjust acquisition of an asset (theft). Slavery was the theft of labour from its owner. Cotton grown in the southern United States before the Civil War was cheaper because of slavery. Those who benefited from the cheaper cotton clothing were exploiting the slaves through the world market. If the Crown Agents had made the colonies of the British Empire pay more than world prices for state purchases, and remitted the difference to the British government, they would have been exploiting the colonials because the monopoly power was conferred by statute, which in turn was ultimately acquired and backed by imperial force. (In fact, according to Davis and Huttenback (1986) the Crown Agents gave the colonies the benefit of experienced bulk buying.)

Factor price equalisation: The tendency under international trade, when transport costs and other barriers to trade are not too high, for the movement of goods to bring about the same result as the free movement of factors would. A corollary of the Hecksher–Ohlin theory, FPE predicts that wages for similar work will be the same in all countries engaging in international trade.

Forward exchange rate: The exchange rate for currency to be traded at a future date, most commonly three months ahead. Under a gold standard that the market believes will persist, there is no forward market because the rate is the same as the spot or current exchange rate. Under a floating rate or an adjustable peg system, if the forward rate is at a discount to (below) the spot then the market expects the rate to fall. The forward market allows traders to avoid exchange rate risk by guaranteeing the rate at which they will be undertaking international transactions in the future.

Gold exchange standard: A modified gold standard in which currencies of other gold standard countries were held as official reserves instead of gold. First recognised in the 1920s, the system has occasionally and wrongly been blamed for the difficulties of the international economy of the 1920s. The system was in operation in the years immediately before the First World War. Japan, for instance, held sterling rather than gold and earned interest on official reserves, which it would not have done upon gold holdings.

Hecksher–Ohlin theory: An explanation for the sources of comparative advantage. According to this theory, countries specialise in the production of goods intensive in their relatively abundant factors. Although not formulated until the twentieth century, the theory best explains nineteenth-century patterns of specialisation. The regions of recent European settlement (Australia, Argentina, mid-USA, Canada) were land abundant and labour scarce. They therefore exported (land-intensive) agricultural produce and imported (labour-intensive) manufactures. Britain possessed substantial coal deposits and therefore exported heat-intensive engineering products, often semi-finished, whereas France without coal concentrated on skilled labour-intensive finished manufactures.

Illiquidity: An organisation is illiquid if it is unable to raise the cash to meet current obligations but is none the less solvent (has a positive net worth, assets greater than liabilities). This is a particular potential problem of banks, which central banks like the Bank of England remedied by assuming the role of 'lender of the last resort'. The USA lacked a central bank for most of the nineteenth century, and therefore its banking system was more vulnerable to financial panics. In effect the Bank of England functioned as lender of the last resort to the US economy, with adverse repercussions on the British economy. In 1890 Argentina abandoned the gold standard and the peso depreciated. Baring held a large volume of Argentine debt which suddenly became hard to dispose of. Baring was therefore illiquid and would have been bankrupted, with adverse consequences for the whole British financial sector, had the Bank of England not organised a rescue package.

Import-substituting industrialisation: Encouraging industrial development by excluding imports, this policy was popular among LDCs in the 1950s and 1960s, grounded in the belief that a country could only develop with manufacturing industry, not by encouraging agriculture. Since these countries were importers of manufactures, they could increase the demand for home-produced manufactures by preventing imports. At the same time, agriculture was taxed heavily to provide funds for industrial development. The upshot was generally a food shortage that had to be rectified by importing. During the 1970s and 1980s, agricultural and open economy development came into favour, but by then world recession was stimulating protection against LDC imports in richer countries.

Infant industries: Industries which need protecting from foreign competition until they grow up. The sources of infancy are a technology that gives rise to economies of scale or learning by doing. The doctrine was popular in Germany and in the United States at the time of British industrial supremacy in the mid-nineteenth century. It contrasted with the opposite doctrine formulated to account for British relative decline from the mid-1870s, the 'costs of an early start'. Here advanced nations need protecting from the less advanced.

International multiplier: The Keynesian income multiplier extended to take account of repercussions on a second nation of a fall in expenditure in a first nation. It is particularly pertinent to the decline of world income 1929–33, but also perhaps to earlier crises such as those of 1907 and 1873. A fall in one economy's demand for imports is a fall in another economy's exports. Higher unemployment and lower spending in the export industries reduces the demand for imports, which are the first nation's exports. And so the contraction of income, expenditure and employment spreads outward like ripples in a pond into which a stone has been thrown. In our case the stone is an initial contractionary shock.

International policy co-ordination: In a world economy, one nation's economic policy can be more effective or less costly if other nations co-operate by co-ordinating their policies. Even under the supposedly automatic gold standard, co-operation between the Bank of France and the Bank of England during the crises of 1890 and 1907 prevented interest rates being driven as high as they might otherwise have been. During the 1920s central bank co-operation between the UK, France, Germany and the USA (Norman, Moreau, Schacht and Strong) may have temporarily held down world interest rates. During the 1931 Credit-Anstalt crisis, Montague Norman attempted to act as lender of the last resort to Austria, but the loan he managed to raise was too small. Had co-operation been more successful, the depression would have been less deep and shorter.

Liberalism: The doctrine, vigorously advanced by Britain in the mid-nineteenth century, that social well-being depended upon individuals being able to do what they liked. If some Chinese wished to buy opium then the Chinese government was tyrannical in trying to stop them. Free trade allowed people to get more of what they wanted and was therefore to be encouraged. Slavery created some difficulty for the doctrine, since the institution was clearly the antithesis of liberty and yet expanding world trade could support slavery by increasing the demand for slave goods. The liberal conscience seems to have been salved by the West Africa squadron of the Royal Navy, maintained entirely to intercept slave ships.

Material balances: A procedure by which centrally planned economies groped towards the volume and distribution of raw materials, intermediate goods and labour necessary to produce the set of goods for final demand.

Most favoured nation clause: The principle that any concession granted by one signatory to the other will be extended to all other trading partners or at least to a wider group. First written into the Cobden–Chevalier treaty between Britain and France in 1860, and the principal plank of GATT, the MFN clause proved a powerful means of liberalising international trade.

Orderly marketing agreements: Means of controlling imports to limit the loss of jobs without contravening GATT rules. The Multi-Fibre Arrangement is the best known and most important example.

Price–specie flow mechanism: An account of how the gold standard worked. If a country lost gold to another, the money supply which was based upon gold would contract, helped by high interest rates raised by the central bank. A smaller money supply meant lower prices, which improved the competitiveness of exports and domestic substitutes for imports. This improved the trade balance and therefore raised the net gold inflow. The mechanism ignores the more important asset arbitrage process. Higher interest rates in one country with fixed exchange rates encourage holders of foreign assets to buy those assets and sell their own. By so doing they increase the demand for currency and encourage a gold inflow much more quickly than the price–specie flow mechanism implies.

Purchasing power parity: A theory of long-run equilibrium exchange rate values, used by Keynes in the 1920s to argue that sterling was overvalued against the dollar. Keynes assumed that the exchange rate was in equilibrium in 1913 and then showed that British prices in the 1920s had risen more than American prices. He maintained that, unless the sterling–dollar exchange rate was lower than in 1913, British goods would be priced out of world markets and the UK would suffer unemployment in the export industries.

Rescheduling: An alternative to default on international debt. Default is the unilateral non-payment of interest and principal. Rescheduling is an agreed reduction or deferment of payments. In the debt crisis from 1982, lending banks have preferred countries to reschedule because otherwise their loans could be worthless and they themselves would become insolvent. This could precipitate a world financial crisis on the scale of 1931–3.

Taut planning: The principle of imposing greater demands than can be met on the resources of a centrally planned economy, in order to ensure that resources are fully utilised. It leads to queues rather than the price inflation of market economies.

Terms of trade: The ratio of export prices to import prices. A favourable movement in the terms of trade is an increase in this ratio, for it means that fewer goods must be exported to pay for a given quantity of exports. The terms of trade between countries which export manufactured goods and those which export primary produce has been the subject of particular investigation. Drawing mainly upon British experience since 1870 (as the exemplar of the manufactures exporter), Prebisch among others maintained that the terms of trade of primary exporters have shown a tendency to deteriorate. He therefore recommended a policy of import-substituting industrialisation for primary exporters, in particular for Latin American countries.

Trade diversion: A customs union diverts trade from low opportunity cost sources to higher opportunity cost sources when it establishes a common external tariff. The displacement of New Zealand butter by French butter in the British market as a consequence of joining the EEC is an example. The related concept is trade creation, the displacement of higher opportunity cost supplies by lower opportunity cost supplies consequent upon the elimination of barriers between member countries. All studies of the trade effects of the Common Market show that trade creation has outweighed trade diversion.

Transfer problem: The problem of transferring resources between two areas that use different currencies. Important historical examples are the payment of war indemnities and reparations after the Franco-Prussian and the First World Wars, and paying for oil imports after the 1973/4 price rise. Transfers tend to drive down the exchange rate of the donor country. If the transfer is denominated in the foreign currency, this depreciation will increase the burden of the transfer and cause a further fall in the exchange rate. When the recipient of the transfer wants neither the donor country's exports nor its assets, the transfer becomes impossible in the absence of third parties, because there is no exchange rate at which the transfer can take place.

References

Abbot, C. C. (1973) 'A re-examination of the 1929 Colonial Development Act', *Economic History Review*, 24, pp. 68–81.
Abramovitz, M. (1989) *Thinking About Growth*, Cambridge: Cambridge University Press.
Adam, J. (1989) *Economic Reforms in the Soviet Union and Eastern Europe since the 1960s*, London: Macmillan.
Aghlevi, B. B. (1975) 'The balance of payments and the money supply under the gold standard regime: The United States 1879–1914', *American Economic Review*, 65, pp. 40–58.
Agraa, A. M. El- (1988) *Japan's Trade Frictions: Realities or misconceptions?*, London: Macmillan.
Ahvenainen, J. (1981) *The Far Eastern Telegraph*, Helsinki: Suomalainen Tiedeakatemia.
Albert, W. (1983) *South America and the World Economy from Independence to 1930*, London: Macmillan.
Aldcroft, D. H. (1977) *From Versailles to Wall Street 1919–1929*, London: Allen Lane.
Aliber, R. Z. (1962) 'Speculation in the foreign exchanges: the European experience 1919–1926', *Yale Economic Essays*, vol. 2 part 1 Spring, pp. 171–245.
Allen, G. C. (1981) *A Short Economic History of Modern Japan*, 4th edn, London: Macmillan.
Allen, G. C., and Donnithorne, A. (1954) *Western Enterprise in Far Eastern Economic Development*, London: Allen and Unwin.
Allen, R. C. (1979) 'International competition in iron and steel, 1850–1913', *Journal of Economic History*, 39, pp. 911–37.
Allen, R. G. D. (1946) 'Mutual aid between the United States and the British empire 1941–45', *Journal of the Royal Statistical Society*, vol. 110, part III, pp. 243–71.
Amdur, R. (1977) 'Rawls' theory of justice: domestic and international perspectives', *World Politics*, 29, pp. 438–61.
American Economic Association (1900) *Essays in Colonial Finance*, New York: Macmillan.
Ames, E., and Rosenberg, N. (1968) 'The Enfield arsenal in theory and history', *Economic Journal*, 78, pp. 827–42.
Ando, A., and Modigliani, F. (1963) 'The "life cycle" hypothesis of saving', *American Economic Review*, 53, pp. 55–84.
Argy, V. (1981) *The Postwar International Money Crisis: An analysis*, London: Allen and Unwin.
Arndt, H. W. (1972) *Economic Lessons of the Nineteen Thirties*, London: F. Cass (first published 1944).

Artus, J. R. (1975) 'The 1967 devaluation of the pound sterling', *IMF Staff Papers*, 22, pp. 595–640.

Ashley, P. (1904) *Modern Tariff History: Germany, the United States, France*, London: John Murray.

Baack, B. D., and Ray, E. J. (1983) 'Political economy of tariff policy: a case study of the United States', *Explorations in Economic History*, 20, pp. 73–93.

Baba, M., and Tatemoto, M. (1968) 'Foreign trade and economic growth in Japan 1858–1937', in L. Klein and K. Ohkawa (eds.), *Economic Growth: The Japanese experience since the Meiji era*, Homewood, Ill.: R. D. Irwin/Yale University.

Baines, D. (1985) *Migration in a Mature Economy: Emigration and internal migration in England and Wales 1861–1900*, Cambridge: Cambridge University Press.

Baines, D. (1991) *Emigration from Europe 1815–1930*, London: Macmillan.

Bairoch, P. (1972) 'Free trade and European development in the nineteenth century', *European Economic Review*, 3, pp. 211–46.

Bairoch, P. (1974) 'Geographical structure and trade balance of European foreign trade, 1800–1970', *Journal of European Economic History*, 3, pp. 557–608.

Bairoch, P. (1976) 'Europe's gross national product, 1800–1975', *Journal of European Economic History*, 5, pp. 273–340.

Bairoch, P. (1989) 'European trade policy 1815–1914', in P. Mathias and S. Pollard (eds.), *Cambridge Economic History of Europe*, vol. 8, Cambridge: Cambridge University Press.

Balassa, B. (1965) 'Trade liberalisation and "revealed" comparative advantage', *The Manchester School*, 33, pp. 99–123.

Balassa, B. (1977) 'Revealed comparative advantage revisited: an analysis of relative export shares of industrial countries 1953–1971', *Manchester School*, 45, pp. 327–44.

Balderston, T. (1981) 'Inflation in Britain and Germany 1908–1923: a comparative study', mimeo, Department of History, University of Manchester.

Baldwin, R. E. (1956) 'Patterns of development in newly settled regions', *Manchester School*, 24, pp. 161–79.

Baldwin, R. E., and Krugman, P. R. (1988a) 'Industrial policy and international competition in wide-bodied jet aircraft', in R. E. Baldwin (ed.), *Trade Policy Issues and Empirical Analysis*, Chicago, Ill.: University of Chicago Press.

Baldwin, R. E., and Krugman, P. R. (1988b) 'Market access and international competition: a simulation study of 16K random access memories', in R. Feenstra (ed.), *Empirical Methods for International Trade*, Cambridge, Mass: MIT Press.

Baldwin, R. E., and Murray T. (1977) 'MFN tariff reductions and developing country benefits under the GSP', *Economic Journal*, 87, pp. 30–46.

Balogh, T., and Graham, A. (1979) 'The transfer problem revisited', *Oxford Bulletin of Economics and Statistics*, 40, pp. 183–92.

Banerji, A. K. (1982) *Aspects of Indo-British Economic Relations 1858–1898*, Bombay: Oxford University Press.

Barbour, D. (1886) *The Theory of Bimetallism and the Effects of the Partial Demonetisation of Silver on England and India*, London: Cassell.

Barrett Brown, M. (1974) *The Economics of Imperialism*, London: Penguin.

Barro, R. J. (1979) 'Money and the price level under the gold standard', *Economic Journal*, 89, pp. 13–33.

Barro, R. J. (1991) 'Economic growth in a cross-section of countries', *Quarterly Journal of Economics*, 106, pp. 407–43.

Barro, R. J., and Sala-i-Martin, X. (1992) 'Convergence', *Journal of Political Economy*, 100, pp. 223–51.

Bartel, R. J. (1974) 'International monetary unions: the 19th century experience', *Journal of European Economic History*, 3, pp. 689–704.

Bastable, C. F. (1922) *The Commerce of Nations*, 9th edn, London: Methuen.

Batchelor, J. (1986) 'The avoidance of catastrophe: two nineteenth century banking crises', in F. Capie and G. E. Wood (eds.), *Financial Crises and the World Banking System*, London: Macmillan.

Batchelor, R. A., Major, R. L., and Morgan, A. D. (1980) *Industrialisation and the Basis for Trade*, Cambridge: Cambridge University Press.

Bauer, P. (1971) *Dissent on Development*, London: Weidenfeld and Nicolson.

Baumol, W. J. (1986) 'Productivity growth, convergence and welfare: what the long run data show', *American Economic Review*, 76, pp. 1072–8.

Beach, W. E. (1935) *British International Gold Movements and Banking Policies 1881–1913*, Cambridge, Mass.: Harvard University Press.

Beachey, R. W. (1976) *The Slave Trade of Eastern Africa*, London: L. R. Collings.

Beasley, W. G. (1955) *Selected Documents on Japanese Foreign Policy 1853–68*, London: Oxford University Press.

Beckerman, A. and associates (1965) *The British Economy in 1975*, Cambridge: Cambridge University Press.

Beenstock, M. (1983) *The World Economy in Transition*, London: Allen and Unwin.

Beenstock, M., and Warburton, P. (1983) 'Long-term trends in economic openness in the United Kingdom and the United States', *Oxford Economic Papers*, 35, pp. 130–5.

Bell, G. (1973) *The Eurodollar Market and the International Financial System*, London: Macmillan.

Bell, P. W. (1956) *The Sterling Area in the Postwar World: Internal mechanism and cohesion 1946–1952*, Oxford: Clarendon Press.

Bell, S. (1940) *Productivity, Wages and National Income*, New York: NBER.

Benjamin, D. K., and Kochin, L. A. (1979) 'Searching for an explanation of unemployment in interwar Britain', *Journal of Political Economy*, 87, pp. 441–78.

Bergman, M., Gerlach, S., and Jonung, L. (1993) 'The rise and fall of the Scandinavian Currency Union 1873–1920', *European Economic Review*, 37, pp. 507–17.

Bernstein, E. M. (1973) 'The evolution of the International Monetary Fund', in A. L. K. Acheson, J. F. Chant and M. J. F. Prachowny (eds.), *Bretton Woods Revisited*, Toronto: University of Toronto Press.

Bevan, V. (1986) *The Development of British Immigration Law*, London: Croom Helm.

Bhagwati, J. N. (ed.) (1976) *The Brain Drain and Taxation*, Amsterdam: North-Holland.

Bhambi, C. P. (1980) *The World Bank and India*, New Delhi: Vikas Publishing House.

Bierman, J. (1989) *Napoleon III and his Carnival Empire*, London: John Murray.

Birnberg, T. B. and Resnick, S. A. (1975) *Colonial Development: An econometric study*, New Haven, Conn., and London: Yale University Press.

Black, J. (1970) 'Trade and the natural growth rate', *Oxford Economic Papers*, 22, pp. 13–23.

Block, F. L. (1977) *The Origins of International Economic Disorder: A study of United States international monetary policy from World War II to the present*, Berkeley, Calif.: University of California Press.

Bloomfield, A. I. (1959) *Monetary Policy under the International Gold Standard 1880–1914*, New York: Federal Reserve Bank.

Board of Trade Journal (1965a) 'Commonwealth preference: tariff duties and preferences on United Kingdom exports', 11 June.

Board of Trade Journal (1965b) 'Commonwealth preferences: United Kingdom customs duties', 31 December.

Borchardt, K. (1991) *Perspectives on Modern German Economic History and Policy*, Cambridge: Cambridge University Press.

Bordo, M. (1975) 'John E. Cairnes on the effects of the Australian gold discoveries 1851–73: an early application of the methodology of positive economics', *History of Political Economy*, 7, pp. 337–59.

Bordo, M. (1981) 'The UK money supply 1870–1914', *Research in Economic History*, 6, pp. 107–25.

Bordo, M. (1993) 'The Bretton Woods international system: a historical overview', in M. Bordo and B. J. Eichengreen (eds.), *A Retrospective on the Bretton Woods System*, Chicago, Ill.: Chicago University Press.

Bordo, M., and Eichengreen, B. J. (eds.) (1993) *A Retrospective on the Bretton Woods System*, Chicago, Ill.: Chicago University Press.

Bordo, M., and Schwartz, A. (1980) 'Money and prices in the nineteenth century', *Journal of Economic History*, 40, pp. 61–7.

Bordo, M., and Schwartz, A. (eds.) (1984) *A Retrospective on the Classical Gold Standard*, Chicago, Ill.: University of Chicago Press.

Borkin, J. (1979) *The Crime and Punishment of I. G. Farben*, London: Andre Deutsch.

Born, K. (1983) *International Banking in the Nineteenth and Twentieth Centuries*, Oxford: Berg.

Bourne, K. (1970) *The Foreign Policy of Victorian England*, London: Oxford University Press.

Boyer, G. R., Hatton, T. J., and O'Rourke, K. (1994) 'The impact of emigration on real wages in Ireland 1850–1914', in T. J. Hatton and J. G. Williamson (eds.), *Migration and World Development 1850–1939*, New York: Routledge.

Bresciani-Turoni, C. (1937) *The Economics of Inflation: A study of currency depreciations in post war Germany 1914–1923*, London: Allen and Unwin.

British Parliamentary Papers (1858) *Select Committee on the Operation of the Bank Act of 1844*, London: HMSO.

British Parliamentary Papers (1867) *Statistical Abstract Relating to British India from 1840 to 1865*, London: HMSO.

British Parliamentary Papers (1886) *Final Report*, London: HMSO.

British Parliamentary Papers (1903) *Royal Commission on Alien Immigration*, IX.

British Parliamentary Papers (1903 and 1914) *Statistical Abstract for the Principal and Other Foreign Countries*, London: HMSO.

Brittain, W. H. B. (1977) 'Developing countries' external debt and the private banks', *Banca Nazionale del Lavoro Quarterly Review*, 30, pp. 365–80.

Broadberry, S. N. (1989) 'Monetary interdependence and deflation in Britain and the United States between the wars', in M. Miller, B. J. Eichengreen and R. Portes (eds.), *Blueprints for Exchange Rate Management*, London: Academic Press.

Broadberry, S. N., and Crafts, N. F. R. (eds.) (1992) *Britain in the International Economy 1870–1939*, Cambridge: Cambridge University Press.

Broadberry, S. N., and Taylor, M. (1992) 'Purchasing power parity and controls during the 1930s', in S. N. Broadberry and N. F. R. Crafts (eds.), *Britain in the International Economy 1870–1939*, Cambridge: Cambridge University Press.

Brown, C. P. (1980) *The Political and Social Economy of Commodity Control*, London: Macmillan.

Brunner, L. P. (1985) 'The effects of trade restrictions on the US economy in the Great Depression', *American Economic Review, Papers and Proceedings*.

Brus, W. (1986a) 'Normalization and conflict', in M. C. Kaser and E. Radice (eds.), *The Economic History of Eastern Europe 1919–1975*, vol. 3, Oxford: Clarendon Press.

Brus, W. (1986b) 'Postwar reconstruction and socio-economic transformation', in M. Kaser and E. Radice (eds.), *The Economic History of Eastern Europe 1919–1975*, vol. 3, Oxford: Clarendon Press.

Brusnikin, A. (1992) 'Aid is good but cooperation is better', *International Affairs*, 11–19 October.

Bulkley, I. G. (1979) *Who Gains from Deep Ocean Mining?*, Berkeley, Calif.: Institute of International Studies, University of California.

Bunde, H. (1931) 'Migratory movements between France and foreign lands', in W. F. Wilcox (ed.), *International Migrations: Interpretations*, New York: NBER.

Burda, M. C. (1993) 'The determinants of East–West German migration: some first results', *European Economic Review*, 37, pp. 452–61.

Burgdorfer, F. (1931) 'Migration across the frontiers of Germany', in W. F. Wilcox (ed.), *International Migrations: Interpretations*, New York: NBER.

Butlin, N. G. (1964) *Investment in Australian Economic Development 1861–1900*, Cambridge: Cambridge University Press.

Cain, P. J. (1979) 'Capitalism, war and internationalism in the thought of Richard Cobden', *British Journal of International Studies*, 5, pp. 112–30.

Cain, P. J. (1980) *Economic Foundations of British Overseas Expansion 1815–1914*, London: Macmillan.

Cain, P. J., and Hopkins, A. G. (1993) *British Imperialism: Innovation and expansion 1688–1914*, Longman: London and New York.

Cairncross, A. (1974) 'The world commodity boom and its implications', *London and Cambridge Economic Bulletin, The Times*, 8 July.

Cairncross, A., and Eichengreen, B. J. (1983) *Sterling in Decline*, Oxford: Basil Blackwell.

Cameron, R. (1961) *France and the Economic Development of Europe 1800–1914*, Princeton, NJ: Princeton University Press.

Cameron, R. (ed.) (1972) *Banking and Economic Development: Some lessons of history*, Oxford: Oxford University Press.

Camps, M. (1959) *The Free Trade Area Negotiations*, London: PEP.

Capie, F. (1981) 'Shaping the British tariff structure in the 1930s', *Explorations in Economic History*, 18, pp. 155–73.

Capie, F. (1983) *Depression and Protectionism: Britain between the Wars*, London: Allen and Unwin.

Capie, F., and Weber, A. (1985) *A Monetary History of the United Kingdom*, vol. 1, London: Allen and Unwin.

Capie, F., Mills, T., and Wood, G. E. (1991) 'Money, interest rates and the Great Depression: Britain from 1870 to 1913', in J. Foreman-Peck (ed.), *New Perspectives on the Late Victorian Economy: Essays in quantitative economic history*, Cambridge: Cambridge University Press.

Cardoso, E. A., and Dornbusch, R. (1989) 'Brazilian debt crises: past and present', in B. J. Eichengreen and P. H. Lindert (eds.), *The International Debt Crisis in Historical Perspective*, Cambridge, Mass.: MIT Press.

Caron, F. (1979) *An Economic History of Modern France*, London: Methuen.

Carr, E. H. (1961) *What is History?*, Trevelyan Lectures, Cambridge University.

Carr-Saunders, A. M. (1936) *World Population: Past growth and present trends*, Oxford: Clarendon Press.

Casson, M. (1979) *Alternatives to the Multinational Enterprise*, New York: Holmes and Meier.
Casson, M. (1986) *Multinationals and World Trade: Vertical integration and the division of labour in world industries*, London: Allen and Unwin.
Cecchini, P. (1988) *The European Challenge 1992: The benefits of a single market*, Aldershot: Wildwood House.
Chambers, E. J., and Gordon, D. (1966) 'Primary products and economic growth', *Journal of Political Economy*, 74, pp. 315–32.
Chandler, A. D. (1977) *The Visible Hand: The Managerial revolution in American business*, Cambridge, Mass.: Harvard University Press.
Chandler, A. D. (1980) 'The growth of the transnational firm in the United States and the United Kingdom', *Economic History Review*, 33, pp. 396–410.
Chandler, A. D. (1990) *Scale and Scope*, Harvard, Mass.: Belknap Press.
Chapman, S. (1984) *The Rise of Merchant Banking*, London: Allen and Unwin.
Chapman, S. (1992) *Merchant Enterprise in Britain: From the Industrial Revolution to World War I*, Cambridge: Cambridge University Press.
Chaudhuri, K. N. (ed.) (1971) *The Economic Development of India under the East India Company 1814–1858*, Cambridge: Cambridge University Press.
Chaudhuri, K. N. (1983) 'Foreign trade and the balance of payments', in D. Kumar (ed.), *The Cambridge Economic History of India*, vol. 2, Cambridge: Cambridge University Press.
Chenery, H. B., and Strout, A. M. (1966) 'Foreign assistance and economic development', *American Economic Review*, 56, pp. 679–33.
Choudri, E. V., and Kochin, L. A. (1980) 'The exchange rate; and the international transmission of business cycle disturbances: some evidence from the Great Depression', *Journal of Money Credit and Banking*, 12, pp. 565–74.
Churchill, W. (1951) *The Second World War: Closing the ring*, Boston, Mass.: Houghton Mifflin.
Clapham, J. H. (1952) *An Economic History of Modern Britain: Free Trade and Steel*, Cambridge: Cambridge University Press.
Clare, G. (1891) *A Money-Market Primer and Key to the Exchanges*, London: Effingham Wilson.
Clements, K. W., and Frenkel, J. A. (1980) 'Flexible exchange rates, money and relative prices: the dollar–pound in the 1920s', *Journal of International Economics*, May, pp. 249–62.
Cloud, P., and Galenson, D. (1987) 'Chinese immigration and contract labour in the late nineteenth century', *Explorations in Economic History*, 24, pp. 22–42.
Coleman, D. C. (1969) *Revisions in Mercantilism*, London: Methuen.
Collins, M. (1981) 'The English banking sector and monetary growth 1844–80', *Discussion Paper 102*, School of Economic Studies, University of Leeds.
Condliffe, J. B. (1930) *The Development of Australia*, London: Collier Macmillan.
Connolly, M., and Taylor, D. (1976) 'Testing the monetary approach to devaluation in developing countries', *Journal of Political Economy*, 84, pp. 849–59.
Conybeare, J. A. C. (1987) *Trade Wars: The theory and practice of international commercial rivalry*, New York: Columbia University Press.
Cooper, R. N. (1977) 'A new international order for mutual gain', *Foreign Policy*, 26, Spring, pp. 66–120.
Cooper, R. N. (1985) 'Economic interdependence and coordination of economic policies', in

R. W. Jones and P. B. Kenen (eds.) *Handbook of International Economics*, vol. 2, Amsterdam: North-Holland.

Coquery-Vidrovitch, C. (1981) 'Industry and empire', in P. Bairoch and M. Levy-Leboyer (eds.), *Disparities in Economic Development since the Industrial Revolution*, London: Macmillan.

Cornwall, J. (1977) *Modern Capitalism*, London: Martin Robertson.

Corti, E. (1928) *The Reign of the House of Rothschild*, London: Gollancz.

Cottrell, P. L. (1992) 'Silver, gold and the international monetary order 1851–96', in S. N. Broadberry and N. F. R. Crafts (eds.) *Britain in the International Economy 1870–1939*, Cambridge: Cambridge University Press.

Council of the Corporation of Foreign Bondholders (1878) *Annual General Report for the Financial year 1877*, London.

Council of the Corporation of Foreign Bondholders (1928) *54th Annual Report for the Year 1927*, London.

Courtney, W. H., and Leipziger, D. M. (1975), 'Multinational corporations in LDCs: the choice of technology', *Oxford Bulletin of Economics and Statistics*, 37, pp. 297–304.

Crafts, N. F. R. (1983) 'Gross national product in Europe 1870–1910: some new new estimates', *Explorations in Economic History*, 20, pp. 387–401.

Crafts, N. F. R. (1989) 'Revealed comparative advantage in manufactures 1899–1950', *Journal of European Economic History*, 18, pp. 127–38.

Crafts, N. F. R., and Thomas, M. (1986) 'Comparative advantage in UK manufacturing trade 1910–1935', *Economic Journal*, 66, pp. 629–45.

Cripps, T. F., and Tarling, R. J. (1973) *Growth in Advanced Capitalist Economies 1950–1970*, Cambridge: Cambridge University Press.

Crockett, A. D. (1978) 'Control over international reserves', *IMF Staff Papers*, 25, pp. 1–24.

Csaba, L. (1990) *Eastern Europe in the World Economy*, Cambridge: Cambridge University Press.

Curzon, G. (1965) *Multilateral Commercial Diplomacy*, New York: Praeger.

Curzon, G., and Curzon, V. (1969) 'Options after the Kennedy Round', in H. G. Johnson (ed.), *New Trade Strategy for the World Economy*, Toronto and Buffalo: University of Toronto Press.

Dakin, D. (1971) *The Unification of Greece 1770–1923*, New York: St. Martin's Press.

Dangerfield, G. (1966) *The Strange Death of Liberal England*, London: MacGibbon and Kee.

David, P. A. (1975) *Technical Choice: Innovation and Economic Growth*, Cambridge: Cambridge University Press.

Davies, R. B. (1976) *Peacefully Working to Conquer the World: Singer sewing machines in foreign markets 1854–1920*, New York: Arno Press.

Davis, J. S. (1975) *The World Between the Wars 1919–39: An economist's view*, Baltimore, Md., and London: Johns Hopkins University Press.

Davis, L., Easterlin, R. A., Parker, W. N., and others (1972) *American Economic Growth: An economist's history of the United States*, New York: Harper and Row.

Davis, L. E., and Huttenback, R. A. (1986) *Mammon and the Pursuit of Empire: The political economy of British imperialism 1860–1912*, Cambridge: Cambridge University Press.

de Cecco, M. (1974) *Gold and Empire*, Oxford: Basil Blackwell.

de Cecco, M. (1990) 'The EMS and other international monetary regimes compared', in P. Feri (ed.), *Prospects for the European Monetary System*, London: Macmillan.

de Grauwe, P. (1991) 'The economic integration of Germany: two tales based on trade theory', in W. Heisenberg (ed.), *German Unification in European Perspective*, London: Brassey's (UK).

de Vries, M. G. (1976) *The IMF 1966–71: The system under stress*, Washington: IMF.

della Paolera, G. (1988) 'How the Argentine economy performed during the international gold standard: a re-examination', unpublished Ph.D., University of Chicago.

DeLong, B. (1989) 'Productivity growth, convergence and welfare: comment', *American Economic Review*, 79, pp. 1138–59.

Denison, E. F. (1962) *The Sources of Economic Growth in the United States and the Alternatives Before Us*, New York: Committee for Economic Development.

Desai, M. (1966) 'An econometric model of the world tin economy 1948–1961', *Econometrica*, 34, pp. 105–34.

Desai, M. (1971) 'Demand for cotton textiles in nineteenth-century India', *Indian Economic and Social History Review*, 8, pp. 337–61.

Di Tella, G. (1985) 'Rents, quasi-rents, normal profits and growth: Argentina and the areas of recent settlement', in D. C. M. Platt and G. di Tella (eds.), *Argentine, Australia and Canada: Studies in comparative development 1870–1965*, London: Macmillan.

Diaz Alejandro, C. F. (1967) 'The Argentine tariff, 1906–1940', *Oxford Economic Papers*, 19, pp. 75–98.

Diaz Alejandro, C. F. (1970) *Essays in Argentine Economic History*, New Haven, Conn.: Yale University Press.

Diebold, F. X., Husted, S., and Rush, M. (1991) 'Real exchange rates under the gold standard', *Journal of Political Economy*, 99, pp. 1252–71.

Dixit, A., and Norman, V. (1980) *Theory of International Trade*, Welwyn, Herts: Nisbet and Cambridge University Press.

Dornbusch, R. (1980) *Open Economy Macroeconomics*, New York: Basic Books.

Dornbusch, R. (1984) 'The overvalued dollar', *Lloyds Bank Review*, pp. 1–12.

Dornstadter, J., and Landsberg (1976) 'The economic background', in R. Vernon (ed.), *The Oil Crisis*, New York: W. W. Norton.

Doyle, M. W. (1986) *Empire*, Ithaca, NY, and London: Cornell University Press.

Driskill, R. A. (1981) 'Exchange rate dynamics: an empirical investigation', *Journal of Political Economy*, 89, pp. 357–71.

Drouin, M. J., and Malmgren, H. B. (1981) 'Canada, the United States and the world economy', *Foreign Affairs*, 60, pp. 393–413.

Drummond, I. (1976) 'The Russian gold standard 1897–1914', *Journal of Economic History*, 36, pp. 663–88.

Drummond, I. (1981) *The Floating Pound and the Sterling Area 1931–1939*, Cambridge: Cambridge University Press.

Drummond, L. (1974) *Imperial Economic Policy 1917–1939*, London: Allen and Unwin.

Dudley, L., and Montmarquette, C. (1976) 'A model of the supply of bilateral foreign aid', *American Economic Review*, 66, pp. 132–42.

Dunning, J. (1970) *Studies in International Investment*, London: Allen and Unwin.

Durand, J. D. (1967) 'The modern expansion of world population', *Proceedings of the American Philosophical Society* (Philadelphia), 111, p. 1.

Dutt, R. C. (1970) *India in the Victorian Age*, 2nd edn, New York: B. Franklin (first published 1904).

Easterlin, R. A. (1961) 'Regional income trends, 1840–1950', in S. E. Hanis (ed.), *American Economic History*, New York: McGraw-Hill.

The Economist (1869) *Commercial and Financial History for 1869*.
The Economist (1982) 'Survey of international banking', 20 March.
The Economist (1993) 'Across the Rio Grande', 9–15 October.
Edelstein, M. (1972) 'Rigidity and bias in the British capital market 1870–1913', in D. N. McCloskey (ed.), *Essays on a Mature Economy*, London: Methuen.
Edelstein, M. (1974) 'The determinants of UK investment abroad', *Journal of Economic History*, 34, pp. 980–1007.
Edelstein, M. (1982) *Overseas Investment in the Age of High Imperialism: The United Kingdom 1850–1914*, London: Methuen.
Ehrlich, C. (1973) 'Building and caretaking: economic policy in British Tropical Africa 1890–1960', *Economic History Review*, 20, pp. 649–63.
Eichengreen, B. J. (1981) *Sterling and the Tariff*, Princeton, NJ: Princeton University Press.
Eichengreen, B. J. (1982) 'Did speculation destabilise the French franc in the 1920s?', *Explorations in Economic History*, 19, pp. 71–100.
Eichengreen, B. J. (1984) 'Central bank cooperation under the interwar gold standard' *Explorations in Economic History*, 21, pp. 63–87.
Eichengreen, B. J. (1987) 'Conducting the international orchestra: Bank of England leadership under the classical gold standard', *Journal of International Money and Finance*, 6, pp. 5–29.
Eichengreen, B. J. (1991) 'Historical research on international lending and debt', *Journal of Economic Perspectives*, 5, pp. 149–69.
Eichengreen, B. J. (1992a) *Golden Fetters: The gold standard and the Great Depresssion 1919–1939*, New York: Oxford University Press.
Eichengreen, B. J. (1992b) 'The gold standard since Alec Ford', in S. N. Broadberry and N. F. R. Crafts (eds.), *Britain in the International Economy*, Cambridge: Cambridge University Press.
Eichengreen, B. J. (1992c) 'The origins and nature of the great slump revisited', *Economic History Review*, 45, pp. 213–39.
Eichengreen, B. J., and Portes, R. (1991) 'After the deluge: default, negotiation and readjustment during the interwar years', in B. J. Eichengreen and P. H. Lindert (eds.), *The International Debt Crisis in Historical Perspective*, Cambridge, Mass.: MIT Press.
Eichengreen, B. J., and Uzan, M. (1992) 'The Marshall Plan: Economic effects and implications for Eastern Europe and the former USSR', *Economic Policy*, 14, pp. 13–76.
Eisner, G. (1961) *Jamaica 1830–1930: A study in economic growth*, Manchester: Manchester University Press.
Ellsworth, P. T. (1956) 'The terms of trade between primary producing and industrial countries', *Inter American Economic Affairs*, 10, pp. 47–65.
Eltis, D. (1987) *Economic Growth and the Ending of the Transatlantic Slave Trade*, Oxford: Oxford University Press.
Eltis, D. (1991) 'Precolonial western Africa and the Atlantic economy', in B. Solow (ed.), *Slavery and the Rise of the Atlantic system*, Cambridge: Cambridge University Press.
Engerman, S. (1986) 'Slavery and emancipation in comparative perspective: a look at some recent debates', *Journal of Economic History*, 46, 317–39.
Erb, R. D. (1979) 'International resource transfers: the international financial system and foreign aid', in R. C. Amacher, G. Haberler and T. D. Willett (eds.), *Challenges to a Liberal International Economic Order*, Washington, DC: American Enterprise Institute.
Erickson, C. (1957) *American Industry and the European Immigrant 1860–85*, Cambridge, Mass.: Harvard University Press.

Erickson, C. (1984) 'Why did contract labour not work in the nineteenth-century United States?', in S. Marks and P. Richardson (eds.), *International Labour Migration: Historical perspectives*, London: Maurice Temple-Smith.
Ethier, S. (1983) *Modern International Economics*, New York: W. W. Norton.
European Commission (1989) *Committee for the Study of Economic and Monetary Union*, Luxemburg: Office for Official Publications of the European Community.
European Commission (1990) *Economic and Monetary Union*, Luxemburg: Office for Official Publications of the European Community.
Faini, R., and Venturini, A. (1993) 'Trade, aid and migration: some basic policy issues', *European Economic Review*, 37, pp. 435–42.
Fairbank, J. K. (1980) 'The creation of the treaty system', in *The Cambridge History of China*, vol. 10, part 2, Cambridge: Cambridge University Press.
Falkus, M. E. (1971) 'United States economic policy and the "dollar gap" of the 1920s', *Economic History Review*, 24, pp. 599–623.
Farnie, D. A. (1979) *The English Cotton Industry and the World Market 1815–1896*, London: Oxford University Press.
Feder, G., and Just, R. E. (1984) 'Debt crisis in an increasingly pessimistic international market: the case of Egyptian credit 1862–1876', *Economic Journal*, 94, pp. 340–56.
Feeny, D. (1989) 'The decline of property rights in man in Thailand 1800–1913', *Journal of Economic History*, 49, pp. 285–96.
Feis, H. (1930) *Europe, the World's Banker 1870–1914*, New Haven, Conn.: Yale University Press.
Ferenczi, I., and Wilcox, W. F. (eds.) (1929), *International Migrations*, New York: NBER.
Feri, P. (1990) 'The EMS, the single market and monetary union: an overview', in P. Feri (ed.), *Prospects for the European Monetary System*, London: Macmillan.
Feuerwerker, A. (1980) 'Economic trends in the late Chi'ing empire 1870–1911', in *The Cambridge History of China*, vol. 1, Cambridge: Cambridge University Press.
Fieldhouse, D. K. (1973) *Economics and Empire 1830–1914*, London: Weidenfeld and Nicolson.
Fieldhouse, D. K. (1978) *Unilever: Anatomy of a multinational*, London: Croom Helm.
Fieldhouse, D. K. (1981) *Colonialism 1870–1945: An introduction*, London: Weidenfeld and Nicolson.
Fieldhouse, D. K. (1982) *The Colonial Empires: A comparative survey from the eighteenth century*, 2nd edn, London: Macmillan.
Findlay Shirras, G. (1931) 'Indian migration', in W. F. Willcox (ed.), *International Migrations: Interpretations*, New York: NBER.
Fisher, I. (1922) *The Making of Index Numbers*, Boston, Mass.: Houghton Mifflin.
Fishlow, A. (1980) 'Brazilian development in long-term perspective', *American Economic Review, Papers and Proceedings*, 70, pp. 102–8.
Flanders, M. June (1963) 'The effects of devaluation on exports, a case study: the United Kingdom 1949–54', *Bulletin of the Oxford University Institute of Statistics*, 25, pp. 165–98.
Flandreau, M. (1993) 'On the inflationary bias of common currencies: the Latin Union puzzle', *European Economic Review*, 37, pp. 501–6.
Flessig, H. (1972) 'The United States and the non-European periphery during the early years of the Great Depression', in H. van der Wee (ed.), *The Great Depression Revisited*, The Hague: M. Nijhoff.
Floud, R., and McCloskey, D. N. (eds.) (1981) *The Economic History of Britain Since 1700*, Cambridge: Cambridge University Press.

Flux, A. W. (1899) 'The flag and trade', *Journal of the Royal Statistical Society*, 62, pp. 489–522.

Fogarty J. (1985) 'Staples, super-staples and the limits of staple theory: the experiences of Argentina, Australia and Canada compared', in D. C. M. Platt and G. di Tella (eds.), *Argentina, Australia and Canada: Studies in comparative development 1870–1965*, London: Macmillan.

Fogel, R. W. (1969) 'The specification problem in economic history', *Journal of Economic History*, 27, pp. 283–308.

Fogel, R. W. (1979) 'Notes on the social saving controversy', *Journal of Economic History*, 39, pp. 1–54.

Food and Agricultural Organisation (1991) *The State of Food and Agriculture*, Rome.

Ford, A. G. (1956) 'Argentina and the Baring Crisis of 1890', *Oxford Economic Papers*, 8, pp. 127–50.

Ford, A. G. (1958) 'Flexible exchange rates and Argentina 1885–1900', *Oxford Economic Papers*, 10, pp. 316–38.

Ford, A. G. (1964) 'Bank rate, the British balance of payments, and the burden of adjustment 1870–1914', *Oxford Economic Papers*, 16, pp. 24–39.

Ford, A. G. (1989) 'International financial policy and the gold standard 1870–1914', in *The Cambridge Economic History of Europe*, vol. 8, Cambridge: Cambridge University Press.

Foreman-Peck, J. (1978) 'Economics of scale and the development of the British motor industry before 1939', unpublished Ph.D. thesis, University of London.

Foreman-Peck, J. (1979) 'Tariff protection and economies of scale: the British motor industry before 1939', *Oxford Economic Papers*, 31, pp. 237–57.

Foreman-Peck, J. S. (1982) 'The American challenge of the twenties: US multinationals and the European motor industry', *Journal of Economic History*, 42, pp. 865–81.

Foreman-Peck, J. S. (1986) 'The motor industry' in M. C. Casson (ed.), *Multinationals and World Trade: Vertical integration and the division of labour in world industries*, London: Allen and Unwin.

Foreman-Peck, J. (1989) 'Foreign investment and imperial exploitation: balance of payments reconstruction for nineteenth century Britain and India', *Economic History Review*, 42, pp. 354–74.

Foreman-Peck, J. (1991a) 'Railways and late Victorian economic growth', in J. Foreman-Peck (ed.), *New Perspectives on the Late Victorian Economy 1860–1914: Essays in quantitative economic history*, Cambridge: Cambridge University Press.

Foreman-Peck, J. (1991b) 'The international transfer of telephone technology 1876–1914', in D. Jeremy (ed.), *The International Technology Transfer 1750–1914*, Aldershot: Edward Elgar.

Foreman-Peck, J. (1992) 'A political economy of international migration 1815–1914', *Manchester School*, 60, pp. 359–76.

Foreman-Peck, J. S (1994) 'A model of later nineteenth century European economic development', *Revista de Historia Economica*, vol. 12.

Foreman-Peck, J., Hughes Hallett, A., Ma, Y. (1992) 'The transmission of the Great Depression in the United States, Britain, France and Germany', *European Economic Review*, 36, pp. 685–94.

Frank, A. Gunder (1967) *Capitalism and Underdevelopment in Latin America: Historical studies of Chile and Brazil*, New York and London: Monthly Review Press.

Frank, A. Gunder (1976) 'Multilateral merchandise trade imbalances and uneven economic development', *Journal of European Economic History*, 5.

Fratianni, M., and Spinelli, F. (1984) 'Italy in the gold standard period', in M. Bordo and A. Schwartz (eds.), *A retrospective on the Classical Gold Standard*, Chicago, Ill.: University of Chicago Press.
Freeman, R. B. (1993) 'Immigration from poor to wealthy countries: experience of the United States', *European Economic Review*, 37, pp. 443–51.
Frey, B. S. (1984) *International Political Economy*, Oxford: Basil Blackwell.
Friedman, M. (1968) 'The role of monetary policy', *American Economic Review*, 58, pp. 1–15.
Friedman, M. (1990) 'The crime of '73', *Journal of Political Economy*, 98, 1159–94.
Friedman, M., and Schwartz, A. (1963) *A Monetary History of the United States 1867–1960*, Princeton, NJ: Princeton University Press.
Friedman, P. (1976) 'The welfare costs of bilateralism: German–Hungarian trade 1933–35', *Explorations in Economic History*, 13, pp. 113–25.
Friedman, P. (1978) 'An econometric model of national income, commercial policy and the level of international trade', *Journal of Economic History*, 28, pp. 148–80.
Fuss, M. A., and Waverman, L. (1992) *Costs and Productivity in Automobile Production: The challenge of Japanese efficiency*, Cambridge: Cambridge University Press.
Gallagher, J., and Robinson, R. (1953) 'The imperialism of free trade 1815–1914', *Economic History Review*, 6, pp. 1–15.
Gallaway, L. E., and Vedder, R. K. (1971) 'Emigration from the UK to the USA 1860–1913', *Journal of Economic History*, 31, pp. 885–97.
Gardner, R. N. (1980) *Sterling–Dollar Diplomacy in Current Perspective*, New York: Columbia University Press.
GATT (1958) *Trends in International Trade*, Geneva.
GATT (1971) *Japan's Expansion and Foreign Trade 1955 to 1970*, Geneva.
GATT (1972) *Trends in United States Merchandise Trade 1953–1970*.
GATT (1979) *International Trade 1978/79*, Geneva.
GATT (1982) *International Trade 1981/82*, Geneva.
GATT (1990) *International Trade 1989/90*, Geneva.
Gavin, M. (1992) 'Intertemporal dimensions of international economic adjustment: evidence from the Franco-Prussian War indemnity', *American Economic Review, Papers and Proceedings*, 80, 174–9.
Giffen, R. (1880a) 'The liquidations of 1873–76', in *Essays in Finance*, London: Bell.
Giffen, R. (1880b) 'Why the depression of trade is so much greater in raw material producing countries', in *Essays in Finance*, London: Bell.
Glynn, S., and Lougheed, A. (1973) 'Comment', *Economic History Review*, 26, pp. 692–4.
Good, D. F. (1992) 'Austria–Hungary', in R. Sylla and G. Toniola (eds.), *Patterns of European Industrialization: The nineteenth century*, London: Routledge.
Goodfellow, D. M. (1931) *A Modern Economic History of South Africa*, London: Routledge.
Goodhart, C. A. E. (1972) *The Business of Banking 1891–1914*, London: Weidenfeld and Nicolson.
Goschen, Rt Hon. Viscount (1866) *The Theory of Foreign Exchanges*, 3rd edn, London: Wilson.
Gould, J. D. (1972) *Economic Growth in History*, London: Methuen.
Gould, J. D. (1979) 'European intercontinental emigration', *Journal of European Economic History*, 8, pp. 593–679.
Graham, F. D. (1930) *Exchange, Prices and Production in Hyperinflation: Germany 1920–23*, Princeton, NJ: Princeton University Press.

Grassman, S. (1980) 'Long-term trends in openness of national economies', *Oxford Economic Papers*, 32, pp. 123–33.

Gravil, R., and Rooth, T. (1978) 'A time of acute dependence: Argentina in the 1930s', *Journal of European Economic History*, 7, pp. 337–78.

Green, A., and Urquhart, M. (1976) 'Factor and commodity flows in the international economy of 1870–1914: a multi-country view', *Journal of Economic History*, 36, pp. 217–52.

Green, N. L. (1985) 'Filling the void: immigration to France before World War I', in D. Hoerder (ed.), *Labor Migration in the Atlantic Economies*, Westport, Conn.: Greenwood Press.

Green, W. A. (1984) 'The West Indies and indentured labour migration: the Jamaican experience', in K. Saunders (ed.), *Indentured Labour in the British Empire 1834–1920*, London: Croom Helm.

Greenhill, R. (1977) 'Shipping 1850–1914', in D. C. M. Platt (ed.), *Business Imperialism 1860–1930*, Oxford: Clarendon Press.

Greenwood, M. J., and McDowell, J. M. (1986) 'The factor market consequences of US immigration', *Journal of Economic Literature*, 24, pp. 1738–72.

Gregory, P. R. (1979) 'The Russian balance of payments, the gold standard and monetary policy: a historical example of foreign capital movements', *Journal of Economic History*, 39, pp. 379–99.

Grennes, T., and Krissof, B. (1993) 'Agricultural trade in a North America Free trade agreement', *World Economy*, 16, pp. 483–502.

Griffin, K. (1970) 'Foreign capital, domestic savings and economic development', *Oxford Bulletin of Economics and Statistics*, 32, pp. 99–112.

Grogan, F. O. (1972) *International Trade in Temperate Zone Products*, Edinburgh: Oliver and Boyd.

Grover Clark, C. (1936) *The Balance Sheet of Imperialism*, New York.

Grubel, H. G., and Lloyd, P. J. (1975) *Intra-Industry Trade*, London: Macmillan.

Habbakuk, H. J. (1962) *American and British Technology in the Nineteenth Century*, Cambridge: Cambridge University Press.

Haberler, G. (1936) *Theory of International Trade*, London: Hodge.

Haberler, G. (1976) *The World Economy, Money and the Great Depression 1919–1939*, Washington, DC: American Institute for Public Policy Research.

Hall, D. (1964) *Ideas and Illustrations in Economic History*, New York: Holt Rinehart.

Hannah, L. (1974) 'Mergers in British manufacturing industry 1880–1918', *Oxford Economic Papers*, 26, pp. 1–20.

Hannah, L., and Kay, J. (1977) *Concentration in Modern Industry*, London: Macmillan.

Hansard (1925) House of Commons, v. 183, 28 April, p. 58.

Hanson II, J. R. (1976) 'The Leff conjecture: some contrary evidence', *Journal of Political Economy*, 84, pp. 401–5.

Hanson II, J. R. (1977a) 'Diversification and concentration of LDC exports: Victorian trends', *Explorations in Economic History*, 14, pp. 44–68.

Hanson II, J. R. (1977b) 'Export instability in historical perspective', *Explorations in Economic History*, 14, pp. 293–310.

Hanson II, J. R. (1986) 'Export shares in the European periphery and the Third World before World War I', *Explorations in Economic History*, 23, pp. 85–99.

Hanson II, J. R. (1991) 'Third World incomes before World War I: further evidence', *Explorations in Economic History*, 28, pp. 367–79.

Hanson, P. (1981) *Trade and Technology in Soviet–Western Relations*, London: Macmillan.

Harcourt, E. (1987) *Taming the Tyrant: The first hundred years of Australia's international telecommunications service*, Sydney: Allen and Unwin.
Hardach, G. (1977) *The First World War 1914–1918*, London: Allen Lane.
Harkness, D. A. E. (1931) 'Irish emigration', in W. F. Willcox (ed.), *International Migration: Interpretations*, New York, NBER.
Harley, C. K. (1971) 'The shift from sailing ships to steamships, 1850–90', in McCloskey, D. N. (ed.), *Essays on a Mature Economy: Britain after 1840*, London: Methuen.
Harley, C. K. (1980) 'Transportation, the world wheat trade and the Kuznets Cycle 1850–1913', *Explorations in Economic History*, 17, pp. 218–50.
Harley, C. K. (1988) 'Ocean freight rates and productivity 1740–1913: the primacy of mechanical invention reaffirmed', *Journal of Economic History*, 48, pp. 851–76.
Harley, C. K. (1992a) 'The antebellum American tariff: food exports and manufacturing', *Explorations in Economic History*, 29, pp. 375–400.
Harley, C. K. (1992b) 'The world food economy and pre-World War I Argentina', in S. N. Broadberry and N. F. R. Crafts (eds.), *Britain in the International Economy 1870–1939*, Cambridge: Cambridge University Press.
Harris, S. E. (1948) *The European Recovery Program*, Cambridge, Mass.: Harvard University Press.
Hartmann, S. M. (1968) *The Marshall Plan*, Columbus, Ohio: C. E. Merrill.
Hasenpflug, H. (1977) 'The stabilization of export earnings in the Lomé Convention', in K. P. Sauvant and H. Hasenpflug (eds.), *The New International Economic Order*, Boulder, Colo.: Westview Press.
Hatton, T. J. (1992) 'Price determination under the gold standard: Britain 1880–1913', in S. N. Broadberry and N. F. R. Crafts (eds.), *Britain in the International Economy 1870–1939*, Cambridge: Cambridge University Press.
Hatton, T. J., and Williamson, J. G. (1992) 'International migration and world development: a historical perspective', *NBER Historical Paper*, no. 41.
Hatton, T. J., and Williamson, J. G. (1994) 'After the famine: emigration from Ireland 1850–1913', *Journal of Economic History* (forthcoming).
Hauser, H. (1917) *Germany's Commercial Grip upon the World*, London: Eveleigh Nash.
Hawtrey, R. (1938) *A Century of Bank Rate*, London: Longmans Green.
Hayek, F. A. (1962) *The Road to Serfdom*, London: Routledge and Kegan Paul (first published 1944).
Hecksher, E. (1949) 'The effects of foreign trade on the distribution of income', in H. Ellis and L. Metzler (eds.), *Readings in the Theory of International Trade*, London: Allen and Unwin.
Hecksher, E. (1955) *Mercantilism*, 2 vols., London: Allen and Unwin.
Heller, H. R., and Khan, M. S. (1978) 'The demand for international reserves under fixed and floating exchange rates', *IMF Staff Papers*, 25, pp. 1–24.
Helliwell, J. (1975) 'Adjustment under fixed and flexible exchange rates', in P. B. Kenen (ed.), *International Trade and Finance*, Cambridge: Cambridge University Press.
Henderson, W. O. (1939) *The Zollverein*, Cambridge: Cambridge University Press.
Hertner, P., and Jones, G. (eds.) (1986) *Multinationals: Theory and practice*, Aldershot: Gower.
Heston, A. (1983) 'National income', in D. Kumar (ed.), *The Cambridge Economic History of India*, vol. 2, Cambridge: Cambridge University Press.
Heston, A., and Summers, R. (1980) 'Comparative Indian economic growth: 1870 to 1970', *American Economic Review, Papers and Proceedings*, 70, pp. 96–101.

Heusser, H. K. (1939) *Control of International Trade*, London: Routledge.
Hidy, R. (1970) *The House of Baring in American Trade and Finance*, New York: Russell and Russell (first published 1949).
Hobson, J. A. (1905) *Imperialism: a study*, revised edn, London: Constable (first published 1902).
Hodgson, J. S. (1972) 'An analysis of floating exchange rates: the dollar–sterling rate 1919–1925, *Southern Economic Journal*, 39, pp. 249–57.
Hoerder, D. (1985) 'An introduction to labour migration in Atlantic economies, 1815–1914', in D. Hoerder (ed.), *Labor Migration in Atlantic Economies: The European and North American working classes during the period of industrialization*, Westport, Conn.: Greenwood Press.
Holmes, C. (1982) 'The impact of immigration on British society 1870–1980', in T. C. Barker and M. Drake (eds.), *Population and Society in Britain 1850–1980*, London: Batsford.
Holtferich, C.-L. (1993) 'Did monetary unification precede or follow political unification of Germany in the nineteenth century?', *European Economic Review*, 37, pp. 518–24.
Holzman, F. D. (1976) *International Trade under Communism: Politics and economics*, London: Macmillan.
Hookham, H. (1972) *A Short History of China*, New York: New American Library.
Hopkins, A. G. (1980) 'Property rights and empire building: Britain's annexation of Lagos', *Journal of Economic History*, 40, pp. 777–98.
Horie, S. (1964) *The International Monetary Fund*, New York: St Martin's Press.
Horsefield, J. K. (1969) *The International Monetary Fund 1945–1965*, 2 vols., Washington DC: IMF.
House of Lords Select Committee on the European Community (1992) *Fraud against the Community*, London: HMSO.
House of Lords Select Committee on European Community (1992) *Ninth Report: Implementation of Reform of the Common Agricultural Policy*, HMSO.
Howson, S. (1980) *Sterling's Managed Float: The operation of the exchange accounts 1932–39*, Princeton, NJ: Princeton University Press.
Huffman, W. E., and Lothian, J. R. (1984) 'The gold standard and the transmission of business cycles 1833–1932', in M. Bordo and A. Schwartz (eds.), *A Retrospective on the Classical Gold Standard, 1821–1931*, Chicago, Ill.: University of Chicago Press.
Hughes Hallett, A., and Ma, Y. (1993) 'East Germany, West Germany and their mezzogiornio problem: a parable for European economic integration', *Economic Journal*, 103, pp. 416–28.
Hughes, J. R. T. (1956) 'The commercial crisis of 1857', *Oxford Economic Papers*, 8, pp. 194–222.
Hughes, J. R. T. (1960) *Fluctuations in Trade, Industry and Finance 1850–1860*, Oxford: Oxford University Press.
Hurd, D. (1967) *The Arrow War: An Anglo-Chinese confusion 1856–1860*, London: Collins.
Hyde, C. K. (1991) 'Iron and steel technologies moving between Europe and the United States before 1914', in D. J. Jeremy (ed.), *International Technology Transfer: Europe, Japan and the USA 1700–1914*, Aldershot: Edward Elgar.
Ichihashi, Y. (1931) 'International migration of the Japanese', in W. F. Willcox (ed.), *International Migrations: Interpretations*, New York: NBER.
Imlah, A. (1958) *Economic Elements in the Pax Britannica: Studies in British foreign trade in the nineteenth century*, Cambridge, Mass: Harvard University Press.

Independent Commission on International Development Issues (1980) *North–South: A programme for survival*, London: Pan.
Ingram, J. C. (1971) *Economic Change in Thailand 1850–1970*, Stanford, Calif.: Stanford University Press.
Inouye, J. (1931) *Problems of Japanese Currency and Exchange 1914–26*, Glasgow: University of Glasgow Press.
Independent Commission on International Development Issues (1980) *North–South: A Programme for survival*, London: Pan.
International Monetary Fund, *Annual Reports*, Washington, DC.
Issani, C. (1961) 'Egypt since 1800: a study in lopsided development', *Journal of Economic History*, 21, pp. 1–26.
James, H. (1992) 'Financial flows across frontiers during the interwar depression', *Economic History Review*, 45, pp. 594–613.
Jenkins, R. O. (1977) *Dependent Industrialization in Latin America: The automotive industry in Argentina, Chile and Mexico*, New York: Praeger.
Jenks, L. H. (1971) *The Migration of British Capital to 1875*, London: Nelson (first published 1927).
Jerome, H. (1926) *Migration and Business Cycles*, New York: NBER.
Jevons, W. S. (1909) 'A serious fall in the value of gold ascertained' in *Investigations in Currency and Finance*, London: Macmillan.
Jevons, W. S. (1906) *The Coal Question*, London: Macmillan.
Johnson, D. Gale (1964) 'Agriculture and foreign economic policy', *Journal of Farm Economics*, 46, pp. 915–29.
Johnson, H. G. (ed.) (1969) *New Trade Strategy for the World Economy*, Toronto and Buffalo, NY: University of Toronto Press.
Johnson, H. G. (1973) 'A general commentary', in A. L. K. Acheson, J. F. Chant and M. J. F. Prachowny (eds.), *Bretton Woods Revisited*, Toronto: University of Toronto Press.
Johnson, H. G. (1975) 'The classical transfer problem: an alternative formulation', *Economica*, 42, pp. 20–31.
Johnson, H. G., and Frenkel, J. A. (eds.) (1976) *The Monetary Approach to the Balance of Payments*, London: Allen and Unwin.
Johnson, S. C. (1913) *A History of Emigration: From the United Kingdom to North America 1763–1912*, London: Routledge.
Jones, C. (1987) *International Business in the Nineteenth Century: The rise and fall of the cosmopolitan bourgeosie*, Brighton: Wheatsheaf.
Jones, J., and Smith, A. D. (1970) *The Economic Impact of Commonwealth Immigration*, Cambridge: Cambridge University Press.
Jones, J. H. (1933) 'The gold standard', *Economic Journal*, 43, pp. 351–74.
Jones, S. (1986) *Two Centuries of Overseas Trading: The origins and growth of the Inchcape Group*, London: Macmillan.
Jonung, L. (1984) 'Swedish experience under the classical gold standard 1873–1914', in M. Bordo and A. Schwartz (eds.), *A Retrospective on the Classical Gold Standard*, Chicago, Ill.: University of Chicago Press.
Jorgensen, E., and Sachs, J. (1989) 'Default and renegotiation of Latin American foreign bonds in the interwar period', in B. J. Eichengreen and P. H. Lindert (eds.), *The International Debt Crisis in Historical Perspective*, Cambridge, Mass.: MIT Press.
Kaldor, N. (1966) *Causes of the Slow Rate of Economic Growth of the United Kingdom*, Cambridge: Cambridge University Press.

Katz, S. I. (1972) 'Devaluation bias and the Bretton Woods system', *Banca Nazionale Del Lavoro Quarterly Review*, 25, pp. 178–98.
Katzenellenbaum, S. S. (1925) *Russian Currency and Banking 1914–1924*, London: P. S. King.
Kelley, A. C. (1965) 'International migration and economic growth: Australia 1865–1935', *Journal of Economic History*, vol. 25, no. 3, pp. 333–54.
Kenen, P. B. (1989) *Exchange Rates and Policy Coordination*, Manchester: Manchester University Press.
Kennedy, P. (1988) *The Rise and Fall of the Great Powers: Economic Change and Military Conflict from 1500 to 2000*, New York: Random House.
Kennedy, W. P. (1976) 'Institutional response to economic growth: capital markets in Britain to 1914', in L. Hannah (ed.), *Management Strategy and Business Development*, London: Macmillan.
Kennedy, W. P. (1987) *Industrial Structure, Capital Markets and the Origins of British Economic Decline*, Cambridge: Cambridge University Press.
Keohane, R. O. (1984) *After Hegemony: Cooperation and discord in the world political economy*, Princeton, NJ: Princeton University Press.
Keynes, J. M. (1913a) *An Essay on Indian Monetary Reform*, London: Macmillan.
Keynes, J. M. (1913b) *Indian Currency and Finance*, London: Macmillan.
Keynes, J. M. (1920) *The Economic Consequences of the Peace*, London: Macmillan.
Keynes, J. M. (1922) *A Revision of the Treaty*, London: Macmillan.
Keynes, J. M. (1925) *The Economic Consequences of Mr Churchill*, London: Hogarth Press.
Keynes, J. M. (1929) 'The German transfer problem', *Economic Journal*, 39, pp. 1–7.
Keynes, J. M. (1930) *Treatise on Money*, vol. 2, London: Macmillan.
Kieve, J. (1973) *The Electric Telegraph: A social and economic history*, Newton Abbot: David and Charles.
Kimura, M. (1989) 'Public finance in Korea under Japanese rule: deficit in the colonial account and colonial taxation', *Explorations in Economic History*, 26, pp. 285–310.
Kindleberger, C. P. (1956) *The Terms of Trade: A European case study*, London: Chapman and Hall.
Kindleberger, C. P. (1967) *Europe's Postwar Growth: The role of labour supply*, London: Oxford University Press.
Kindleberger, C. P. (1973) *The World in Depression 1929–1939*, London: Allen Lane.
Kindleberger, C. P. (1975) 'The rise of free trade in Western Europe 1820–1875', *Journal of Economic History*, 35, pp. 20–55.
Kindleberger, C. P. (1987) *Marshall Plan Days*, London: Allen and Unwin.
Kindleberger, C. P. (1989) *Manias, Panics and Crashes*, 2nd edn, London: Macmillan.
King, M. A. (1974) 'Economic growth and social development: a statistical investigation', *Review of Income and Wealth*, 20, pp. 251–72.
King, W. T. C. (1972) *History of the London Discount Market*, London: F. Cass.
Kirby, M. C. (1981) *The Decline of British Economic Power Since 1870*, London: Allen and Unwin.
Kissinger, H. (1982) *Years of Upheaval*, Boston, Mass.: Little, Brown.
Kitson, M., and Solomou, S. (1990) *Protectionism and Economic Revival: The British interwar economy*, Cambridge: Cambridge University Press.
Kleiman, E. (1976) 'Trade and the decline of colonialism', *Economic Journal*, 86, pp. 459–80.

Kleiman, E. (1978) 'Metropolitan exports lost through decolonisation', *Bulletin of the Oxford University Institute of Economics and Statistics*, 40, pp. 273–8.
Klein, H. S. (1978) *The Middle Passage: Comparative studies in the Atlantic slave trade*, Princeton, NJ: Princeton University Press.
Klug, A. (1993) *The German Buybacks, 1932–1939: A cure for overhang?*, Princeton Studies in International Finance, no. 75.
Kock, K. (1969) *International Trade Policy and the Gatt 1947–1967*, Stockholm: Almquist and Wiksell.
Krantz, O. (1988) 'New estimates of Swedish historical GDP since the beginning of the nineteenth century', *Review of Income and Wealth*, vol. 34, no. 2, pp. 215–42.
Kravis, I. (1970) 'Trade as a handmaiden of growth: similarities between the nineteenth and twentieth centuries', *Economic Journal*, 80, pp. 850–72.
Kravis, I., Heston, A. W. and Summers, R. (1978) 'Real GDP per capita for more than one hundred countries', *Economic Journal*, 88, pp. 215–42.
Kreinen, M. (1975) *International Economics*, 2nd edn, New York: Harcourt Brace.
Krueger, A. O. (1975) *The Benefits and Costs of Import Substitution in India: A microeconomic study*, Minneapolis, Minn.: University of Minnesota Press.
Krueger, A. O. (1992) 'Government, trade and economic integration', *American Economic Review, Papers and Proceedings*, 82, pp. 109–14.
Kucynski, P. P. (1988) *Latin American Debt*, Baltimore, Md: Johns Hopkins University Press.
Kuznets, S. (1971) 'The contribution of immigration to the growth of the labour force', in R. W. Fogel and S. L. Engerman (eds.), *The Reinterpretation of American Economic History*, New York: Harper and Row.
Landes, D. S. (1958) *Bankers and Pashas*, Cambridge, Mass.: Harvard University Press.
Landes, D. S. (1969) *The Unbound Prometheus*, Cambridge: Cambridge University Press.
Latham, A. J. H. (1978a) 'Merchandise trade imbalances and uneven economic development in India and China', *Journal of European Economic History*, 7, pp. 33–60.
Latham, A. J. H. (1978b) *The International Economy and the Undeveloped World 1865–1914*, London: Croom Helm.
Latham, A. J. H. (1986) 'Ethnic Chinese multinationals in the international grain trade before the Second World War', *South African Journal of Economic History*, 1, pp. 4–18.
Latham, A. J. H. (1988) 'From competition to constraint: the international rice trade in the nineteenth and twentieth centuries', *Business and Economic History*, 2nd Series, 17, pp. 91–102.
Latham, A. J. H., and Neal, L. (1983) 'The international market in rice and wheat 1868–1914', *Economic History Review*, 36, pp. 260–80.
Laughlin, J. L. (1897) *The History of Bimetallism in the United States*, New York: D. Appleton.
Laughlin, J. L. (1919) *Money and Prices*, London: P. S. King.
Lawrence, R. Z., and Schultze, C. L. (1987) 'Overview' in R. Z. Lawrence and C. L. Schultze (eds.), *Barriers to European Growth: A transatlantic view*, Washington, DC: Brookings Institution.
Le Roy Bennett, A. (1980) *International Organizations*, 2nd edn, New York: Prentice Hall.
League of Nations (1927) *Cartels and Combines* (by K. Wiedenfeld), Geneva.
League of Nations (1935) *Commercial Banks 1929–34*, Geneva.
League of Nations (1945) *Industrialization and Foreign Trade* (by F. Hilgerdt), Geneva.
Leff, N. H. (1982) *Underdevelopment and Development in Brazil*, vol. 1: *Economic Structure and Change 1822–1947*, London: Allen and Unwin.

Lenin, V. I. (1934) *Imperialism, the Highest Stage of Capitalism*, London: Lawrence and Wishart (first published 1916).
Leontief, W. (1969) 'Domestic production and foreign trade: the American position re-examined', in J. N. Bhagwati (ed.), *International Trade: Selected readings*, Baltimore, Md: Penguin.
Leveen, Phillip E. (1975) 'A quantitative analysis of the impact of British suppression policies on the volume of the nineteenth-century African slave trade', in S. L. Engerman and E. D. Genovese (eds.), *Race and Slavery in the Western Hemisphere*, Princeton, NJ: Princeton University Press.
Levi, P. (1987) *If This is a Man*, London: Abacus.
Levin, J. V. (1960) *The Export Economies*, Cambridge, Mass.: Harvard University Press.
Levine, A. L. (1967) *Industrial Retardation in Britain 1880–1914*, New York: Basic Books.
Levy-Leboyer, M., and Bourgignon, F. (1990) *The French Economy in the Nineteenth Century: An essay in econometric analysis*, Cambridge: Cambridge University Press.
Lewis, W. A. (1949) *Economic Survey 1919–1939*, London: Allen and Unwin.
Lewis, W. A. (1978a) *The Evolution of the International Economic Order*, Princeton, NJ: Princeton University Press.
Lewis, W. A. (1978b) *Growth and Fluctuations 1870–1913*, London: Allen and Unwin.
Lewis, W. A. (1981) 'The rate of growth of world trade 1830–1973', in S. Grassman and E. Lundberg (eds.), *The World Economic Order, Past and Prospects*, New York: St. Martin's Press.
Liberal Industrial Inquiry (1928) *Britain's Industrial Future*, London.
Liepmann, H. (1938) *Tariff Levels and the Economic Unity of Europe*, London: Allen and Unwin.
Lindert, P. H. (1969) *Key currencies and Gold 1900–1914*, Princeton, NJ: Princeton University Press.
Lindert, P. H. (1989) 'Response to debt crisis: what is different about the 1980s?, in B. J. Eichengreen and P. H. Lindert (eds.), *The International Debt Crisis in Historical Perspective*, Cambridge, Mass.: MIT Press.
Lindert, P. H. (1984) 'Comment', in M. Bordo and A. Schwartz (eds.), *A Retrospective on the Classical Gold Standard*, Chicago, Ill.: University of Chicago Press.
Lindert, P. H., and Morton, P. J. (1988) 'How sovereign debt has worked', in J. D. Sachs (ed.), *Developing Country Debt and Economic Performance: The international financial system*, Chicago, NBER: University of Chicago Press.
Lindert, P. H., and Trace, K. (1971) 'Yardsticks for Victorian entrepreneurs', in D. N. McCloskey (ed.), *Essays on a Mature Economy: Britain after 1840*, London: Methuen.
Lipson, C. (1989) 'International debt and national security: comparing Victorian Britain and postwar America', in B. J. Eichengreen and P. H. Lindert (eds.), *The International Debt Crisis in Historical Perspective*, Cambridge, Mass.: MIT Press.
List, F. (1856) *National System of Political Economy*, (translated by G. A. Matile, preliminary essay by S. Colwell), Philadelphia, PA: J. B. Lippincott & Co.
Little, I. M. D., Scitovsky, T., and Scott, M. Fg. (1970) *Industry and Trade in Some Developing Countries*, London: Oxford University Press.
Livingstone, I. (1978) 'Metropolitan exports lost through decolonization: a comment', *Bulletin of the Oxford University Institute of Economics and Statistics*, 40, pp. 279–80.
Logue, D. E., and Sweeney, R. J. (1977) *The Economics of the Law of the Sea Negotiations*, Los Angeles, Calif.: International Institute for Economic Research.
Longrigg, S. H. (1961) *Oil in the Middle East*, 2nd edn, Oxford: Oxford University Press.

Lundberg, E. (1968) *Instability and Economic Growth*, New Haven, Conn., and London: Yale University Press.
Luxemburg, R. (1972) *The Accumulation of Capital*, London: Allen Lane (first published 1913).
MacDougall, D., and Hutt, R. (1954) 'Imperial preference: a quantitative analysis', *Economic Journal*, 64, pp. 233–57.
Macinnes, C. M. (1934) *England and Slavery*, Bristol: Arrowsmith.
MacKenzie, C. (1954) *Realms of Silver*, London: Routledge and Kegan Paul.
Macpherson, W. J. (1972) 'Economic development in India under the British Crown 1858–1947', in A. J. Youngson (ed.), *Economic Development in the Long Run*, London: Allen and Unwin.
Macrosty, H. (1907) *The Trust Movement in British Industry*, London: Macmillan.
Maddison, A. (1969) *Economic Growth in Japan and the USSR*, London: Allen and Unwin.
Maddison, A. (1971) *Class Structure and Economic Growth: India and Pakistan since the Moghuls*, London: Allen and Unwin.
Maddison, A. (1977) 'Phases of capitalist development', *Banca Nazionale del Lavoro Quarterly Review*, 39, pp. 103–38.
Maddison, A. (1982) *Phases of Capitalist Development*, Oxford: Oxford University Press.
Maddison, A. (1983) 'A comparison of levels of GDP per capita in developed and developing countries 1700–1980', *Journal of Economic History*, 43, pp. 27–42.
Maddison, A. (1989) *The World Economy in the Twentieth Century*, Paris: OECD.
Maddison, A. (1991) *Dynamic Forces in Capitalist Development*, Oxford: Oxford University Press.
Maizels, A. (1963) *Industrial Growth and World Trade*, Cambridge: Cambridge University Press.
Majumdar, R. C., Raychaudhuri, H. C., and Datta, K. (1958) *An Advanced History of India*, London: Macmillan.
Malchow, H. L. (1979) *Population Pressures: Emigration and government in late nineteenth-century Britain*, Palo Alto, California: Society for the Promotion of Science and Scholarship.
Mannix, D. P., and Cowley, M. (1976) *Black Cargoes: A history of the Atlantic slave trade*, London: Penguin.
Mantoux, E. (1946) *The Carthaginian Peace; or the economic consequences of Mr Keynes*, London: G. Cumberledge.
Marshall, A. (1925) 'Remedies for fluctuations of general prices', in A. C. Pigou (ed.), *Memorials of Alfred Marshall*, London: Macmillan.
Marshall, A. (1926) *Official Papers*, London: Macmillan.
Martin-Acena, P. (1993) 'Spain during the classic gold standard years 1880–1914', in M. Bordo and F. Capie (eds.), *Monetary Regimes in Transition*, Cambridge: Cambridge University Press.
Martinez, A. B., and Lewandowski, M. (1911) *The Argentine in the Twentieth Century*, London: T. Fisher Unwin.
Marx, K., and Engels, F. (1893) *Manifesto of the Communist Party*, Moscow: Progress Publishers (first published 1848).
Mason, E. S., and Asher, R. E. (1973) *The World Bank Since Bretton Woods*, Washington, DC: The Brookings Institution.
Mathew, W. M. (1977) 'Antony Gibbs & Sons, the guano trade and the Peruvian government 1842–61', in D. C. M. Platt (ed.), *Business Imperialism 1860–1930*, Oxford: Clarendon Press.

Mathews, J. T. (1990) *Preserving the Global Environment: The challenge of shared leadership*, New York: Norton.
Matthews, K. G. P. (1986a) *The Interwar Economy: An equilibrium analysis*, Aldershot: Gower.
Matthews, K. G. P. (1986b) 'Was sterling overvalued in 1925?', *Economic History Review*, 39, pp. 572–87.
McClain, C. J. (1990) 'Chinese immigration: a comment on Cloud and Galenson', *Explorations in Economic History*, 27, pp. 363–78.
McCloskey, D. N. (1980) 'Magnanimous Albion: free trade and British national income, 1841–81', *Explorations in Economic History*, 17, pp. 303–20.
McCloskey, D. N., and Zecher, J. R. (1975) 'How the gold standard worked, 1880–1913', in J. Frenkel and H. G. Johnson (eds.), *The Monetary Theory of the Balance of Payments*, London: Allen and Unwin.
McDiarmid, O. J. (1948) *Commercial Policy in the Canadian Economy*, Cambridge, Mass.: Harvard University Press.
McInnis, R. M. (1986) 'The emergence of a world economy in the latter half of the nineteenth century', in *Ninth International Economic History Congress Bern 1986: Debates and Controversies*, Zurich: Verlag der Fachvereine.
McIntosh, D. C. (1977) 'Mantoux versus Keynes: a note on German income and the reparations controversy', *Economic Journal*, 87, pp. 765–7.
McKie, J. W. (1976) 'The United States', in R. Vernon (ed.), *The Oil Crisis*, New York: Norton.
Mendershausen, H. (1941) *The Economics of War*, New York: Prentice Hall.
Meredith, D. (1975) 'The British government and colonial economic policy 1919–39', *Economic History Review*, 28, pp. 484–99.
Mill, J. S. (1929) *Principles of Political Economy*, 7th edn, vol. 2, book 3, London: Longman (first published 1848).
Mill, J. S. (1972) *Utilitarianism, On Liberty, and Considerations on Representative Government* (edited by H. B. Acton), London: Dent.
Milward, A. S. (1972) *The Fascist Economy in Norway*, Oxford: Clarendon Press.
Milward, A. S. (1977) *War Economy and Society 1939–1945*, London: Allen Lane.
Milward, A. S. (1984) *The Reconstruction of Western Europe 1945–51*, London: Methuen.
Milward, A. S., and Saul, S. B. (1973) *The Economic Development of Continental Europe 1780–1870*, London: Allen and Unwin.
Milward, A. S., and Saul, S. B. (1978) *The development of the Economies of Continental Europe 1850–1914*, London: Allen and Unwin.
Mishan, E. J. (1977) *The Economic Growth Debate: An assessment*, London: Allen and Unwin.
Mitchell, B. R. (1975) *Abstract of European Historical Statistics*, Cambridge: Cambridge University Press.
Mitrany, D. (1936) *The Effects of the War on South Eastern Europe*, New Haven, Conn.: Yale University Press.
Moggridge, D. E. (1972) *British Monetary Policy 1924–1931: The Norman Conquest of $4.86*, Cambridge: Cambridge University Press.
Moggridge, D. E. (1989) 'The gold standard and national financial policies 1919–1939', in *The Cambridge Economic History of Europe*, vol. 8, Cambridge: Cambridge University Press.
Mokyr, J., and Savin, R. E. (1976) 'Stagflation in historical perspective: the Napoleonic wars revisited', *Research in Economic History*, 1, pp. 198–259.

Morgan, E. V. (1952) *Studies in British Financial Policy 1916–1925*, London: Macmillan.
Morris, M. D. (1963) 'Towards a reinterpretation of nineteenth-century Indian economic history', *Journal of Economic History*, 23, pp. 612–13.
Morris, M. D. (1968) 'Trends and tendencies in Indian economic history', *Indian Economic and Social History Review*, 5, pp. 377–84.
Morris, M. D., Matsue, T., Chandra, B., and Raychaudhuri, T. (eds.) (1969) *The Indian Economy in the Nineteenth Century: A symposium*, Delhi: Delhi School of Economics.
Mouré, K. (1991) *Managing the Franc Poincaré: Economic understanding and political constraint in French monetary policy 1928–1936*, Cambridge: Cambridge University Press.
Mukerjee, T. (1972) 'Theory of economic drain: impact of British rule on the Indian economy, 1840–1900', in K. E. Boulding and T. Mukerjee (eds.), *Economic Imperialism: A book of readings*, Ann Arbor, Mich.: University of Michigan Press.
Mulhall, M. G. (1892) *Dictionary of Statistics*, London: Routledge.
Mulhall, M. G. (1899) *The Dictionary of Statistics*, London: Routledge.
Mundell, R. A. (1981) 'International trade and factor mobility', in J. N. Bhagwati (ed.), *International Trade Selected Readings*, Cambridge, Mass.: MIT Press.
Mundell, R. A., and Swoboda, A. (eds.) (1969) *Monetary Problems of the International Economy*, Chicago, Ill.: Chicago University Press.
Musson, A. E. (1976) 'Industrial motive power in the United Kingdom 1800–1870', *Economic History Review*, 29, 415–39.
Musson, A. E. (1980) 'The engineering industry', in R. A. Church (ed.), *The Dynamics of Victorian Business*, London: Allen and Unwin.
Myint, H. (1977) 'Adam Smith's theory of international trade in the perspective of economic development', *Economica*, 44, pp. 231–48.
Nagle, J. C. (1976) *Agricultural Trade Policies*, Farnborough: Saxon House.
Narain, B. (1926) 'Exchange and prices in India 1873–1924', *Weltwirtschaftliches Archiv*, 23, pp. 246–92.
Neal, L. (1979) 'The economics and finance of bilateral clearing arrangements: Germany 1934–8', *Economic History Review*, 32, pp. 391–404.
Nelson, R. R., and Wright, G. (1992) 'American technological leadership', *Journal of Economic Literature*, 30, pp. 1931–64.
Neuberger, H. M., and Stokes, H. H. (1979) 'The relationship between interest rates and gold flows under the gold standard', *Economica*, 46, pp. 261–79.
Nevins, A., and Hill, F. (1957) *Ford: Expansion and Challenge 1915–1933*, New York: Scribner.
Newmarch, W. (1860) 'Reports of the character and results of the trade of the United Kingdom during the year 1859', *Journal of the Statistical Society*.
Newmarch, W. (1878) 'On the progress of the foreign trade of the United Kingdom since 1856. . .', *Journal of the (Royal) Statistical Society*, 41, pp. 187–282.
Nicholas, S. (ed.) (1988) *Convict Workers: Reinterpreting Australia's past*, Cambridge: Cambridge University Press.
Nicholas, S. (1991) 'The expansion of British multinational companies: testing for managerial failure', in J. Foreman-Peck (ed.), *New Perspectives on the Late Victorian Economy*, Cambridge: Cambridge University Press.
Nishimura, S. (1971) *The Decline of Inland Bills of Exchange in the London Money Market 1855–1913*, Cambridge: Cambridge University Press.
North, D. C. (1960) 'The US balance of payments 1790–1860', *Studies in Income and Wealth*, 24, pp. 573–628.

Nozick, R. (1974) *Anarchy, State and Utopia*, New York: Basic Books.
Nugent, J. B. (1973) 'Exchange rate movements and economic development in the late nineteenth century', *Journal of Political Economy*, 81, pp. 1110–35.
Nurkse, R. (1944) *International Currency Experience*, Princeton, NJ: League of Nations.
Nurkse, R. (1961) *Equilibrium and Growth in the World Economy* (edited by G. Haberler and R. M. Stern), Cambridge, Mass.: Harvard University Press.
Nye, J. V. (1991a) 'Changing French trade conditions, national welfare and the 1860 Anglo-French Treaty of Commerce', *Explorations in Economic History*, 28, pp. 460–77.
Nye, J. V. (1991b) 'The myth of free-trade Britain and fortress France: tariffs and trade in the nineteenth century', *Journal of Economic History*, 51, pp. 23–46.
O'Brien, P. K. (1977) *The New Economic History of Railways*, London: Croom Helm.
O'Brien, P. K. (1982) 'European economic development: the contribution of the periphery', *Economic History Review*, 35, pp. 1–18.
O'Brien, P. K. (1983) *Railways and Economic Development, 1830–1914*, London: Macmillan.
O'Brien, P. K. (1988) 'The costs and benefits of British imperialism 1846–1914', *Past and Present*, 120, pp. 163–200.
O'Brien, P. K. (1989) 'Debate', *Past and Present*, 121, pp. 192–99.
O'Brien, P. K. (1991) *Power with Profit: The state and the economy 1688–1815*, inaugural lecture, University of London.
O'Brien, P. K., and Engerman, S. (1991) 'Exports and the growth of the British economy', in B. Solow (ed.) *Slavery and the Rise of the Atlantic System*, Cambridge: Cambridge University Press.
O'Brien, P. K., and Keyder, C. (1978) *Economic Growth in Britain and France 1780–1914*, London: Allen and Unwin.
Odell, P. R. (1975) 'Oil', in C. Payer (ed.), *The Commodity Trade of the Third World*, London: Macmillan.
OECD (1969) *Development Assistance: 1969 review*, Paris.
OECD (1970) *Gaps in Technology: Comparisons between member countries*, Paris.
OECD (1977) 'The adjustment process since the oil crisis', *Economic Outlook*, 21, pp. 86–97.
OECD (1979) *Balance of Payments of OECD Countries 1960-1977*, Paris.
OECD (1980) *Technical Change and Economic Policy*, Paris.
OECD (1986) *OECD Economics Outlook*, 39, Paris.
OECD (1987) *National Policy and Agricultural Trade*, Paris.
OECD (1991) *Agricultural Policies, Markets and Trade: Monitoring and outlook*, Paris.
Offer, A. (1993) 'The British Empire 1870–1914: a waste of money?', *Economic History Review*, 46, 215–38.
Ohlin, B. (1931) *The Course and Phases of the World Depression*, Geneva: League of Nations.
Ohlin, B. (1933) *Inter-regional and International Trade*, Cambridge, Mass.: Harvard University Press.
Olson, M. (1974) 'The UK and the world market for wheat and other primary products 1885–1914', *Explorations in Economic History*, 11, pp. 352–6.
O'Rourke, K., and Williamson, J. G. (1992) 'Were Hecksher and Ohlin right? Putting history back into factor-price-equalization theorem', *Harvard Discussion Papers*, 1593.
O'Rourke, K., Taylor, A. M., and Williamson, J. G. (1993) 'Land, labour and the wage–rental ratio: factor price convergence in the late nineteenth century', *Harvard Economics Discussion Papers*, 1629.

Page, S. A. B. (1979) 'The revival of protectionism and its consequences for Europe', *Journal of Common Market Studies*, 20.
Paine, S. (1974) *Exporting Workers: The Turkish case*, Cambridge: Cambridge University Press.
Palmer, R. (1977) *Land and Racial Domination in Rhodesia*, Berkeley, Calif.: University of California Press.
Pamuk, S. (1986) 'The decline and resistance of Ottoman cotton textile 1820–1913', *Explorations in Economic History*, 23, pp. 205–25.
Papanek, G. (1973) 'Aid, foreign private investment, savings and growth in less developed countries', *Journal of Political Economy*, 81, pp. 120-30.
Parker Willis, H. (1968) *A History of the Latin Monetary Union: A Study in International Monetary Action*, New York: Greenwood Press.
Patterson, G. (1966) *Discrimination in International Trade: The policy issues 1945–1965*, Princeton, NJ: Princeton University Press.
Payer, C. (1974) *The Debt Trap: The IMF and the Third World*, New York and London: Monthly Review Press.
Pelaez, C. M. (1976) 'A comparison of long-term monetary behaviour and institutions in Brazil, Europe and the United States', *Journal of European Economic History*, 5, pp. 439–50.
Penrose, E. T. (1968) *The Large International Firm in Developing Countries: The international petroleum industry*, London: Allen and Unwin.
Phelps-Brown, E. H., and Ogza, S. A. (1955) 'Economic growth and the price level', *Economic Journal*, 65, pp. 1–18.
Piore, M. J. (1979) *Birds of Passage: Migrant labour and industrial societies*, Cambridge: Cambridge University Press.
Piramal, G., and Herdeck, M. (1986) *India's Industrialists*, vol. 1, Washington, DC: Three Continents Press.
Platt, D. C. M. (ed.) (1977) *Business Imperialism 1860–1930*, Oxford: Clarendon Press.
Platt, D. C. M. (1984) *Foreign Finance in Continental Europe and the USA 1815–1870*, London: Allen and Unwin.
Pollard, S. (1962) *The Development of the British Economy 1914–1950*, London: Edward Arnold.
Pontecorvo, G. (1986) *The New Order of the Oceans: The advent of a managed environment*, New York: Columbia University Press.
Posner, M. V. (1961) 'International trade and technical change', *Oxford Economic Papers*, 13, pp. 323–41.
Prebisch, R. (1962), 'The economic development of Latin America and its principal problems', *Economic Bulletin for Latin America*, 7, pp. 1-22.
Pressnell, L. S. (1968) 'Gold reserves, banking reserves and the Baring crisis of 1890', in C. R. Whittesley and J. S. G. Wilson (eds.), *Essays in Money and Banking*, Oxford: Oxford University Press.
Pressnell, L. S. (1978) '1925: the burden of sterling', *Economic History Review*, 31, pp. 67–88.
Pressnell, L. S. (1982) 'The sterling system and financial crises before 1914', in C. P. Kindleberger and J. P. Lafargue (eds.), *Financial Crises: Theory, history and policy*, Cambridge: Cambridge University Press.
Preston, S. H. (1980) 'Causes and consequences of mortality declines in less developed countries during the twentieth century', in R. A. Easterlin (ed.), *Population and*

Economic Change in Developing Countries, Chicago, Ill., and London: University of Chicago Press.
Price, H. B. (1955) *The Marshall Plan and its Meaning*, Ithaca, NY: Cornell University Press.
Prout, C. (1976) 'Finance for developing countries: an essay', in A. Shonfield (ed.), *International Economic Relations of the Western World 1959–71*, vol. 2, London: Oxford University Press.
Putnam, R. D., and Bayne, N. (1987) *Hanging Together: Cooperation and conflict in seven power summits*, revised edn, London: Sage.
Quigley, M. M. (1972) 'A model of Swedish emigration', *Quarterly Journal of Economics*, 86, pp. 111–26.
Radice, E. A. (1986) 'The German economic programme in Eastern Europe', in M. C. Kaser and E. A. Radice (eds.) *The Economic History of Eastern Europe 1919–1975*, vol. 2, Oxford: Clarendon Press.
Ranki, G. (1983) 'On the development of the Austro-Hungarian monarchy', in J. Komlos (ed.) *Economic Development in the Hapsburg Monarchy*, New York: Columbia University Press.
Ravallion, M. (1987) 'Trade and stabilisation: another look at British India's controversial foodgrain exports', *Explorations in Economic History*, 24, pp. 354–70.
Ravallion, M. (1989) *Markets and Famines*, Oxford: Clarendon Press.
Rawls, J. (1971) *A Theory of Justice*, Cambridge, Mass.: Belknap Press.
Raychaudhuri, T. (1969) 'A reinterpretation of nineteenth-century Indian economic history?', in M. D. Morris, T. Matsue, B. Chandra and T. Raychaudhuri (eds.), *The Indian Economy in the Nineteenth Century: A symposium*, Delhi: Delhi School of Economics.
Raychaudhuri, T. (1983) 'The mid-eighteenth century background', in D. Kumar (ed.), *The Cambridge Economic History of India*, vol. 2, Cambridge: Cambridge University Press.
Raychaudhuri, T., and Datta, K. (1958) *An Advanced History of India*, London: Macmillan.
Reader, W. J. (1970) *Imperial Chemical Industries: A history*, vol. 1, London: Oxford University Press.
Reddaway, W. B., Potter, S. J., and Taylor, C. T. (1967) *Effects of UK Direct Investment Overseas*, Cambridge: Cambridge University Press.
Redlich, F. (1967) 'Two nineteenth-century financiers and autobiographers: a comparative study in creative destructiveness and business failure', *Economy and History*, 10, pp. 38–66.
Redmond, J. (1980) 'An indicator of the effective exchange rate of the pound in the nineteen thirties', *Economic History Review*, 33, pp. 83–91.
Redmond, J. (1984) 'The sterling overvaluation in 1925: a multilateral approach', *Economic History Review*, 37, 530–2.
Remer, C. F. (1926) 'International trade between gold and silver countries: China 1885–1913', *Quarterly Journal of Economics*, 40, pp. 597–643.
Ricardo, D. (1817) *On the Principles of Political Economy and Taxation*, London: John Murray.
Richardson, H. W. (1972) 'British emigration and overseas investment 1870–1914', *Economic History Review*, 25, pp. 99–113.
Richardson, J. H. (1938) 'Tariffs, preferences and other forms of protection', in British Association for the Advancement of Science, *Britain in Recovery*, London: Pitman.
Rieben, H. (1971) 'Intra-European migration of labour and the migration of high-level

manpower from Europe to North America, in C. P. Kindleberger and A. Shonfield (eds.), *North American and Western European Economic Policies*, London: Macmillan.

Robbins, L. (1939) *The Economic Causes of War*, London: Jonathan Cape.

Robson, P. (1980) *The Economics of International Integration*, London: Allen and Unwin.

Rockoff, H. (1984) 'Some evidence on the real price of gold, its costs of production and commodity prices', in M. Bordo and A. Schartz (eds.), *A Retrospective on the Classical Gold Standard*, Chicago, Ill.: University of Chicago Press.

Rodney, W. (1972) *How Europe Underdeveloped Africa*, London: Bogle L'Ouverture.

Rodrigues, C. A. (1981) 'The non-equivalence of tariffs and quotas under retaliation', in J. N. Bhagwati (ed.), *International Trade: Selected readings*, Cambridge, Mass.: MIT Press.

Roemer, J. E. (1976) 'The effect of sphere of influence and economic distance on the commodity composition of trade in manufactures', *Review of Economics and Statistics*, 56, pp. 318–27.

Roemer, J. E. (1982) 'Exploitation, alternatives and socialism', *Economic Journal*, 92, pp. 87–107.

Rosenraad, C. (1900) 'The international money market: discussion', *Journal of the Royal Statistical Society*, 63, p. 33.

Rosovsky, H. (1968) 'Japan's transition to modern economic growth', in H. Rosovsky (ed.), *Industrialisation in Two Systems*, New York: Wiley.

Rostow, W. W. (1948) *The British Economy of the Nineteenth Century*, London: Oxford University Press.

Rostow, W. W. (1972) *The Stages of Economic Growth*, Cambridge: Cambridge University Press.

Rostow, W.W. (1978) *The World Economy: Theory, history, prospects*, London: Macmillan.

Rowe, J. W. F. (1965) *Primary Commodities in International Trade*, Cambridge: Cambridge University Press.

Rowthorne, R. E. (1971) *International Big Business 1957–1967: A study of comparative growth*, Cambridge: Cambridge University Press.

Rowthorne, R. E., and Solomou, S. (1991) 'The macroeconomic effects of overseas investment on the UK balance of trade 1870–1913', *Economic History Review*, 44, pp. 654–64.

Royal Institute of International Affairs (1931) *The International Gold Problem*, London: Oxford University Press.

Royal Institute of International Affairs (1953) *World Production of Raw Materials*, London.

Rubinstein, W. D. (1981) *Men of Property: The very wealthy in Britain since the Industrial Revolution*, London: Croom Helm.

Sampson, A. (1973) *The Sovereign State of ITT*, New York: Stein and Day.

Sampson, A. (1975) *The Seven Sisters: The great oil companies and the world they made*, New York: Viking.

Sampson, A. (1982) *The Money Lenders: Bankers and a world in turmoil*, New York: Viking.

Samuelson, P. A. (1949) 'International factor price equalisation once again', *Economic Journal*, 59, pp. 181–97.

Sandberg, L. G. (1969) 'American rings and English mules', *Quarterly Journal of Economics*, 83, pp. 25–43.

Sapsford, D. (1985) 'The statistical debate on the net barter terms of trade between commodities and manufactures', *Economic Journal*, 95, pp. 781–8.

Sato, K. (1966) 'On the adjustment time in neo-classical growth models', *Review of Economic Studies*, 33, pp. 263–8.

Saul, S. B. (1960) *Studies in British Overseas Trade 1870–1914*, Liverpool: Liverpool University Press.
Sayers, R. S. (1933) 'The question of the standard in the eighteen fifties', *Economic History*, 2, pp. 575–601.
Sayers, R. S. (1976) *The Bank of England 1891–1944*, Cambridge: Cambridge University Press.
Scammell, W. M. (1980) *The International Economy Since 1945*, London: Macmillan.
Schedvin, C. B. (1990) 'Staples and the regions of the Pax Britannica', *Economic History Review*, 43, pp. 233–559.
Schlesinger, Jr., A. M. (1965) *A Thousand Days: John F. Kennedy in the White House*, Boston: Houghton Mifflin.
Schlesinger, S., and Kinzer, S. (1982) *Bitter Fruit*, New York: Doubleday.
Schmid, G. C. (1974) 'The politics of currency stabilization: the French franc 1926', *Journal of European Economic History*, 3, pp. 359–77.
Schubert, A. (1991) *The Credit-Anstalt Crisis of 1931*, Cambridge: Cambridge University Press.
Schumpeter, J. (1939) *Business Cycles*, New York and London: McGraw-Hill.
Schwartz, A. (1984) 'Introduction', in M. Bordo and A. Schwartz (eds.), *A Retrospective on the Classical Gold Standard*, Chicago, Ill.: University of Chicago Press.
Scitovsky, T. (1976) *The Joyless Economy*, London: Oxford University Press.
Scott, A. (1971) 'Translantic and North American international migration', in C. P. Kindleberger and A. Shonfield (eds.), *North American and Western European Economic Policies*, London: Macmillan.
Seabourne, T. (1986) 'The summer of 1914', in F. Capie and G. E. Wood (eds.), *Financial Crises and the World Banking System*, London: Macmillan.
Servan-Schreiber, J.-J. (1968) *The American Challenge*, New York: Athenaeum.
Seyd, E. (1870) 'On international coinage and the variation of the foreign exchanges during recent years', *Journal of the Statistical Society*, 33, pp. 42–73.
Sheridan, R. B. (1976) 'Sweet malefactor: the social costs of slavery and sugar in Jamaica and Cuba 1804-54', *Economic History Review*, 29, pp. 236–57.
Shonfield, A. (ed.) (1976) *International Economic Relations of the Western World 1959-71*, London: Oxford University Press.
Sicsic, P. (1992) 'Was the franc Poincaré deliberately undervalued?', *Explorations in Economic History*, 29, pp. 69–92.
Siemens, G. von (1957) *History of the House of Siemens*, 2 vols., Freiberg/Munich: Alber.
Singer, H. W. (1950) 'The distribution of the gains between investing and borrowing countries', *American Economic Review, Papers and Proceedings*, 40, pp. 473–85.
Smith, G. W., and Schink, G. R. (1976) 'The international tin agreement: a reassessment', *Economic Journal*, 86, pp. 715–28.
Smith, M. A. M. (1977) 'Capital accumulation in the open two sector model', *Economic Journal*, 87, pp. 273–82.
Smith, R. L. (1979) 'Australian immigration 1945–1975', in P. J. Brain, R. L. Smith and G. P. Schuyers, *Population, Immigration and the Australian Economy*, London: Croom Helm.
Smith, R. S. (1979) *The Lagos Consulate 1851–1861*, Berkeley, Calif.: University of California Press.
Smithsonian Institute (1927) *World Weather Records*, Washington, DC.
Snowden, P. N. (1985) *Emerging Risk in International Banking: Origins of financial vulnerability in the 1980s*, London: Allen and Unwin.

Solberg, C. (1985) 'Land tenure and land settlement; policy and patterns in the Canadian prairies and the Argentine pampas, 1880–1930', in D. C. M. Platt and G. di Tella (eds.), *Argentine, Australia and Canada: Studies in comparative development 1870–1965*, London: Macmillan.

Solow, B. L., and Engerman, S. (eds.) (1987) *British Capitalism and Caribbean Slavery: The legacy of Eric Williams*, Cambridge: Cambridge University Press.

Spear, P. (1978) *The Oxford History of Modern India 1740–1975*, 2nd edn, Delhi: Oxford University Press.

Spraos, J. (1980) 'The statistical debate on the net barter terms of trade between primary commodities and manufactures', *Economic Journal*, 90, pp. 107–28.

Staley, M. (1935) *War and the Private Investor*, Chicago, Ill.: University of Chicago Press.

Stead, W. T. (1901) *The Americanisation of the World*, London: Reviews Annual.

Stein, L. (1977) 'Export instability and development: a review of some recent findings', *Banca Nazionale Del Lavoro Quarterly Review*, 30, pp. 277–90.

Stern, F. (1977) *Gold and Iron*, New York: Alfred Knopf.

Stern, R. M. (1960) 'A century of food exports', *Kyklos*, 13, pp. 44–64.

Strange, S. (1976) *International Monetary Relations*, vol. 2: *International Economic Relations of the Western World 1959–71* (edited by A. Shonfield), London: Oxford University Press.

Sturmey, S. G. (1962) *British Shipping and World Competition*, London: University of London/Athlone Press.

Sugiyama, S. (1988) *Japan's Industrialization in the World Economy 1859–1899: Export trade and overseas competition*, London: Athlone Press.

Suret-Canale, J. (1971) *French Colonialism in Tropical Africa 1900–1945*, London: C. Hurst.

Svedberg, P. (1978) 'The portfolio-direct composition of private foreign investment in 1914 revisited', *Economic Journal*, 88, pp. 763–77.

Svedberg, P. (1981) 'Colonial enforcement of foreign direct investment', *Manchester School*, 39, pp. 21–38.

Svennilson, I. (1954) *Growth and Stagnation in the European Economy*, Geneva: United Nations Economic Commission for Europe.

Swann, D. (1972) *The Economics of the Common Market*, 2nd edn, Harmondsworth: Penguin.

Swoboda, A. (ed.) (1969) *Monetary Problems of the International Economy*, Chicago, Ill.: Chicago University Press.

Taussig, F. W. (1931) *The Tariff History of the United States*, 8th edn, New York and London: Putnam.

Taylor, A. J. P. (1967) *Europe: Grandeur and decline*, Harmondsworth: Penguin.

Taylor, A. M., and Williamson, J. G. (1991) 'Capital flows to the New World as an intergenerational transfer', *Harvard Economics Discussion Paper*, no. 1579.

Taylor, P. (1971) *The Distant Magnet: European migration to the United States*, London: Eyre and Spottiswoode.

Temin, P. (1966) 'Labour scarcity and the problem of American industrial efficiency in the 1850s', *Journal of Economic History*, 26, pp. 361–79.

Temin, P. (1971) 'Labour scarcity in America', *Journal of Interdisciplinary History*, 1, pp. 251–61.

Temin, P. (1976a) *Did Monetary Factors Cause the Great Depression?*, New York: Norton.

Temin, P. (1976b) 'Lessons for the present from the Great Depression', *American Economic Review, Papers and Proceedings*, 66, pp. 40–5.

Temin, P. (1989) *Lessons from the Great Depression*, Cambridge, Mass.: MIT Press.

Tew, J. H. B. (1978) 'Policies directed towards the balance of payments', in F. T. Blackaby (ed.), *British Economic Policy 1960–74*, Cambridge: Cambridge University Press.
Thiel, E. (1990) 'Changing patterns of monetary independence', in W. Wallace (ed.), *The Dynamics of European Integration*, London: Pinter for RIIA.
Thomas, B. (1973) *Migration and Economic Growth: A study of Great Britain and the Atlantic economy*, 2nd edn, Cambridge: Cambridge University Press.
Thomas, L. B. (1973) 'Behaviour of flexible exchange rates: additional tests from the post-World War 1 episode', *Southern Economic Journal*, 40, pp. 167–82.
Thomas, M. (1992) 'Institutional rigidity in the British labour market', in S. N. Broadberry and N. F. R. Crafts (eds.), *Britain in the International Economy 1870–1939*, Cambridge: Cambridge University Press.
Tomaske, M. J. A. (1971) 'The determinants of inter-country differences in European migration 1881–1900', *Journal of Economic History*, 31, pp. 840–53.
Toniolo, G. (1990) *An Economic History of Liberal Italy 1850–1918*, London: Routledge.
Tooke, T., and Newmarch, W. (1928) *A History of Prices*, London: P. S. King.
Tracy, M. (1989) *Government and Agriculture in western Europe 1880–1988*, 3rd edn, Hemel Hempstead: Harvester Wheatsheaf.
Triffin, R. (1960) *Gold and the Dollar Crisis*, New Haven: Yale University Press.
Triffin, R. (1968) *Our International Monetary System*, New York: Random House.
Tugendhat, C., and Hamilton, A. (1975) *Oil: The biggest business*, London: Eyre Methuen.
Twomey, M. J. (1983) 'Employment in nineteenth century Indian textiles', *Explorations in Economic History*, 20, pp. 37-57.
United Nations (1948a) *Postwar Shortages of Food and Coal*, New York: United Nations Department of Economic Affairs.
United Nations (1948b) *The Foreign Exchange Position of the Devastated Countries*, New York: United Nations Department of Economic Affairs.
United Nations (1949a) *Economic Survey of Asia and the Far East*, New York: United Nations Department of Economic Affairs.
United Nations (1949b) *Economic Survey of Europe*, New York: United Nations Department of Economic Affairs.
United Nations (1949c) *Economic Survey of Latin America*, New York: United Nations Department of Economic Affairs.
United Nations (1953) *Commodity Trade and Economic Development*, New York: United Nations Department of Economic Affairs.
United Nations (1974) *The World Population Situation in 1970–75*, New York: United Nations Department of Economic and Social Affairs.
United Nations (1980) *World Industry in 1980*, New York: United Nations Industrial Development Organisation.
United Nations (1993) *Economic Survey of Europe in 1992–3*, New York: United Nations Economic Commission for Europe.
Urquhart, M. C., and Buckley, K. A. H. (1965) *Historical Statistics of Canada*, Toronto: Macmillan.
Vanek, J. (1963) *The Natural Resource Content of United States Foreign Trade 1870–1955*, Cambridge, Mass.: MIT Press.
Vanous, J., and Marrese, M. (1982) 'Soviet subsidies to eastern economies', *Wall Street Journal*, 15 January.
Vernon, R. (1966) 'International investment and international trade in the product cycle', *Quarterly Journal of Economics*, 80, pp. 190–207.

Vernon, R. (1976) 'An interpretation', in R. Vernon (ed.) *The Oil Crisis*, New York: W. W. Norton.

Visaria, L., and Visaria, P. (1983) 'Population 1757–1947', in D. Kumar (ed.), *The Cambridge Economic History of India*, vol. 2, Cambridge: Cambridge University Press.

Vives, J. C. (1969) *An Economic History of Spain*, Princeton, NJ: Princeton University Press.

Vogel, E. F. (1991) *The Four Little Dragons: The spread of industrialization in East Asia*, Cambridge, Mass.: Harvard University Press.

Wakeman, F. (1980) 'The Canton trade and the Opium War', in *The Cambridge Economic History of China*, vol. 10, part 2, Cambridge: Cambridge University Press.

Wallerstein, I. (1979) *The Capitalist World Economy*, Cambridge: Cambridge University Press.

Wang, J. S. L. (1993) 'The profitability of Anglo-Chinese trade 1861–1913', *Business History*, 35, pp. 39–65.

Watkins, M. H. (1963) 'A staple theory of economic growth', *Canadian Journal of Economics and Politics*, 24, pp. 41–58.

Weiller, J. S. (1971) 'Long run tendencies in foreign trade', *Journal of Economic History*, 21, pp. 804–21.

Wells, L. T. (ed.) (1972) *The Product Life Cycle and International Trade*, Boston, Mass.: Harvard Business School.

Welsh, F. (1993) *A History of Hong-Kong*, London: Harper Collins.

Wexler, I. (1983) *The Marshall Plan Revisited: The European recovery programme in economic perspective*, Westport, Conn.: Greenwood Press.

Wheelock, D. C. (1991) *The Strategy and Consistency of Federal Reserve Monetary Policy 1924–1933*, Cambridge: Cambridge University Press.

Whitaker, J. K. and Hudgins, Jr, M. W. (1977) 'The floating pound sterling of the 1930s', *Southern Economic Journal*, 43, pp. 1478–85.

Wijkman, P. M. (1990) 'Patterns of production and trade', in W. Wallace (ed.), *The Dynamics of European Integration*, London: Pinter for RIIA.

Wilczynski, J. (1969) *The Economics and Politics of East–West Trade*, New York: Praeger.

Wilczynski, J. (1974) *Technology in Comecon*, London: Macmillan.

Wilkins, M. (1970) *The Emergence of Multinational Enterprise: American business abroad from the colonial era to 1914*, Cambridge, Mass.: Harvard University Press.

Wilkins, M. (1989) *A History of Foreign Investment in the United States to 1914*, Cambridge, Mass.: Harvard University Press.

Wilkinson, M. (1971) 'European migration to the United States', *Review of Economics and Statistics*, 52, pp. 272–9.

Willett (1979) 'Major challenges to the international economic system', in R. C. Amacher, G. Haberler and T. D. Willett (eds.), *Challenges to a Liberal International Economic Order*, Washington D.C.: American Enterprise Institute.

Williams, E. (1964) *Capitalism and Slavery*, London: Deutsch.

Williams, E. E. (1896) *Made in Germany*, London: Heinemann.

Williamson, J. (1973) 'International liquidity: a survey', *Economic Journal*, 83, pp. 685–746.

Williamson, J. (1977) *The Failure of World Monetary Reform*, Sunbury-on-Thames, Middlesex: Nelson.

Williamson, J. G. (1974) *Late Nineteenth Century American Development*, Cambridge: Cambridge University Press.

Williamson, J. G. (1980) 'Greasing the wheels of sputtering export engines: midwestern grains and American growth', *Explorations in Economic History*, 18, pp. 109–217.

Williamson, J. G., and Lindert, P. H. (1980) 'Long term trends in American wealth inequality', in J. D. Smith (ed.) *Modelling the Distribution and Intergenerational Transfer of Wealth*, Chicago, Ill.: and London: University of Chicago Press.

Wilson, D. (1992) *Rothschild: A story of wealth and power*, London: Mandarin.

Wimmer, L. T. (1975) 'The gold crises of 1869: stabilising or destablishing speculation under floating exchange rates?', *Explorations in Economic History*, 12, pp. 105–22.

Winston, A. P. (1927) 'Does trade follow the dollar?', *American Economic Review*, pp. 458–77.

Winters, L. A. and Venables, A. (eds.) (1991) *European Integration: Trade and industry*, Cambridge: Cambridge University Press.

Wolf, E. R. (1982) *Europe and the People Without History*, Berkeley, Calif., and London: University of California Press.

Wolfe, M. (1951) *The French Franc between the Wars 1919–1939*, New York: Columbia University Press.

Wong, J. Y. (1974) 'The Arrow Incident: a re-appraisal', *Modern Asian Studies*, 8, pp. 373–89.

Woytinsky, W. S., and Woytinsky, E. S. (1953) *World Population and Production: Trends and outlook*, New York: Twentieth Century Fund.

Wright, G. (1990) 'The origins of American industrial success 1879–1940', *American Economic Review*, 80, pp. 651–68.

Wright, J. F. (1981) 'Britain's inter-war experience', *Oxford Economic Papers*, Supplement, pp. 282–305.

Yamazawa, I. (1990) *Economic Development and International Trade: The Japanese Model*, Honolulu: Resource Systems Institute.

Yates, P. L. (1959) *Forty Years of Foreign Trade*, London: Allen and Unwin.

Yeager, L. B. (1969) 'Fluctuating exchange rates in the nineteenth century: the experiences of Austria and Russia', in R. A. Mundell and A. Swoboda (eds.), *Monetary Problems of the International Economy*, Chicago, Ill.: Chicago University Press.

Yeager, L. B. (1976) *International Monetary Relations: Theory, history and policy*, New York: Harper and Row.

Yergin, D. (1991) *The Prize: The epic quest for oil, money and power*, New York: Simon and Schuster.

Yotopoulos, P. A., and Nugent, J. B. (1976) *Economics of Development: Empirical investigations*, New York: Harper and Row.

Zauberman, A. (1955) *Economic Imperialism: The lessons of Eastern Europe*, London: Ampersand.

Zauberman, A. (1964) *Industrial Progress in Poland, Czechoslovakia and East Germany 1937–1962*, London: Oxford University Press.

Zimmerman, L. J. (1962) 'The distribution of world income 1860–1960', in E. de Vries (ed.), *Essays on Unbalanced Growth*, S. Gravenage: Mouton & Co.

Index

Acceptance credit 66
Achnacarry Agreement 192
Africa,
  colonisation 108, 109 Fig
  debt rescheduling 349
  devaluation of currencies 1980s 348
  economic situation in 1970s and 1980s 324, 332, 333
  motor vehicle registrations 1920s 183
  slave trade 7
Agriculture industry xv
  modernisation 187
  price falls between 1929 and 1931 199
  production fall in 1947/8 243
  protection 187, 271–2
Aid, official *see* Official capital (international) flows
Anglo-French commercial treaty 45
Anglo-Iranian Oil Co. (BP Oil) 313
Argentina,
  average growth rates of GNP per capita 1913–50 183, 184 Table
  beef exports 1875–1914 99
  devaluation 332
  foreign debt 166–7, 217, 318
  gold standard 1885–1900 165
  immigration policy 149
  loan 349
  national income 19th century 100
  tariffs late 19th century 115
Armenia 339
ASEAN 329
Asia, economic situation in 1970s 333–5
Atlantic Ocean, passenger transport time taken in late 19th century 144
Australia,
  average growth rates of GNP per capita 1913–50 183, 184 Table
  beef exports 1875–1914 99
  convict transportation 57
  gold mining 75, 99
  immigration 55, 120, 141, 149, 283, 285
  imports from UK in 19th century 23

  international flow as percentage of GNP 1870–1910 120, 121 Fig
  tariffs 47, 114
  trade agreements with New Zealand 324
Austro-German Union 79, 87
Austro-Hungarian Empire, division 178
  economic development 53–4
  emigration 141
  exchange rate policy 77, 78
Austro-Prussian War 1866 84

Balance of payments,
  definition 20–1
  effect of devaluation on 300, 301 Table
  mid 19th century,
    India 26, 27 Table, 28, 29 Table
    measurement 19–21
    UK 21–4
    USA 24–6
  under Bretton Woods 299–305
Bank of England 88, 169, 170–2, 221
Bank of France 169
Bank for International Settlements xvii, 221, 223, 248
Banks,
  international debts 318–19
  rescheduling 319, 320 Fig, 321–2
Baring Brothers 66, 166–7
Basle Agreement 297
Beef 98 Fig, 99
Belgian Congo (Zaire) 193, 239
Belgium 32, 33 Table, 40, 42, 53, 80, 177, 227, 230, 231
Bilateral exchange, definition xiv
Bills of Exchange 66–7, 87
Bimetallism 80, 158–9
Birnberg and Resnick's model of colonial development 188, 189 Fig, 190
Boer War 100
Bolivia,
  debt-equity swaps 349
  devaluation 332
Bond arbitrage 72

Index

Brazil 252
  average growth rates of GNP per capita 1913–50 183, 194 Table
  coffee industry 1923 192, 193
  exchange rate policy 19th century 77
  immigration policy 149
  import duties 332
  protection of manufacturing 276–7
Bretton Woods Agreement 131, 240, 299–305, 321
British East India Company 4, 5
British Empire,
  colonies tariff policy 47
  see also UK, colonial activity
British Railmakers' Association 117
Brown Boveri 137
Brussels convention 1902 xvii

Canada,
  average growth rates of GNP per capita 1913–50 183, 184 Table
  economic development late 19th century 100
  emigration 1890s 141
  floating exchange rates 304
  Free Trade Agreement with USA 47, 324
  immigration 120, 149, 286
  international flow as percentage of GNP 1870–1910 120, 121 Fig
  motor industry 331
  tariffs late 19th century 114
  wheat pool 1928 193–4
Canadian-United States Automotive Agreement 269
Capital movements,
  1875–1914 120–39
    capital exporters 123–6
    determinants of demand 122
    foreign direct investment 136–8
    gains from foreign investment 133–6
    political impact of foreign investment 126–33
  and debt 1980s 348–50
  mid to late 19th century 58–9
Cartels, late 19th century 116–18
Central bank policies in 1920s 219–21
CFCs abolition 352
Chile 11, 55, 62, 252, 349
China 16
  agriculture reforms 1980s 334–5
  Boxer Uprising 129
  direct investment by Hong Kong 334
  foreign investment and Chinese trade 1882–1916 128, 129 Fig
  foreign trade 19th century 5, 51
  opium imports 5–6, 8
  railway loans and exports 135, 136 Table
  silver standard 227
  Special Economic Zones 335

Taiping Rebellion (1851–64) 10
tariff policy 19th century 47
Treaty of Tientsin 129
Ching, T.K. 51
Churchill, Winston 226, 236–7
City of Glasgow Bank, failure 166
Clothing trade expansion 1980s 325
CMEA 281
Coal industry 34, 42, 92, 93 Table, 191–2, 243, 273
Cocom 338
Code of Trade Liberalisation 248
Coffee 192, 193
Cold War 248–50, 355
Collective bargaining 218
Colonial development, Birnberg and Resnick's model 188, 189 Fig, 190
Colonial Development Act (UK) 1929 188
Colonial trade, compared with total trade 1892–6 110 Table, 111
Comecon 338–9
Commercial policy 19th century 43–8
Commodity control schemes interwar years 192–4
Commodity price equalisation 52–3
Common Agricultural Policy 290, 325, 342, 344–6
Commonwealth of Independent States (CIS) 339
  see also USSR
Communist economies 291
Communist-system trade policy 281–2
Comparative advantage
  definition 36–7
  Hecksher–Ohlin theory 40–3, 265, 266, 330–2
  manufactures 1875–1914 91–7
  mid 19th century 36–48
  UK, France and Germany in 1899 and 1913 97 Table
Comptoi-d'Escompte rescue 168
Convergence of world economy 330–2
Convict transportation 56, 57
Copper 193
Corn Laws, repealed 22, 52, 115
Cotton industry xiii, xv, 9–10, 16, 21–2, 39, 273
Council for Mutual Economic Assistance (CMEA) 250
Crédit Mobilier 59, 84
Credit-Anstalt Bank, failure 216, 223
Cuba, restriction of sugar output 193
Czechoslovakia 229–30, 249, 282

Danatbank, failure 223
Debt,
  national default 139
  rescheduling 349
  see also individual countries
Debt-equity swaps 349

Index

Depression 1873,
    cartels 116–18
    tariffs 113–16
Depression (1929–32) 176, 233
    world-wide nature 222–8
Depression (1974–5) 307–11
Direct foreign investment 312–13
    transferred technology 314–15
    see also Foreign investment
Dollar diplomacy 128
Du Pont 117

EC 324–5
    Common Agricultural Policy 271, 342, 344–6
    completion of internal market 340, 341 Fig
    Delors Report 342–3
    food stocks 346
    formation 268, 269
    Maastricht Treaty 343, 356
    monetary policy,
        ECU 342
        Exchange Rate Mechanism (ERM) 342
        monetary union 340, 342, 343
        snake 342
Economic activity 1850–75, international fluctuations 81–6
Economic growth 1875–1914 90–119
    commodity terms of trade 106–8
    comparative advantage in manufactures 91–7
    temperate zone primary product exporters 97–101
    trade and colonisation 108–13
    tropical trade and LDCs 102–6
Economic growth 1960–70, direct contribution to exports 287, 288 Fig
Economic growth, mid to late 19th century 61–3
Economic performance 1913–50 182–5
Economic policy co-ordination 1970s and 1980s 350–1, 354
Economic policy collaboration 350
Economic policy consultation 350
Edison General Electric 138
EEC *see* EC
Egypt 122, 127, 130, 131–2
Empire Cotton Growing Corporation 188
Eurocurrency system 312
Europe,
    deficit with USA 1940s 244
    economic integration 340–6
    see also EC
    immigration 1950s onwards 283
    trade with China in 19th century 5
Europe (Eastern),
    1980s 339
    reconstruction after Second World War 248–50
Europe (West) reconstruction after Second World War 245–8

European Coal and Steel Community 245, 256
European Free Trade Area 268, 269
European Payments Union (EPU) 247–8
Exchange Rate Mechanism (ERM) xviii, 342, 353, 355
Exchange rates,
    1845–59 76, 77 Table
    1850–75 71–2
    effect of price inflation 307
    fixed rate regime 1970s 305–7
    floating,
        after First World War 209–12
        managed exchange rates 1930s 228–32
    regimes 1850–75 77–9
    swings 1979 onwards 346–8
Exploitation 15
Exports 18, 135–6
    1899–1913 95 Table, 96
    1913–37 185 Table
Extractive industries xvi

Factor mobility, international (1850–75) 50–64
    capital movements 58–9
    economic growth and economic relations 61–3
    Far East 51
    foreign merchants 50–2
    income and welfare 60–1
    trade, factor price equalisation and income gaps between countries 52–3
    worker migration 53, 54–8
Factor price equalisation 52–3
Fair Trading League 1881 114
Fee simple xiv
Finance bills 67
Finland 213
First World War,
    death toll 177
    European economic relations 176–80
Ford 197–8
Foreign investment,
    1875–1914 136–8
        defaulting countries 126–7, 129–31
        return on state bonds issued 133, 134 Table
    mid to late 19th century 58–9
    private capital mobility 1960s and 1970s 311–15
    tying exports to 135–6
    see also Capital movements; Direct foreign investment
France,
    abolition of slavery 31
    average growth rates of GNP per capita 1913–50 183, 184 Table
    bimetallic standard 23, 75, 76, 80, 81
    central bank policies 1920s 220–1
    colonial activity and expenditure 108, 112, 188

colonial trade   111, 201
economic position after First World War   211–12
emigration   141
Exchange Rate Mechanism (ERM)   342
foreign investment   58–9, 124–5
free trade and tariffs 19th century   45, 113–14
gold reserves   222–3
immigration   141, 149, 285
international flow as percentage of GNP 1870–1910   121 Fig
manufacturing trade 1913–37   195
Meline Tariff 1892   199
military expenditure 1905–9   165
price levels 1875–1914   156 Table
receipt of Marshall Aid   246, 247 Table
trade with UK in 19th century   42–3
trade with USA in 19th century   42
Free trade and liberalisation 1850–75   7–8, 31–49
  absolute advantage   36
  commercial policy   43–8
  comparative advantage   36–48
  integration of markets   35–6
  reduction in transport costs   313–15
Fuel,
  trade 1913–37   191–2
  see also Coal industry; Oil industry

GATT,
  Dillon Round   268, 269
  Kennedy Round   268–9, 290
  origins   242
  principles   242
  Tokyo Round   269, 326, 346
  Uruguay Round   328
General Agreement on Tariffs and Trade see GATT
General Arrangements to Borrow   297, 298
General Motors   197–8
Germany,
  army and navy expenditure 1909   165
  average growth rates of GNP per capita 1913–50   183, 184 Table
  Bruening's deflationary policy   223
  chemical industry   92
  colonial activity and expenditure   108, 111, 112
  economic position after First World War and hyperinflation   210
  establishment of trading agreements with central Europe and South America in 1930s   202–3, 205–6
  Exchange Rate Mechanism (ERM)   342, 343
  free trade   39, 45–6, 79
  immigration 1875–1914   141, 143
  international flow as percentage of GNP 1870–1910   120, 121 Fig
  introduction of old-age pensions and unemployment insurance in 1888   165
  iron and steel industry   91–2
  manufacturing trade 1913–37   194, 196
  motor industry   331
  price levels 1875–1914   156 Table
  Reparations Commission   216
  revaluation of mark in 1961   297
  tariffs late 19th century   113
  unification (1990)   xiii–xiv, 343
  Versailles Peace Settlement   178
  Zollverein (customs union)   39, 45, 46, 79
Germany (West),
  immigration   285, 330
  receipt of Marshall Aid   246, 247 Table
Gibbs, H.   15
GNP, average growth rates per capita 1913–50   183, 194 Table
Gold   30
  ratio of price to silver   73 Table, 158, 159 Fig, 173
  transactions in mid 19th century   21
Gold mining, Australia   99
Gold Rush (1849)   76
Gold standard   154–74
  benefits   165, 172–3
  disintegration   208–34
  fluctuations in economic activity   166–9
  monetary policy   169–73, 219–21, 233
  money and international price level   155–9
  public finances   164–6
  reasons for failure after First World War,
    impact of war on monetary system   215
    international debt redistribution   215–17
    monetary policy   219–21, 233
    price levels   217–19
    return of fixed exchange rates in 1920s   212–14
  workings   161–4
    reserve currencies   162–4
  see also International reserves
Gold supplies 1850–75   72–5
Grain prices,
  1850's   xiii
  see also Wheat
Great Exhibition 1851   31
Greece   130–1, 132–3, 283–4
Guano trade in Peru   13–15, 17, 62

Havana Charter   242
Hecksher-Ohlin theory   265, 266, 330–2
  definition   40–3
Hoechst   138
Holland   112, 285
Hong Kong   4, 6, 51, 334
Hungary   282, 339

I G Farben, use of *slave labour* during Second World War   238
IATA formation   xvii

Import-substituting industrialisation (ISI) 275, 276
Income gaps between countries, mid to late 19th century 52–3
Income and welfare, 19th century 60–1
Indenture system 55–6, 143
India 13, 255–6
 agriculture production 277
 average growth rates of GNP per capita 1913–50 183, 194 Table
 balance of payments 1858/9 26, 27 Table, 28, 29 Table
 division into India and Pakistan 238
 economy during Second World War 237–8
 emigration 1871–1915 143
 grain exports 1875–1914 102–3
 introduction of British rule 4–5
 mutiny and revolt (1857) 5
 opium exports to China 5–6, 8
 payment of Home Charges 160
 per capita income 1870 61
 price levels 1875–1914 156 Table
 railway system in 19th century 33
 silver imports 1855–66 76
 silver prices 1890s 161
 tariff policy 19th century 48
 trade with UK 9–10, 16, 104
  see also UK, economic links with India and USA in mid 19th century
Indonesia, foreign debt 318
Inflation, after First World War 209–12
Influenza epidemic 1918 177
Integration of markets 19th century 35–6
Interest Equalisation Tax 312
International Bank of Reconstruction and Development (IBRD) 240, 241
 see also World Bank
International debt 1960–80 318–19
 rescheduling 319, 320 Fig, 321–2
International Energy Agency (IEA) 336
International Monetary Fund (IMF) 240–1, 254, 303, 349
International multiplier 224
International Office of Telegraphy, establishment xvii
International policy co-ordination and co-operation 221, 227, 350–1
International Rail Syndicate 117
International reserves 1950s onwards, composition 295 Table, 296–7
 demand 294–5
 growth 294–9
 measures to increase 297–9
International Tin Agreement 275
International trade 1914–44 180–2
Ireland 54
Iron and steel industry 91–2, 327
Italy,
 average growth rates of GNP per capita 1913–50 183, 184 Table
 economic development in mid 19th century 54
 Exchange Rate Mechanism (ERM) 342
 international flow as percentage of GNP 1870–1910 120, 121 Fig
 migration 141, 283
 military expenditure 1905–9 165
 monetary policy 1894–1913 164
 receipt of Marshall Aid 246, 247 Table
 silver coins mid 19th century 80
 use of quotas as reprisals 199
ITT, Chile 313

Jamaica 102
Japan 16
 Action Programme 327
 average growth rates of GNP per capita 1913–50 183, 184 Table
 collapse of silk exports in 1930s 191
 competitiveness 196–7, 266
 cost of 1905–6 Russian war 130
 economic plight 1940s 245
 economic position after First World War 209
 emigration policy 19th century 56
 expansion in Far East 237
 exports 204, 325–6
 manufacturing trade 1913–37 194, 196
 Meiji Restoration (1868) 6
 Ministry for International Trade and Industry (MITI) 270
 motor industry 331
 privatisation of NTT 348
 tariff policies 47, 327
 trade policy 1960 269–70
 utilisation of foreign R & D 266
 western trade in 19th century 6
 yen–dollar exchange rate 1970–92 326

Keynes, J.M. 106, 214, 216, 239
Kikkoman 138
Knickerbocker Trust 167
Knoop, J. 51
Korean war 249, 252, 256

Labour mobility (international), 1950s onwards 283–6
 see also Migration
Labour productivity and trade in 1860, USA 37, 38 Table
Latin America,
 debt rescheduling 349
 economic situation in 1970s and 1980s 332
 immigration policy 149
 tariff policy 19th century 47
Latin Monetary Union 79–80, 87

LDCs, rising manufacturing exports 1960s and 1970s   265–6
Lend-lease   243, 256
Lenin, V.I.   111, 176–7
Liberalism, definition   7–9
Life expectancy in 19th century   60
List, F.   9
LM Ericsson   137

Manufacturing industries   xvi
  1913–37   194–8
  decline in world trade   205
  export volumes 1937–50   253 Table
  rates of protection   275 Table, 276
Market structure and gains from trade in 19th century   13–15
Marshall Plan   244, 245–8, 256, 350
Mexico,
  average growth rates of GNP per capita 1913–50   183, 184 Table
  debt default   348–9, 353
  foreign direct investment   328–9
  railway system in 19th century   33
Michelin   138
Middle East   13
Migration,
  1875–1914   140–53
    causes   143–6
    pattern and measurement   140–3
    transport costs   144
  1960s and 1970s   284 Table, 291
  1980s   329–30
  effects on indigenous population   285–6
  fall in 1920s and 1930s   182
  gains and losses from migration   147–8
  intercontinental rates 1851–80   55 Table
  mid to late 19th century   53, 54–8
  political economy   148–52
  relevance of Rybczynski's theorem   58
  see also Labour mobility (international)
Mill, J.S., *Essay on Liberty*   8
Mint par exchange rate   71
Monetary unions   79–81, 87
  see also EC
Most favoured nation clause   44, 199, 242
Motor industry 1970s, comparison of efficiencies   331
Motor industry, interwar years   197–8
Motor vehicle registrations   183
Multi-Fibre Agreement (MFA)   270, 325
Multilateral trade liberalisation   328
  see also Tariffs and quotas
Multinational companies   137, 197–8, 247, 260–1
  see also Direct foreign investment

NAFTA   xviii, 328–9, 353, 356
NATO, formation   249

Navigation Act   115
Nestlé   137
Netherlands   45, 137, 246, 247 Table
New Russian Company   59
New Zealand   100, 141, 272, 283, 324
Nobel Dynamite Trust Company   116–17

OEEC   247
Official capital (international) flows (official aid)   315–18, 322
Oil industry   252
  1913–37   191–2
  1974 crisis   335
  control by OPEC   335–8
  Iranian shutdown 1978   336
  Iraq attack on Iran 1980   336
  Iraq invasion of Kuwait   337–8, 354
  price increases 1970s   262, 274, 310, 321
  prices and world price level 1970–92   337 Fig
  West African Supply Agreement   274
OPEC   xiii, 274, 323, 335–8
Opium   5–6, 8, 30
Orenstein-Arthur Koppel   138
Ottawa Conference 1932   201, 205
Ozone layer depletion   352

Passport requirements in Europe late 19th century   115
Peru   13–15, 17, 32, 62
Philips   137
Poland   282, 319
Population,
  growth   258, 325 Table
  world   2 Fig, 356–7
Portugal   45, 111
Precious metals,
  transactions in mid 19th century   21
  see also Gold; Silver
Prices,
  balance of supply and demand   61
  explosion 1970s   280–1
  inflation and exchange rates   307
  levels 1875–1914 for various countries   156 Table
Primary commodities   186 Table, 187–92, 205, 270–4
Property rights   xiv
Protectionism *see* Tariffs
Purchasing power parity calculations   36, 212–13, 228, 229 Table

Quotas   199

R&D (research and development)   259–60, 266, 340
Railways   32, 33 Table, 34, 85–6, 117
Reichsbank   169, 171
Reserve currencies   162–4

Resources, international policy 351–3
Ricardo, D. 36
Romania 282
Roosevelt, Theodore 194, 226, 236–7
Rothschilds 59, 65–6, 124, 167
Rubber, output restriction scheme 193
Russia,
  abolishment of serfdom 59
  budgetary problems after First World War 209
  collapse of empire in 1917 177
  exchange rate policy mid 19th century 77
  foreign investment in 124
  gold standard 172
  immigration 1875–1914 141
  inflation from 1919 212
  lend-lease arrangements with USA 236
  military expenditure 1905–9 165
  railways mid-to-late 19th century 86
  tariffs late 19th century 114
  see also USSR
Russia/USSR, average growth rates of GNP per capita 1913–50 183, 194 Table
Rybczynski's theorem, relevance to worker migration 58

San Remo Conference 1920 192
Scandinavia, free trade 45
Scandinavian Monetary Union 87
Schacht, Hjalmar 181, 221
Scotland, worker migration 1860's 54
Sea transport 34–5
Second World War,
  economic consequences 238–9
  economic relations 236–9
    lend-lease arrangements 236–7
  post-war trading system 239–40
Service industries xvi
Settlements, world pattern in 1910 91, 93 Fig
Shell 137, 192
Siam *see* Thailand
Siemens 51, 68
Silver 30, 80
  Indian imports 1855–66 76
  ratio of price to gold 73 Table, 158, 159 Fig, 173
Silver standard economies 1875–1914 158, 160–1
Singer–Prebisch thesis 106–7
Slavery xiv, 12–13, 16
South Africa 114–15, 141
Soviet Union *see* USSR
Spain,
  colonial trade late 19th century 111
  emigration 1890s 141
  Exchange Rate Mechanism (ERM) 342
  floating exchange rate 1929–32 227
  foreign residents 330

  gold standard 1876–1883 172
  money supply 228
Special drawing rights (SDR) 297, 298, 305
Specialisation xv
STABEX 318
Standard Oil of New Jersey 192
Staple Theory 63
Steel, non-tariff barriers 1970s 327
Sterling Area 254, 256
Stock market,
  boom 220
  crash 1974 310
  crash 1987 347
Stolper–Samuelson theorem 43
Suez Canal 34, 85
Sugar xvii, 22
Sweden 137, 172
Switzerland 45, 137

Tariffs,
  1920s and 1930s 198–201
  19th century 43–8, 113–16
  cuts agreed Tokyo Round 326
  effective rate formula for protection 292
  potential as instruments of foreign policy 43
  reductions 1959–71 269
  Stolper–Samuelson theorem 43
Taut planning 338
Technology,
  generation and international transfer 259–65
    trade patterns 262–5
  imports of technology and economic performance 261–2
  international flow 289
  new in 1920s 183
Telegraph communication 87
  international development 67–8, 69–70 Figs
Temperate zone primary product exporters, international trade and economic growth 1875–1914 97–101
Terms of trade, between primary and manufactured products 1970s 280–1
Thailand 7, 106, 188, 334, 346
Tin, output control scheme 194
Trade,
  1875–1914, European domination 90–119
  based on voluntary exchange 7–9
  and colonisation 1875–1914 108–13
  and growth rate 61–2
  patterns 251–5, 278–80
  policy 1960s and 1970s,
    industrial countries 268–74
    non-industrial countries 274–7
  summary of regional international economic experience during 1980s 324, 325 Table
  use of sanctions as political weapon 203
  world,
    1850 2 Fig

Index

1875–1913   103, 104 Table
1913   104, 105 Fig
1937–55   252, 253 Table
1963–78   263 Table
1979   263, 264 Table
Transferred technology, direct foreign investment   314–15
Transport costs and trade (1850–75)   31–5
Treaty of Rome 1957   268
Tripartite Agreement 1936   232, 233
Tropical trade and LDCs, 1875–1914   102–6
Turkey, default on servicing foreign borrowing   127, 130–1
Turkish Empire,
  division in 1918   178
  migrants   286

Uganda, devaluation of currency 1980s   348
UNCTAD   274–5
Undersea mining   352–3
Unemployment,
  1976   311
  1992   354
  comparison before and after 1914   218
Unilever   137, 197
United Fruit Company, Guatemala   313
United Kingdom,
  aid, main recipients 1974   317 Table
  average growth rates of GNP per capita 1913–50   183, 194 Table
  balance of payments in mid 19th century   21–4, 22 Table, 22, 23, 24
  central bank policies see Bank of England
  coal industry in 19th century   34
  colonial activity late 19th century   108, 110, 111
  competitiveness in international manufacturing markets   252, 267
  corn and sugar imports in mid 19th century   22
  cotton industry   xv, 9–10, 16, 21–2, 39
  de Gaulle's veto of EC membership   268
  decline in monetary growth in 1870s   74–5
  the Depression   224
  devaluation of sterling   195, 300–1
  direct foreign investment   137
  domination of international economy in 19th century   3–5
  economic links with India and USA in mid 19th century   19–30
  economic position after First World War   209
  Exchange Rate Mechanism (ERM)   342
  exports to Australia in mid 19th century   23
  exports to Australia, New Zealand and South Africa 1913–37   196
  exports to India 1913–37   196
  exports to Japan 1913–37   195–6
  financial crisis 1914   168
  floating exchange rates in 1930s   230–2
    Exchange Equalisation Account   230–1
  foreign investment,
    19th century   58, 59
    income in 1913   125–6
  free trade   43–4
  gold standard   208, 212–14, 224
  growth of money, national income and prices 1846–80   74 Table
  growth in money supply after 1896   157
  Import Duties Act 1932   195, 200
  international flow as percentage of GNP 1870–1910   121 Fig
  introduction of old-age pensions and unemployment insurance in 1888   165
  iron and steel industry   42, 92
  lend-lease arrangements with USA   236, 237
  manufacturing competitiveness and economic growth 1899–1913   96–7
  manufacturing trade 1913–37   195, 196
  motor vehicle registrations 1919–28   183
  Mutual Aid Agreement with USA   236–7
  price levels 1875–1914   156 Table
  railway system in 19th century   33 Table, 34
  ratio of R & D to GDP in 1960s   259
  receipt of Marshall Aid   246, 247 Table
  Safeguarding of Industry Act 1921   182
  sugar-beet industry   187
  tariff protection   200, 201
  tariffs mid to late 19th century   43–5
  trade with empire in 1920s   201–2
  trade with France 19th century   42–3
  trade with India   9–10, 16, 104
  unemployment in 1920s   214
  urbanisation 19th century   60
  wheat   101
Universal Postal Union   xvii
UNRRA   243
US-Canada Free Trade Agreement   328
US-Canadian Automotive Trade Agreement (1965)   278
USA,
  Agricultural Adjustment Act   271
  average growth rates of GNP per capita 1913–50   183, 184 Table
  balance of payments in 1855   24, 25 Table, 26
  bimetallism   75–6
  Bland Act 1878   158
  central bank   170, 220, 221
  depressions,
    1893   167
    comparison of 1929 and 1974   309–10
    The Depression   225–6
  direct foreign investment in   138
  direct investment in Canada   137
  dollar exchange rate 1979–80s   346–8
  EC trade policy   354

economic links with UK, *see* UK, economic links with India and USA in mid 19th century
economy during Second World War 236
European deficit 1940s 244
fall in imports 1929–33 224
floating exchange rates mid 19th century 77–8, 82
foreign investment 19th century 59
Free Trade Agreement with Canada 324
free trade and tariffs 19th century 46, 47
Gadsden Purchase 26
gold exports mid 19th century 24–5
gold reserves 1920s and 1930s 222
gold standard 225
high-technology industries 331
immigration 23, 25, 56–7, 141, 144, 145, 146, 149, 150, 151–2, 283, 330
Immigration Act 1965 330
import controls,
   cars from Japan 327
   cotton and wheat 270–1
import duties mid 19th century 24
international flow as percentage of GNP 1870–1910 120, 121 Fig
intervention in Haiti (1915) 128
intervention in Nicaragua (1909) 128
labour productivity and trade in 1860 37, 38 Table
manufacturing trade 1913–37 194
Morrell Tariff Act 1861 46
motor industry 197–8, 331
motor vehicle registrations 1919–28 183
Mutual Aid Agreement with UK 236–7
net manufacturing exports 1960–70 266 Table
net receipts from war debts 1931 216–17
per capita income 1870 61
price levels 1875–1914 156 Table
production of manufactured goods in late 1940s 243
productivity growth after 1973 330
Quota Act 1921 152
quota system for oil imports 273

railway system in 19th century 32, 33 Table, 34, 86
Reciprocal Trade Agreement Act 201, 268
Silver Purchase Act 1880 158–9
Smoot–Hawley Tariff Act 1930 199
specialisation 1875–1914 94
tariffs 114, 181
technology leadership until 1960s 259
Trade Expansion Act 1962 268
trade with France 19th century 42
yen–dollar exchange rate 1970–92 326 Fig
USSR,
   imports of western machinery 1955–78 262
   price fixing 1955–67 281
   reparations 250
   taut planning 338
   trade with West after Second World War 249
   see also Commonwealth of Independent States (CIS)

Venezuela, oil industry 192
Versailles Peace Treaty 178, 179 Fig, 204, 235
Voluntary Export Restraints (VER) 270, 324, 327, 346
Voluntary Restraint Agreements (VRA) 270, 324, 327, 346

Welfare and economic growth 1950s and 1960s 287–9
West African Supply Agreement, oil 274
Western Electric 139
Wheat 41, 42 Table, 52, 98 Fig, 101, 194
Wholesale prices 1928 compared with 1914 217, 218 Table, 219
World Bank (IBRD) 241, 251, 256, 349

Yen,
   depreciation 1870s 130
   depreciation 1930s 230
Yen–dollar exchange rate 1970–92 326 Fig
Yugoslavia, economy after Second World War 250